Progress in Molecular and Subcellular Biology

15

Series Editors

Ph. Jeanteur, Y. Kuchino,
W.E.G. Müller (*Managing Editor*)
P.L. Paine

Springer
Berlin
Heidelberg
New York
Barcelona
Budapest
Hong Kong
London
Milan
Paris
Santa Clara
Singapore
Tokyo

B. Rinkevich W.E.G. Müller (Eds.)

Invertebrate Immunology

With 40 Figures

 Springer

Prof. Dr. B. RINKEVICH
Israel Oceanographic and Limnological Research
National Institute of Oceanography
Tel Shikmona, P.O.B. 8030
Haifa 31080
Israel

Prof. Dr. W.E.G. MÜLLER
Institut für Physiologische Chemie
Abteilung Angewandte Molekularbiologie
Universität Mainz
Duesbergweg 6
55099 Mainz
Germany

ISBN 3-540-59239-3 Springer-Verlag Berlin Heidelberg New York

Library of Congress Cataloging-in-Publication Data. Rinkevich. B. (Baruch), 1948–   . Invertebrate immunology/B. Rinkevich, W.E.G. Müller,   p.   cm. – (Progress in molecular and
subcellular biology; 15) Includes bibliographical references and index. ISBN 3-540-59239-3
1. Invertebrates – Immunology.   I. Müller, W.E.G. (Werner E.G.). 1942–   . II. Title. III. Series.
QL362.85. R56   1996   592'.029 – dc20   95-24586

© Springer-Verlag Berlin Heidelberg 1996
Printed in Germany

Cover design: Springer-Verlag, Design & Production

Typesetting: Thomson Press (India) Ltd., New Delhi

SPIN: 10476871   39/3132/SPS – 5 4 3 2 1 0 – Printed on acid-free paper

Preface

The biological bases of invertebrate immune responses have interested scientists for decades, from the first relevant observation by E. Metchnikoff in 1882, who discovered phagocytosis while studying starfish larvae. Invertebrate immunology first began to be appreciated as an important field in the late 1960s and 1970s. However, in the following years there was much controversy regarding the question: do invertebrates offer insight into the origin of the sophisticated immune responses of the vertebrates? There are several reasons why progress in research on invertebrate immune competence has been painfully slow. One of the main impediments to the progress, as compared to the fast development of knowledge in the vertebrate systems, was the fact that most of the studies concentrated on "whole organism" assays, mainly on grafting tissues between allogeneic partners. Only in the last few years have more and more aspects of invertebrate immunity been investigated on the cellular, biochemical and molecular levels. These studies led to discoveries of novel defense reactions, new pathways of effector mechanisms which are elicited after recognition of "nonself", and complex, sometimes highly polymorphic genetic elements that control invertebrate immune reactions. The importance of invertebrate immunity for understanding "immunology" as a whole, despite the conflicting models and hypotheses, is now much more recognized than before. Although most of the 20 phyla belonging to the invertebrates have different modes of life, body organizations, habitats occupied, and biochemical patterns, they show striking aspects of exceptional precision for discriminating between self and nonself. This book summarizes aspects of invertebrate immune responses on the cellular, biochemical and molecular levels. Some chapters try to find unified patterns between phyla; others concentrate on one specific group of organisms. Concurrently, these review chapters present some of the rapid progress recently achieved in invertebrate immunity, placing this discipline as a fast-growing and important field in biological sciences.

Haifa, Israel   B. RINKEVICH
Mainz, Germany   W.E.G. MÜLLER

Contents

Humoral Factors in Marine Invertebrates 1
M. LECLERC

References... 7

Earthworm Immunity 10
E.L. COOPER

1	Introduction.................................	10
2	Cells of the Immune System	10
3	Natural Immunity and Nonspecific Cellular Responses	13
4	Immune Reactions to Transplants: Coelomocyte Responses	15
4.1	Histopathologic Alterations	16
4.1.1	Will Pollution Affect Wound Healing and Graft Rejection?..........................	16
4.1.2	Where Do Coelomocytes Come From?	19
4.1.3	Adoptive Transfer of Accelerated Responses........	19
5	Proliferative Response of Coelomocytes to Transplantation Antigens.....................	20
6	Membrane Components Allied to the Ig Superfamily.	24
6.1	Phylogeny of the Immunoglobulin Joining (J) Chain .	26
7	Communication Between Immune, Nervous and Endocrine Systems	26
8	Cell-Mediated Cytotoxicity......................	27
9	Mechanisms of Cellular Defense: Adaptive Cellular Response	28
10	Lysozyme-Like Substances	30
10.1	Hemolytic and Hemagglutinating Systems of the Coelomic Fluid	31
10.1.1	Cytolytic Substances	32
10.1.2	Adaptively Formed Substances	33
10.1.3	Tumorostatic Activity of Coelomic Fluid	34
11	Evolution of Immune Responses and Where Earthworms Fit In	35

References... 37

The Prophenoloxidase Activating System
and Associated Proteins in Invertebrates   46
M.W. Johansson and K. Söderhäll

1   Introduction................................   46
2   Components of the Prophenoloxidase
         Activating System   48
2.1   Prophenoloxidase   48
2.2   Prophenoloxidase Activating Enzyme   50
2.3   Prophenoloxidase Activation Inhibitors...........   50
2.4   Elicitors and Elicitor-Binding Proteins.............   52
2.5   Associated Proteins   55
2.5.1   The Cell-Adhesion Protein   55
2.5.2   The Receptor for the β-1,3-Glucan Binding Protein
         and the Cell-Adhesion Protein   57
3   Complement-like Proteins.......................   58
4   Clotting Proteins   59
5   Summary   60
References..   60

Inducible Humoral Immune Defense Responses in Insects ...   67
R.D. Karp

1   Introduction................................   67
2   Antibacterial Responses   68
2.1   Lysozyme...................................   69
2.2   Lectins   69
2.3   Cecropins and Allied Factors   71
2.4   Attacins and Allied Factors   72
2.5   Defensins   73
2.6   The Antibacterial Response
         in the American Cockroach   73
2.7   Hemolin....................................   75
2.8   Unclassified Antibacterial Factors   76
3   The Response to Soluble Proteins   76
4   Concluding Remarks...........................   80
References..   82

Blood Clotting in Invertebrates........................   88
S. Srimal

1   Introduction................................   88
1.1   Hemostasis and Immunity – a Marriage
         of Convenience   89
2   The Limulus Clotting System   89
2.1   Amebocytes and Clotting   89
2.1.1   The Proteolytic Cascade   90

3   Clotting in Crustaceans and Other Arthropods   94
4   Conclusion   96
References   97

Immune Function of α_2-Macroglobulin in Invertebrates   101
P.B. ARMSTRONG and J.P. QUIGLEY

1   Introduction   101
2   Structure of α_2-Macroglobulin   102
3   The α_2-Macroglobulin Protein Family   104
4   α_2-Macroglobulin in Invertebrates   104
5   Receptor-Mediated Clearance
     of α_2-Macroglobulin-Protease Complex   112
6   α_2-Macroglobulin in the Blood Cells   114
7   Interaction of α_2-Macroglobulin with the Proteases
     of the Clotting System of *Limulus*   115
8   Cytokine-Binding Activities of α_2-Macroglobulin   116
9   The Plasma Based Cytolytic System of *Limulus*   116
10   α_2-Macroglobulin and the Inactivation
     of the Proteases of Invading Parasites   119
11   Interaction of the α_2-Macroglobulin
     with Other Systems of Immunity in Invertebrates   119
12   Evolutionary Considerations   121
References   122

Host–Parasite Interactions in Molluscs   131
S.E. FRYER and C.J. BAYNE

1   Introduction   131
2   Molluscan Internal Defenses   133
3   The Fates of Invading Molluscan Parasites   135
4   How Do Molluscs Respond to Parasite Infections? . .   139
5   Are Susceptible Snails Simply Incompetent?   141
6   How Does a Parasite Escape Destruction
     by a Molluscan Host?   142
7   Conclusions   147
References   148

Clotting and Immune Defense in Limulidae   154
T. MUTA and S. IWANAGA

1   Introduction   154
2   Hemolymph   156
3   Hemocytes   156
4   Hemocyte Granules   156
5   Clotting Cascade   158
5.1 Coagulogen and Coagulin   160

5.2   Proclotting Enzyme and Clotting Enzyme   163
5.3   Factor B and Factor $\bar{B}$ 164
5.4   Factor C and Factor $\bar{C}$ 165
5.5   Factor G and Factor $\bar{G}$.   167
6   Protease Inhibitors   171
6.1   Anticoagulants   171
6.1.1   *Limulus* Intracellular Coagulation Inhibitor,
           Types 1 and 2 (LICI-1 and -2)   171
6.2   Trypsin Inhibitor in the Hemocytes   173
6.3   Inhibitor in Plasma   173
6.3.1 α_2-Macroglobulin   173
7   Transglutaminase and Its Substrates   174
7.1   Transglutaminase   174
7.2   Intracellular Substrates of Transglutaminase   176
8   Antibacterial Substances   176
8.1   Antilipopolysaccharide Factor (ALF)   177
8.2   Tachyplesin Family   177
8.3   Big Defensin   178
9   Agglutinins/Lectins   179
9.1   Plasma Agglutinins   179
9.1.1   Limulin/C-Reactive Protein/Carcinoscorpin   179
9.1.2   Polyphemin   180
9.1.3   Lectins from Japanese Horseshoe Crab
           (*Tachypleus tridentatus*)   180
9.2   Intracellular Agglutinins   181
9.2.1   LPS-Binding Protein (L6)   181
9.2.2   Limunectin   182
9.2.3   *Limulus* 18K Agglutination-Aggregation Factor
           (18K-LAF)   182
10   Summary   182
References.   183

Cytotoxic Activity of Tunicate Hemocytes   190
N. PARRINELLO

1   Introduction.   190
2   Ascidian Hemocytes   191
3   Multiple Modes of Self/Nonself Recognition
           and Cytotoxicity   191
3.1   Graft Rejection in Solitary Ascidians.
           Specific Immunorecognition
           and Hemocyte Cytocidal Responses to Foreignness . .   191
3.2   Tunic Reaction to Erythrocyte Subcuticular Injection . .   193
3.3   Nonfusion Reaction in Colonial Ascidians   194
3.4   Natural Cytocidal Response.
           In Vitro "Contact Reaction"   196

3.5   In Vitro Spontaneous Hemocyte Cytotoxic Activity
        Against Mammalian Target Cells   197
3.5.1   Erythrocyte Targets.   197
3.5.2   Tumor Cells   197
4   The Hemocytes Involved in Cytotoxic Reactions   198
4.1   Are Immunocompetent Lymphocyte-Like Cells
        Cytotoxic?   198
4.2   Are Lymphocyte-Like Cells Involved in NFR?   201
4.3   Stem-Cell Function of Lymphocyte-Like Cells   201
4.4   Hemocytes Involved in Botryllid Cytotoxic
        Reactions   203
5   Hemocytes Involved in the Contact Reaction   205
6   Globular Granulocytes (Morula Cells):
        Are They Involved in Cytotoxic Reactions?.   206
7   Hemocytes with NK-Like Activity   208
7.1   Antierythrocytes   208
7.2   Antitumor   209
8   Conclusions   210
References.   213

Humoral Factors in Tunicates   218
Y. SAITO

1   Introduction.   218
2   Hemagglutinins (Lectins)   219
3   Antimicrobial Substances   224
4   Other Humoral Factors   225
5   Humoral Factors Involved
        in Allogeneic Recognition   227
6   Concluding Remarks.   230
References.   231

Molecular Aspects of Immune Reactions in Echinodermata. .   235
V. MATRANGA

1   Introduction.   235
2   Response to Allogeneic Transplants.   236
3   Clearance Studies   236
4   Humoral Molecules.   237
5   Cytokines.   239
6   Immune Effector Cells.   240
7   Summary and Concluding Remarks   243
References.   245

Subject Index   249

List of Contributors

Addresses are given at the beginning of the respective contribution

Armstrong, P.B.   101
Bayne, C.J.   131
Cooper, E.L.   10
Fryer, S.E.   131
Iwanaga, S.   154
Johansson, M.W.   46
Karp, R.D.   67
Leclerc, M.   1

Matranga, V.   235
Muta, T.   154
Parrinello, N.   190
Quigley, J.P.   101
Saito, Y.   218
Söderhäll, K.   46
Srimal, S.   88

Humoral Factors in Marine Invertebrates

M. LECLERC

Lymphoid organs in Crustacea and Insecta were first discovered by L. Cuénot between 1890 and 1910. For many years, however, little research was done into the immune processes of invertebrates (which represent 95% of animal species), as it was assumed that with a short life-span and a high rate of reproduction, a complex and highly efficient immune system was not required.

It is only in the last 20 years, with the development of comparative immunology, that there has been a renewal of interest in the study of immunity in invertebrates. The reasons for this development are twofold.

Firstly, in the field of pure research, comparative study of the various immune systems encountered in the invertebrates may lead to a better understanding of the highly complex phenomena observed in the vertebrates. Moreover, the discovery of analogies or homologies, or even of previously unknown systems, such as the prophenoloxidase system in Arthropoda, yields useful information about the phylogenesis of immunity. The second reason is socio-economic: understanding the physiology of the defence mechanism in invertebrates (and in particular, the host–parasite relationship) could allow more efficient control of those organisms which are carriers of infectious disease or which cause damage in agriculture and forestry, as well as ensuring better protection during the intensive breeding of the economically important invertebrates.

Invertebrates, like vertebrates, possess a natural immunity consisting, first of all, of the presence of various physico-chemical barriers which act as highly efficient obstacles to outside invasion. Most invertebrates have an external skeleton such as, for example, the cockle of many molluscs, the test of sea-urchins or the cuticle of arthropods. In the absence of this mechanical barrier, and in some cases in addition to it, there are anti-microbial substances such as mucus, melanin, or agglutinins (to ward off bacteria, viruses or other pasasites), which coat the body of most invertebrates either permanently, or after stimulation of epidermic and hypodermic glands.

The second line of defence brings into play cells of various kinds. These cells fall into two main categories: circulating cells, called haemocytes or coelomocytes, and fixed cells which are found in all tissues or located in the lymphoid

Immunologie des Invertébrés, UFR Sciences, Université d'Orléans, France and phagocytic organs. These cells are able to trigger a wide range of defence mechanisms, either directly or in collaboration with humoral factors.

For example, after sustaining injury, most invertebrates develop an inflammatory-type reaction, which calls into play hemocytes/coelomocytes and humoral factors such as lymphokins. During the cicatrization process, cell aggregation, coagulation and/or muscular contraction take place. In addition, certain cells at this level are able to eliminate any foreign body which has penetrated the organism (this is called the phenomenon of clearance). If the intruder is small in size, it is eliminated by phagocytosis, a process which involves various stages: chimiotactism, fixation, ingestion and destruction, by means of substances such as opsonins (agglutinins), complement factors and, in certain species, activation products of the ProPo system. When the intruders are numerous or large, there is the formation of nodules or capsules, either cellular or humoral (multicellular mantle or melanin deposit), which capture the foreign bodies, enabling them to be eliminated. This process of encapsulation involves cellular adherence factors (Kobayashi et al. 1990) or components of the prophenoloxidase system (Schmit and Ratcliffe 1977).

Invertebrates have developed many different faculties in order to kill any microorganism or parasite which succeeds in passing this initial physico-chemical barrier. First of all, associated with phagocytosis, we find hydrolytic activity due to enzymes, such as esterases, peroxidases and lysozymes, or due to toxic factors such as quinones, the production of which is linked to the melanization process. Cytotoxic cells are also found in Spongeae, Coelenterata and Prochordata, where their role is to preserve the integrity of adjacent colonies, and in other invertebrates where this cytotoxicity, which requires close contact with the target cell and is probably achieved by agglutinins, is not unlike the "natural killer" process in vertebrates.

In addition to agglutinins and other toxic factors, many other humoral factors are involved in defence reactions (Cooper 1985):

1. Lymphokin factors, found in Tunicata and Echinodermata, some of which are able to induce immunological reactions in vertebrates and present close functional resemblances to Interleukin I in vertebrates (Beck et al. 1989a,b)

2. Inductible factors, found specifically in Echinodermata (Leclerc and Brillouet 1981)

3. Complement factors, which have been shown to be present mainly in Echinodermata (Day et al. 1972; Bertheusen 1982, 1983, 1984; Leonard et al. 1990) and which are involved in a process which may be considered similar to the primitive alternative pathway of the complement system in vertebrates (Leonard et al. 1990).

Clearly, these two types of response, cellular and humoral, which exist and interact closely in invertebrates, are capable of non-specific binary self/not-self recognition (Du Pasquier and Duprat 1968). However, there is some debate as to their ability for specific recognition, or, in other words, their ability to develop an adaptative defence reaction (Hildemann 1974; Cooper

1979; Voisin 1983; Klein 1989; Reinisch and Litman 1989; Marchalonis and Schluter 1990).

The first level of self/not-self recognition plus elimination of the foreign body or of the modified self, is present in all invertebrates (Theodor 1970), as are phagocytosis and encapsulation through the activity of "scout molecules". Among the most likely candidates for this role are lectins, proteins or glyco-proteins, which react specifically with certain glucidic motives, and produce agglutinating or opsonizing effects (Sminia and Van der Knaap 1987).

By carrying out transplantation experiments (in some cases with graft versus host reaction, or GvHr; Panijel et al. 1977) and in vitro cytotoxicity experiments on mixed leucocyte cultures (MLC), some authors have demonstrated the existence of both xenogenic and allogenic recognition. Moreover, certain inverte-brates go a step further and are capable of an accelerated reaction of secondary allograft or xenograft rejection (Hildeman and Dix 1972). Likewise, the specific reactivity of certain sub-populations of immunocompetent invertebrate cells to certain mitogens Con A (Concanavlin A), PHA (Phytohaemagglutin) PWM (Pokeweed mitogen), LPS (Lipopolysaccharide of *S. typhimurium*) for example (Brillouet et al. 1981), as well as the production of mixed leucocyte reactions (Tanaka 1975; Warr et al. 1977; Luquet et al. 1984) demonstrate the highly discriminatory capacity of these cells, not unlike the lymphocytic subpopulations in the vertebrates.

Lastly, the existence of molecules in the complement system and of specifically induced humoral substances (sometimes with antibody-like activity) suggests that it is not only vertebrates which are endowed with complex, specific, memo-rized defence reactions. Similarly, molecules related to the superfamily of immu-noglobulins (Williams and Barclay 1988) have been identified in certain groups

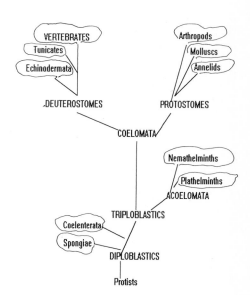

Fig. 1. Simplified genealogical tree

such as β2-microglobulins in Crustacea (Shalev et al. 1981), or Thy-1 antigen in Mollusca (Williams and Gagnon 1982) and Tunicata (Mansour and Cooper 1984, 1987; Mansour et al. 1985). Moreover, the identification of these interleukin (IL)-1-type molecules in certain invertebrates presupposes the presence of receptors for this molecule, and, in vertebrates, this receptor belongs to the superfamily of immunoglobulins (Sims et al. 1988). It would seem, therefore, that the ancestral genes of immunoglobulins and of the major histocompatibility complex in vertebrates originally appeared in invertebrates (Hildeman et al. 1977; Scofield et al. 1982; Pla and Shalev 1983; Marchalonis et al. 1984; Buss and Green 1985; Rosenshein et al. 1985; Warr 1985; Cooper 1986; Edelman 1987; Williams and Barclay 1988). The final section of this chapter gives further details about the study of humoral factors carried out on Echinodermata in our laboratory and, in particular, those organisms at the junction point between invertebrates and vertebrates. Our work specifically concerns the axial organ (AO) of Asterida, sea stars, which is an ancestral lymphoid organ. The study of immunity in Echinodermata is of particular interest given the phylogenetic position of this branch (Fig. 1). Indeed, whereas all other invertebrates are protostomes, Echinodermata are deuterostomes like vertebrates, and possess the same common ancestor from which they are believed to have diverged about 600 million years ago, in the early Cambrian.

The earliest immune reaction observed in Echinodermata was phagocytosis, by Durham (1888) and Metchnikoff (1891). However, it was not until the 1960s that the immune response of these invertebrates was studied in detail. Most work examined the phenomena of graft rejection and the role of certain cells and molecules found in the coelomic cavity (Bang 1982). The coelomic liquid contains various cell types involved in preserving the organism's integrity. Some of these cells are involved in coagulation and cicatrization of injuries, while others are responsible for phagocytosis and cytoxicity, thus enabling clearance and graft rejection. Bertheussen and Seljelid (1978), for instance, have shown that 67% of the coelomocytes in *Strongylocentrotus droebachiensis* are phagocytes. An approach to the study of lytic activity in Echinodermata consisted of the search for lysing substances in the coelomic fluid (Ryoyama 1973, 1974). Ryoyama has demonstrated, in various sea urchins, not only agglutination but also lytic activity against erythrocytes of several species of vertebrates. Such haemagglutinins and haemolysis have also been detected in the coelomic liquid of a holothuria (Parrinello et al. 1979), another sea urchin (Canicatti 1987), and a sea star (Canicatti 1989). An exhaustive study of the lytic system of *Holothuria polii* has been carried out by the team of Canicatti (1989). They have shown that two categories of haemolysins are produced by two different sub-categories of amibocytes. The first, called haemolysin 1, or Hel, Ca-dependent and thermosensitive, is a natural constituent of the coelomic liquid. The second, called haemolysin 2, or He2, Ca-independent and not thermo-sensitive, is only produced after antigenic stimulation. These haemolysins have a dual function: firstly, they are responsible for opsonizing activity via which the not-self is recognized and phagocytized; secondly, they can interact with the membrane of the target cell and damage it, leading to lysis of the cell. The latter process is reminiscent of the cytolytic effect mediated by the constituents of the complement system and by the perforins in vertebrates with which a phylogenetic affinity has been established by Bertheussen (1984) and Canicatti (1988, 1990). It should be mentioned that anti-bacterial activity due to lysosomal enzymes has also been revealed in *Holothuria polii* (Canicatti and Roch 1989). We have found no anti-bacterial activity after immunization of sea stars with Gram-positive and Gram-negative bacteria, either in the coelomic liquid or in the axial organ, nor have we detected any anti-bacterial peptides. Day et al. (1972) isolated, in the coelomic liquid of a sea star, a factor which when mixed with CVF (Cobra venom factor, an analogue of C3b in Mammifera) is capable of activating the complement system of amphibia. Subsequently, Bertheussen discovered a complement-type activity, as well as the existence of receptors of the C3 component of the complement system, on the surface of the phagocytes of sea urchins (Bertheussen 1982). Finally, Leonard et al. (1990) showed a complement-type activity in the coelomic liquid of an asterid, and suggested the existence of several constituents of an alternative primitive pathway of the complement. These molecules remain to be characterized.

Another category of molecules which have considerable importance in the immune response of vertebrates is cytokins (lymphokins, monokins and interleukins). They are polypeptides which are synthesized by a wide range of cells such as monocytes/macrophages, lymphocytes and fibroblasts, and they enable the transmission of information between cells, information about both the level of haematopoiesis and about the development of various responses by an organism to attack. In this connection, the work of Prendergast et al. (1974) should be mentioned. They isolated, from the coelomocytes of the sea star *Asterias forbesi*, a 39-kDa protein which they called sea star factor (SSF) and which produces many effects on Mammifera, in a similar manner to the lymphokins of vertebrates. For example, SSF can induce chimiotactism against monocytes, inhibit migration, and activate macrophages in vertebrates (Prendergast and Suzuki 1970; Prendergast and Unanue 1970). It can also inhibit the response of T-lymphocytes to bacterial infection and stimulation by mitogens (ConA, PHA) of other lymphocytes in MLC (Willenborg and Prendergast 1974; Prendergast et al. 1974; Prendergast and Liu 1976). A last point to mention is that SSF can both induce the adherence and aggregation of coelomocytes around a foreign body introduced into the coelomic cavity, and inhibit, on the surface of macrophages of vertebrates, the expression of Ia molecules (class II molecules) of the major histocompatibility complex MHC, which play a role in the stimulation of helper T-lymphocytes (Donelly et al. 1985). Prendergast suggests therefore that SSF could be a primitive interleukin-1-type molecule. In 1981, an interleukin-II type was first characterized by Leclerc et al. from *A. rubens*. Cells (T- and B-like cells) were separated by nylon wool and were cultured for 24 h at 10 °C in the presence of PWM-coated Sepharose beads. Supernatants from these cultures were centrifuged to eliminate beads and cell debris, sterilized by membrane filtration (0.22 μm Millipore) and concentrated five- or ten-fold. The supernatants were added to cultures of total AO (axial organ) cells (T-like, B-like, phagocyte cells) which had been started 24 h before, and 24 h later, the cells were harvested and mitogenic stimulation was evaluated by measuring 3H-thymidine incorporation.

It was found that the supernatant from the T-like cells cultured in the presence of PWM beads stimulated the AO cells. This result indicates that T-like cells produce an interleukin-II-like substance which is able to stimulate the cells. In 1986, Beck and Habicht isolated and characterized, from the coelomic liquid of *A. forbesi*, a protein which they assimilate to a primitive IL 1, given the numerous analogies between this molecule and the IL 1 of Mammifera. Although the affinity between this protein, with a molecular weight of 29 kDa, and SSF, remains to be determined, the authors suggest that SSF could be the precursor of the IL-1-like molecule secreted in the coelomic liquid (Beck and Habicht 1991). The last point to mention concerns ALF (antibody-like factor) which we have characterized and isolated in our laboratory (Leclerc et al. 1981; Delmotte et al. 1986).

The general idea that emerges from the experiments made in our laboratory is that Echinodermata, as exemplified by the sea star, possess an immune system able to mount cellular and humoral-specific responses after stimulation with a foreign antigen. First experiments used proteins as antigens (Leclerc 1973). Second experiments used hapten-carrier conjugates (Leclerc and Brillouet 1981).

It appears that all the elements necessary to these responses are present in the AO which can be considered as an ancestral lymphoid organ. Two main populations of cells are present in the organ; phagocytic cells and cells which are morphologically and functionally similar to the lymphocytes of vertebrates.

The immunocompetence of the AO was clearly demonstrated by the production of a soluble substance which appears to be similar in many respects to the humoral antibodies of the vertebrates (complement-dependent). There are a few important comments to be made if we accept the hypothesis of the existence of an immune system in the sea star. The AO has about 10^6 to 10^7 cells. the number of clones available to respond to any particular antigenic determinant must therefore be small, and it could be expected that the resulting specificity would not be very fine. Indeed, we have observed that sea stars immunized with either rat IgG or human lambda Bence-Jones protein react indiscriminately to both proteins, indicating extensive cross-reactivity. Rodent antisera, on the other hand, never show cross-reaction between these two proteins in spite of the many homologies in their sequences (there is a great evolutionary distance between Echinodermata and mammals).

The main acquisition of Echinodermata seems to be cellular differentiation into two subpopulations, perhaps ancestral to B- and T-lymphocytes, and their interplay with macrophage-like cells (Leclerc et al. 1986, 1993), resulting in the synthesis of specific humoral antibody-like factors.

In the near future we shall be able to analyze these factors by the use of molecular sonds.

We hope that the present review will help in suggesting provocative and imaginative ideas on humoral factors in invertebrates, especially in marine invertebrates.

References

Bang FB (1982) Disease processes in seastars: a metchnikovian challenge. Biol Bull 162: 135–148

Beck G, Habicht GS (1986) Isolation and characterization of a primitive interleukin-1-like protein from an invertebrate *Asterias forbesi*. Proc Natl Acad Sci USA 83: 1–5

Beck G, Habicht GS (1991) Primitive cytokines habingers of vertebrate defense. Immunol Today 87: 180–183

Beck G, Vasta GR, Marchalonis JJ, Habicht GS (1989a) Characterization of interleukin-1 activity in tunicates. Comp Biochem Physiol 92B: 93–98

Beck G, O'Brien R, Habicht GS (1989b) Invertebrate cytokines: the phylogenetic emergence of interleukin-1. Bio Essays 11: 62–67

Bertheussen K (1982) Receptors for complement on echinoid phagocytes. II. Purified human complement mediates echinoid phagocytosis. Dev Comp Immunol 6: 635–642

Bertheussen K (1983) Complement-like activity in sea urchin coelomic fluid. Dev Comp Immunol 7: 21–31

Bertheussen K (1984) Complement-like activity in sea urchin coelomic fluid. Dev Comp Immunol 7: 21–31

Bertheussen K, Seljelid R (1978) Echinoid phagocytes in vitro. Exp Cell Res 11: 401–412

Brillouet C, Leclerc M, Panijel J, Binaghi RA (1981) In vitro effect of various mitogens on starfish (*Asterias rubens*) axial organ cells. Cell Immunol 57: 136–144

Buss LW, Green DR (1985) Histocompatibility in vertebrates: the relict hypothesis. Dev Comp Immunol 9: 191–201

Canicatti C (1987) Evolution of the lytic system in echinoderms. Naturally occurring hemolytic activity in *Paracentrotus lividus* (Echinoidea) coelomic fluid. Boll Zool 4: 325–329

Canicatti C (1988) The lytic system of *Holothuria polii* (Echinodermata): a review. Boll Zool 55: 139–144

Canicatti C (1989) Evolution of the lytic system in echinoderms. II. Naturally occurring hemolytic activity in *Marthasterias glacialis* (Asteroidea) coelomicfluid. Comp Biochem Physiol 93A: 587–591

Canicatti C (1990) Hemolysins: pore-forming proteins in invertebrates. Experientia 46: 239–244

Canicatti C, Roch PH (1989) Studies on *Holothuria polii* (Echinoderma) antibacterial proteins. I. Evidence for and activity of coelomocyte lysozyme. Experientia 45: 756–759

Cooper EL (1979) L'évolution de l'immunité. Recherche 103: 824–833

Cooper EL (1985) Overview of humoral factors in invertebrates. Dev Comp Immunol 9: 577–583

Cooper EL (1986) Evolution of histoincompatibility. In: Brehélin M (ed) Immunity in invertebrates. Springer, Berlin Heidelberg New York, pp 139–149

Day NKB, Geiger H, Finstad H, Good RA (1972) A starfish hemolymph factor which activates vertebrate complement in the presence of cobra venom factor. J Immunol 109: 164–167

Delmotte F, Brillouet C, Leclerc M, Luquet G, Kader JC (1986) Purification of an antibody-like protein from the sea star *Asterias rubens* (L.). Eur J Immunol 16: 1325–1330

Donnely JJ, Vogel SN, Prendergast RA (1985) Down regulation of Ia expression on macrophages by the sea star factor. Cell Immunol 90: 408–415

Du Pasquier L, Duprat P (1968) Aspects humoraux et cellulaires d'une immunité naturelle non spécifique chez l'oligochète *Eisenia foetida* Sav. (Lumbricinae). CR Acad Sci Paris 266: 538–541

Durham F (1988) On the emigration of ameboid corpuscles in the starfish. Proc R Soc Lond B43: 328–330

Edelman GM (1987) CAMs and Igs: cell adhesion and the evolutionary origins of immunity. Immubol Rev 100: 11–45

Hildemann WH (1974) Some new concepts in immunological phylogeny. Nature 250: 116–120

Hildemann WH, Dix TG (1972) Transplantation reactions of tropical Australian echinoderms. Transplantation 15: 624–633

Hildemann WH, Raison RL, Cheung G, Hull CJ, Akaka L, Okamoto J (1977) Immunological specificity and memory in a scleractinian coral. Nature 270: 219–223

Klein J (1989) Are invertebrates capable of anticipatory immune responses? Scand J Immunol 29: 499–505

Kobayashi M, Johansson MW, Söderhäll K (1990) The 76 kD cell-adhesion factor from crayfish haemocytes promotes encapsulation in vitro. Cell Tissue Res 260: 13–18

Leclerc M (1973) Etude ultrastructurale des reactions d'*Asterina Gilbbosa* (Echinoderme Asteride) au niveau de l'organe axial après injection de protéines. Ann Immunol 124 C: 363–374

Leclerc M, Brillouet C (1981) Evidence of antibody-like substances secreted by axial organ cells of the starfish *Asterias rubens*. Immunol Lett 2: 279–281

Leclerc M, Brillouet C, Luquet G, Agogué P, Binaghi RA (1981) Properties of cell subpopulations of starfish axial organ: in vitro effect of pokeweed mitogen and evidence of lymphokine-like substances. Scand J Immunol 14: 281–284

Leclerc M, Brillouet C, Luquet G, Binaghi RA (1986) Production of an antibody-like factor in the seastar *Asterias rubens*: involvement of at least three cellular populations. Immunology 57: 479–482

Leclerc M, Arneodo V, Legac E, Bajelan M, Vaugier G (1983) Identification of T like and B like lymphocyte subsets in sea star *Asterias rubens* by monoclonal antibodies to human leucocytes. Thymus 21: 133–139

Leonard LA, Stranberg JD, Winkelstein JA (1990) Complement-like activity in the sea star, *Asterias forbesi*. Dev Comp Immunol 14: 19–30

Luquet G, Brillouet C, Leclerc M (1984) M.L.R.-like reaction between axial organ cells from asterids. Immunol Lett 7: 235–238

Mansour MH, Cooper EL (1984) Serological and partial molecular characterization of a Thy-1 homolog, in tunicates. Eur J Immunol 14: 1031–1039

Mansour MH, Cooper EL (1987) Tunicate Thy-1. An invertebrate member of the Ig superfamily. In: Cooper EL, Langlet C, Bierne J (eds) Progress in clinical and biological research, vol 233. Developmental and comparative immunology. Alan R Liss, New York, pp 33–42

Mansour MH, De Lange R, Cooper EL (1985) Isolation, purification and amino acid composition of the tunicate hemocyte Thy-1 homolog. J Biol Chem 260: 2681–2686

Marchalonis JJ, Schluter SF (1990) On the relevance of invertebrate recognition and defence mechanisms to the emergence of the immune response of vertebrates. Scand J Immunol 32: 13–20

Marchalonis JJ, Vasta GR, Warr GW, Barker WC (1984) Probing the boundaries of the extended immunoglobulin family of recognition molecules: jumping domains, convergence and minigenes. Immunol Today 5: 133–142

Metchnikoff E (1891) Lectures on the comparative pathology of inflammation. Dover Publ., New York (transl 1968 by Starling FA, Starling EH, Keagan, Trench, Trubuer, London)

Panijel J, Leclerc M, Redziniak G, El Lababidi M (1977) Specific reactions induced in vertebrates by sea star axial organ cells. In: Solomon JB, Horton JD (eds) Developmental immunobiology. Elsevier, Amsterdam, pp 91–97

Parrinello N, Rindone D, Canicatti C (1979) Naturally occurring hemolysins in the coelomic fluid of *Holuthuria polii* Delle Chiaje (Echinodermata). Dev Comp Immunol 3: 45–54

Pla M, Shalev A (1983) Déterminants chez les invertébrés, analogues aux antigènes du complexe majeur d'histocompatibilité. Bull Inst Pasteur (Paris) 81: 273–275

Prendergast RA (1970) Macrophage activating and chemotactic factor from the sea star *Asterias forbesi*. Fed Proc 29: 647

Prendergast RA, Liu SH (1976) Isolation and characterization of sea star factor. Scand J Immunol 5: 873–880

Prendergast RA, Suzuki M (1970) Invertebrate protein simulating mediators of delayed hypersensitivity. Nature 227: 277–279

Prendergast RA, Unanue ER (1970) Macrophage activation and migration inhibition factor from the sea star *Asterias forbesi*. Fed Proc Fed Am Soc Exp Biol 29: 771

Prendergast RA, Cole GA, Henney CS (1974) Marinein vertebrate origin of a reactant to mammalian T cells. Ann NY Acad Sci 234: 7–16

Reinisch GL, Litman GW (1989) Evolutionary immunobiology. Immunol Today 10: 278–281

Rosenshein IL, Schluter SF, Vasta GR, Marchalonis JJ (1985) Phylogenetic conservation of heavy chain determinants of vertebrates and protochordates. Dev Comp Immunol 9: 783–795

Ryoyama K (1973) Studies on the biological properties of coelomic fluid of sea urchin. I. Naturally occurring hemolysin in sea urchin. Biochim Biophys Acta 320: 157–165

Ryoyama K (1974) Studies on the biological properties of coelomic fluid of sea urchin II. Naturally occurring hemagglutinin in sea urchin. Biol Bull 146: 404–414

Schmit AR, Ratcliffe NA (1977) The encapsulation of foreign tissue implants in *Galleria mellonella* larvae. Insect Physiol 23: 175–184

Scofield VL, Schlumberger JM, West LA, Weissman IL (1982) Protochordate allorecognition is controlled by a MHC-like gene system. Nature 295: 499–502

Shalev A, Greenberg AH, Lögdberg L, Björck L (1981) β2-Microglobulin-like molecules in low vertebrates and invertebrates. Immunology 127: 1186–1191

Sims JE, March CJ, Cosman D, Widmer MB, MacDonald HR, McMahan CJ, Grubin CE, Wignall JM, Jackson JL, Call SM, Friend D, Alpaert AR, Gillis S, Urdal DL, Dower SK (1988) cDNA expression and cloning of the IL-1 receptor, a member of the immunoglobulin superfamily. Science 241: 585–589

Sminia T, Van der Knaap WPW (1987) The forgotten creatures: immunity in invertebrates. In: Cooper EL, Langlet C, Bierne J (eds) Progress in clinical and biological research, vol 233. Developmental and comparative immunology. Alan R Liss, New York, pp 157–162

Tanaka K (1975) Allogeneic distinction in *Botryllus primigenus* and in other colonial ascidians. Adv Exp Med Biol 64: 115–124

Theodor JL (1970) Dinstinction between "self" and "not-self" in lower invertebrates. Nature 227: 690–692

Voisin GA (1983) Réflexions sur l'immunologie comparée. Bull Inst Pasteur (Paris) 81: 277–279

Warr GW, Decker JM, Mandel TE, De Luca D, Hudson R, Marchalonis JJ (1977) Lymphocyte-like cells of the tunicate *Pyura stolonifera*: binding of lectins, morphological and functional studies. Aust J Exp Biol Med Sci 55: 151–164

Warr GW (1985) Beta 2 microglobulins: a brief comparative review. Dev Comp Immunol 9: 769–775

Willenborg DO, Prendergast RA (1974) The effect of sea star coelomocyte extract on cell-mediated resistance to *Listeria monocytogenes* in mice. J Exp Med 139: 820–833

Williams AF, Barclay AN (1988) The immunoglobulin superfamily – domains for cell surface recognition. Annu Rev Immunol 6: 381–406

Williams AF, Gagnon J (1982) Neuronal cell Thy-1 glycoprotein: homology with immunoglobulin. Science 216: 696–703

Earthworm Immunity

E.L. COOPER

1 Introduction

The purpose of this chapter is to draw attention to the mass of information that has accumulated on the earthworm immune system, since about 1963. This work has emanated from several laboratories, primarily those of Cooper (USA), Duprat, Valembois, Roch (France), Bilej, Rejnek, Šíma, Tucková, Větvička (Czech Republic), and recently Goven (USA). To my knowledge, these are the laboratories that have been most concerned with major studies of earthworm immunity. Goven's work represents a unique departure, a "clinical" approach, since he has proposed that earthworms be used as sentinels, and that components of their immune system serve as surrogate biomarkers indicative of changes caused by environmental pollution (Rodriguez-Grau et al. 1989; Fitzpatrick et al. 1990; Chen et al. 1991; Eyambe et al. 1991; Venables et al. 1992; Goven et al. 1993; Goven and Kennedy 1995). This will not be an elaborate review of earthworm immunity nor will it include information on polychaete annelids found in the book by Vetvicka et al. (1994), for this has been done in several recent reviews including another book: Tucková and Bilej (1995); Roch (1995). My purpose will be to present certain areas that, although reviewed earlier, surely require further investigation, since, in my opinion, they are critical subjects.

2 Cells of the Immune System

The cellular armamentarium of the earthworm is indeed complex. There are at least six major cell types whose function is only partially known (Tables 1–3). These cell types have been classified by Cooper and Stein (1981) and Valembois et al. (1982a). Stein et al. (1977) examined morphology and phagocytic properties of earthworm coelomocytes, and later, Stein and Cooper (1978) analyzed their cytochemical properties. Intermittently, there have been studies by Valembois (1971b), and Linthicum et al. (1977a,b) concerning their fine structure in relation to transplants. Thus, a significant contribution that could be made, would be the

Laboratory of Comparative Immunology, Department of Neurobiology, School of Medicine, University of California, Los Angeles, California 90024, USA

Table 1. Coelomocyte characteristics – *Lumbricus terrestris*[a]. (Stein et al. 1977)

Cell type	Frequency[b]	Cell size in μm	Nucleus[c]	Color of cytoplasm	Granules[d]
Basophil	63.5 ± 6.1	5–30	4–8 Central or eccentric dark blue-violet	Blue	0.5, few dark blue
Neutrophil	18.0 ± 5.9	12–15	8–10 Central or eccentric rose	Pale blue, pink or lavender	0.5–1, few to moderate light blue, pink or lavender
Acidophil Type I	6.2 ± 3.4	10–30	5–9 Usually eccentric dark red-violet	Blue	1–2, many pink to red
Acidophil Type II	0.7 ± 0.5	10–15	6–8 Central or eccentric dark red-violet	Blue	2–4; moderate to many pink to red
Granulocyte	8.1 ± 6.6	8–40	5–9 Often eccentric dark violet	Blue	0.5–2, variable pink, lavender, blue
Chloragogen Type I	see Type II (below)	10–25 by 30–60	6–7 Usually eccentric rose	Pale blue	1–2, many medium blue
Chloragogen Type II	1.4 ± 0.9[e]	12–20	7–8 Usually eccentric rose	Blue	0.5–2, many dark blue-violet
Transitional	1.9 ± 1.0	Variable	Variable	Variable	Variable

[a]Cells stained with Wright's stain. [b]Percent of the total population, expressed as mean value ± SD. [c]Size in μm, position in cell and colour. [d]Approximate size in μm, number, color. [e]Combined frequency for chloragogen types I and II.

Table 2. Summary of various cell types present in *Lumbricus terrestris*

General name	Light microscopy designation (Wright's stain)	Electron microscopy designation
Hyaline amoebocytes	1. Basophils	Lymphocytic coelomocytes of 2 types
	2. Neutrophils	Type I granulocytes
	3. Acidophils (2 types)	Inclusion-containing coelomocytes
Granular amoebocytes	4. Granulocytes	Not identified
	5. Not identified	Type II granulocytes
Eleocytes	6. Chloragogen cells (eleocytes) (2 types)	Chloragogen cells (eleocytes) (1 type only)

Table 3. Coelomocyte functions (see text for mitogen responses, adoptive transfer and rosette formation)

Function	Coelomocyte type	Species studied
Phagocytosis	Amoebocytes	Species from many families
	Eleocytes	*Eisenia foetida*
Encapsulation and formation of "brown bodies"	Amoebocytes	Many species
	Eleocytes[a]	(various families)
Wound closure (formation of wound plug)	All coelomocytes	Lumbricidae
Wound healing and regeneration (general)	Amoebocytes	*Eisenia foetida*
	Eleocytes	*Eisenia foetida*
Formation of substratum for regenerating epidermis during wound healing	Hyaline amoebocytes	*Eisenia foetida*
Graft rejection	Amoebocytes	*Lumbricus terrestris*
	"Splanchnopleural macrophage"[b]	*Eisenia foetida*
	Type I granulocyte[b]	*Lumbricus terrestris*
	Hyaline amoebocytes	*Eisenia foetida*
	"Small basophil"[b]	*Eisenia foetida*
Cytotoxicity	Unspecified	*Lumbricus terrestris*
		Eisenia foetida
	Eleocytes[c]	*Eisenia foetida*
	Amoebocytes[d]	
Possible blood cell precursors	Amoebocytes	Lumbricidae (several species)
		Pheretima indica
Possible precursor cells for epidermal basal cells and inter-muscular granule-containing cells	Hyaline amoebocytes	*Eisenia foetida*
Hemoglobin production	Hemoglobin producing cell	*Eisenia foetida*
Hemopoiesis	Eleocytes	*Eisenia foetida*
Nutritive ("trephocytic")	Eleocytes	Species from many different families
Association with developing eggs (possible nutritive function	Lamprocytes	*Enchytraeus fragmentosus*
Excretory	Eleocytes	Species from many families

[a]Probably passively involved.
[b]Subtype of hyaline amoebocytes.
[c]Nonspecific, naturally occurring.
[d]Specific, acquired.

development of techniques for the successful isolation and purification of coelomocytes with the view of cultivating them in vitro. There is a definite need to analyze earthworm leukocytes more specifically according to each cell population in relation to function (Cooper et al. 1979).

3 Natural Immunity and Nonspecific Cellular Responses

The nonspecific responses have consisted primarily of those devoted to understanding chemotaxis, (Marks et al. 1979), phagocytosis (Bilej et al. 1990c; Dales and Kalac 1992), wound healing (Cameron 1932; Burke 1974a,b,c) encapsulation and other events that may not involve memory and specificity. These defense reactions are considered to be basic, inherent in all responses in complex animals. This raises two new questions concerning (1) the evolution of natural killer cells and when this function might have dissociated from the phagocytic response during evolution, and (2) the role of molecules that mediate communication (e.g. the cytokines) (Raftos et al. 1991a,b, 1992; Beck et al. 1993). Assuming a close kinship in evolution between phagocytosis and natural killing (NK), then once an NK response is initiated, suppose cells are phagocytic, can they kill? What is the time course of cytotoxicity and phagocytosis?

What happens when phagocytosis does not occur? The fate of foreign particles in the coelomic cavity of annelids is influenced by the size of the particle. Small ones, such as bacteria, are usually phagocytosed, whereas larger ones are eliminated by a process of encapsulation. There is interesting evidence supporting the view that phagocytosis may be influenced by various opsonins such as IgG and C3b (Laulan et al. 1983, 1988). Cameron (1932) was probably the first person in more recent times to analyze encapsulation by observing coelomocyte reactions to cotton thread inserted into the earthworm's (*Lumbricus terrestris*) coelomic cavity. All coelomocyte types surround individual fibers within the first 24 h, leading to cell proliferation in septal lining cells. Later, a dense capsule composed of flattened cells is formed within several days, and small vessels growing outwards from the body wall are surrounded by proliferating cells. After seven days, the encapsulating cells are closely packed producing thin strands of fibrous tissue. Finally, a fibrous capsule (sometimes called "brown body" owing to the content of brown pigment is produced). As a model, this response to inert material can serve to reveal what may transpire in responses against foreign, living cells, since similar observations have been described in the case of defense against parasites. In summary and to add credence to these early observations, some 40 years later, Bang (1973) postulated that encapsulation begins with recognition but that phagocytosis, which would normally follow, cannot be triggered if a foreign object is too large. Although encapsulation has often been analyzed as the mechanism for eliminating large inert bodies that have been artificially introduced into the coelomic cavity, it represents one of the main natural defense mechanisms against potentially invading parasites and other foreign cells.

More recently, encapsulation has been analyzed thoroughly in *Eisenia foetida andrei* by Valembois et al. (1992). They examined more than 1000 worms, which were either kept in the laboratory or collected from natural farm mature, in order to follow mechanisms and the role that coelomocytes play in brown body formation. Brown body formation generally requires several months and results in a convex, disk-shaped mass. When it reaches a diameter of 1–2 mm, its external cells flatten and lose their adhesiveness toward free coelomocytes or waste particles. Once formed, a brown body can then migrate to the posterior segments where it can be eliminated by autonomy (Keilen 1925). During the last weeks of maturation, melanin in brown bodies (Ratcliffe et al. 1985), causes darkening, proceeding from a yellowish color to red-brown to distinctly brown. Because its color depends on the stage, the term "brown body" has been proposed for use with only large structures found in posterior segments; the terms "nodule" or "aggregate" seem to be more suitable as generic terms.

Histologically, nodules smaller than 1 mm strongly adhere to the coelomic epithelium, exhibiting strands of cellular material binding them to the coelomic wall. With larger nodules, cellular binding to the coelomic epithelium is more and more narrow, and eventually completely disappears. Examination of small nodules (150–250 μm) has revealed that most nodules (85%) contain tissue wastes, especially necrotic muscle cells and setae. Agglutinated bacteria (isolated bacteria cannot initiate encapsulation and are engulfed by phagocytosis) are present in 29% of small nodules. Gregarines and nematodes have been observed less frequently (8 and 1%) and in some cases, two or three types of foreign particles or even altered self tissues have been found in the same nodule. When considered in the larger context of hosts eliminating foreignness, encapsulation and phagocytosis are crucial.

Going a step further, by means of histochemical and electron-microscope techniques, Valembois et al. (1994) found evidence of lipofuscin and melanin in brown bodies of *Eisenia fetida andrei*. Lipofuscin, often called the aging pigment, is the most abundant and the first to be synthesized, whereas melanin seemed to be present in mature brown bodies. A computer-assisted image analysis was used to quantify the relative abundance of the two pigments at different stages of brown body formation. With the results of a previous study showing phenoloxidase activity in coelomocytes, and the present demonstration of their ability to produce reactive oxygen systems when agglutinated, we can now correlate the simultaneous presence of the two pigments in brown bodies. More recently, as an extension of this work, Seymour et al. (1995) have searched for phenoloxidase activity in coelomic fluid by quantitatively assaying the two substrates, L-DOPA and dopamine, using high pressure liquid chromatography coupled with electrochemical detection; the activity is partly inhibited by 1-phenyl-2-thiourea. Oxidase activity exerted by coelomic fluid seems inoperative on tyrosine (Valembois et al. 1991). The coelomic fluid factor responsible for phenoloxidase activity, has been found to be a protein with a mass of about 38 kDa, which is likely to be devoid of peroxidase activity. Whether the non-self material is natural or introduced experimentally, these two defense responses, phagocytosis and encapsulation, serve to protect a host whose life may be threatened. At the same time, these two mechanisms reveal more specifically aspects of earthworm immunodefense responses that are unrelated to mechanisms of adaptive immunity. This has been referred to as anticipatory, in contrast to nonadaptive or nonanticipatory (Klein 1989).

4 Immune Reactions to Transplants: Coelomocyte Responses

Moving from nonspecific reactions, earthworms have been investigated extensively with respect to their capacity to destroy transplants of body wall (Cooper 1971). The body wall consists of an outer epithelium composed of several types of cells, one of which actively secretes mucus, about which nothing is known concerning its protective properties. A well-developed basement membrane separates the epithelium from longitudinal and circular muscles layers. These, in turn, are separated, only tenuously, by a simple squamous epithelium from the fluid-filled coelomic cavity which contains the immunocompetent coelomocytes. By diapedesis, coelomocytes can readily move into the body wall which easily accommodates their entrance into transplants. Both light and electron-microscope observations have been utilized to analyze histopathologic changes that occur during healing and the subsequent rejection process. By light microscopy observations, the first major change that occurs after wounds are healed is coelomocyte accumulation near graft sites and infiltration into the matrix at 24 h post transplantation, a response that occurs in association with autografts, allografts, and xenografts. This movement of coelomocytes is initially somewhat nonspecific; it is a necessary "cleansing activity", since it ceases in relation to autografts once damaged tissue is removed – an inflammatory response to injury (Cooper 1969a,b; Hostetter and Cooper 1972). In some instances, coelomocytes may respond to xenografts by completely walling them off in an encapsulation-like reaction (Hostetter and Cooper 1972; Parry 1976), the extent of which may vary depending upon the size of the graft bed (Hostetter and Cooper 1972). In contrast to this nonspecific encapsulation response, observation of finer detail using electron microscopy, has revealed that within 1–3 days posttransplantation, the graft matrix is infiltrated by coelomocytes referred to as type-I granulocytes (neutrophils by light microscopy), which destroy the viable inner muscle layers (Linthicum et al. 1977b). In this instance, there is a major distinction between reactions to *self* and *nonself* grafts. Only nonspecific, nonadaptive, transient, inflammatory reactions occur in response to autografts. In contrast, at 5 days these same coelomocytes, after a response to a xenograft, contain numerous phagosomes, residual bodies, and glycogen granules. Later, by 11–13 days, type-I granulocytes have partially destroyed the outer muscle layer of xenografts, leaving the outer epithelium mostly intact. Small lymphocytic (basophilic) coelomocytes are often visible within xenografts and they seem to function primarily as minor scavengers, i.e., they phagocytose small cellular debris, but not viable muscle fragments. This latter observation indicates the primary role of two cell types as effectors in graft rejection.

4.1 Histopathologic Alterations

Analyzing the mechanism of graft rejection reveals somewhat different results when using *E. foetida* as host (Valembois 1974). According to this interpretation, in *E. foetida*, destruction of first-set grafts apparently occurs as a result of autolysis, involving synthesis of lysosomal enzymes by the transplanted tissues. Second-set grafts are, by contrast, destroyed as a result of coelomocyte invasion, as occurs in first-set xenografts of *L. terrestris*. Such reacting coelomocytes which are stimulated and infiltrate the graft are referred to as small leukocytes. Coelomocytes also infiltrate autografts and wounds of sham-operated worms as a nonspecific, general inflammatory response, related to wound healing and removal of damaged tissue, without visible phagocytosis of viable muscle (Linthicum et al. 1977b).

4.1.1 Will Pollution Affect Wound Healing and Graft Rejection?

Goven's group has recently shown that the earthworm is rapidly becoming an excellent model for testing the effects of soil contamination on the immune system (Rodriguez-Grau et al. 1989; Chen et al. 1991; Eyambe et al. 1991; Venables et al. 1992; Goven et al. 1993, 1994; Goven and Kennedy 1995), and similar collaborations have emerged from Cooper's laboratory. Recently, we have exposed *L. terrestris* to a nominal concentration of 10 μg/cm^2 Aroclor 1254 at 15 °C, producing gross pathologic effects, i.e. defects in wound healing and inflammation (Cooper and Roch 1992). The pattern of wound healing could reflect an inhibitory effect of Aroclor on cells that are involved in wound closure and then inflammation; processes that may go on simultaneously. Analyses of host coelomocyte subpopulations stimulated after wounding or grafting, reveals

──────────────────────────────▶

Fig. 1. SEM observations of *E.f. andrei* coelomocytes from unexposed earthworms (*1, 3, 5*) compared to cells from earthworms exposed over 5 days to a concentration of 3 μg/cm^2 of PCBs (*2, 4, 6*). The *bar* corresponds to 1 μm. **1, 2** Macrophages. *1* Note the presence of numerous pseudopodia spreading onto the surface and the contact with a chlorogosome, coming from the degranulation of chloragocytes (*arrowhead*) in the process of phagocytosis. *2* The cytoplasm appears condensed in a round, smooth, cellular body. Note the less extended pseudopodia *arrowhead* when compared to *1*. **3, 4** Leukocytes. *3* Note the presence of numerous microvilli on the cell surface (*arrowheads*). The small granules on the background are chloragosomes coming from the spontaneous degranulation of chloragocytes. *4* After exposure to Aroclor 1254, the cell surface presents flat large short villi (*arrowheads*), less numerous than in *3*. **5, 6** Phagocytosis by macrophages after being in contact with yeast. *5* The macrophage is full of phagocytosed yeasts modifying the cell shape (*arrowheads*). *6* A free yeast (*arrowhead*) can be seen near a nonphagocytosing macrophage.

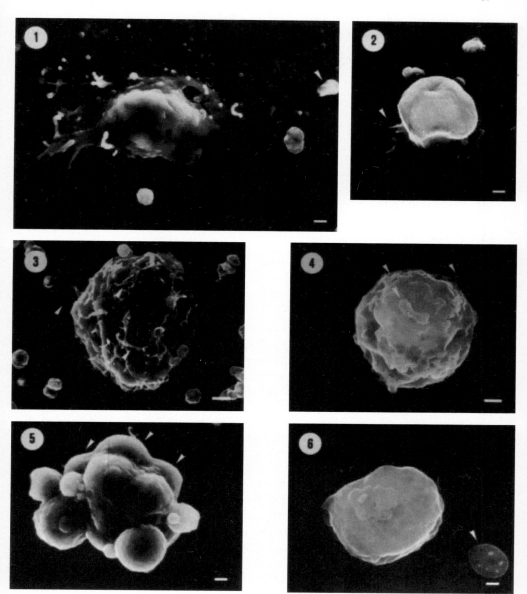

Table 4. Modulation of cellular and humoral-related immunodefense activities in earthworms induced by exposure to PCBs ($\nearrow$ = increased activity; $\searrow$ = decreased activity)

Cellular immunity		Active cells
Wound healing[a]	$\searrow$	Macrophages
Graft rejection[a]	$\nearrow$	Leukocytes
Phagocytosis of yeasts	$\searrow$	Macrophages
elimination of non pathogenic bacteria	$\searrow$	Mainly macrophages

Humoral immunity		Origin of activity
in vivo antibacterial activity (LD50) against pathogenic bacteria	$\nearrow$	Mainly Chloragocytes
in vitro inhibition of bacterial growth	$\nearrow$	Coelomic fluid
Hemolysis	$\nearrow$	Chloragocytes
Lysozyme activity	$\nearrow$	Leukocytes
Proteolysis	$\nearrow$	Chloragocytes

[a]From Cooper and Roch (1992).

two major points: (1) leukocytes are mainly involved in recognition of diverse antigens, including those of grafted tissues; (2) macrophages participate in inflammatory reactions and wound healing (Valembois and Roch 1977; Cooper and Roch 1984). From these results, we suggest that Aroclor exerts a suppressive effect on macrophage activity, as revealed by defects in healing, inflammation, followed by graft rejection (Table 4, Fig. 1).

Quite surprisingly, the highest percentage of allografts showing severe signs of rejection, occurred earlier in PCB exposed worms than in unexposed ones. This contradiction in the usual response was found to be obvious from the first day postgrafting, when 66.6% of allografts were swollen. Thus it is suggested that exposure of worms to PCB accelerates, by one mechanism or another, the allograft rejection process (Cooper and Roch 1986). Signs of autograft rejection were never observed and there was only early, transient inflammation in exposed earthworms. Thus, Aroclor might exert a stimulatory effect on leukocytes that recognize and process foreign or allograft antigens. Alternatively, exposure may perturb an early, essential mechanism that is manifest during the first 24 hours post-transplantation. According to our previous research, such an early mechanism consists of the interaction/cooperation between leukocyte subpopulations (Valembois et al. 1980a,b; Cooper and Roch 1984).

Two interpretations emerge when viewed together. First, we induced defects in the healing process by exposure to Aroclor, but graft rejection in surviving exposed worms was accelerated. Second, from the results of previous experiments, injecting worms at the time of grafting with erythrocytes, which induces agglutinin synthesis (Stein et al. 1982), depresses healing, but normal rejection occurs (Cooper and Roch 1984). In turn, we present two alternative hypotheses.

First, if the healing of wounds is interfered with by an antigen that will induce phagocytosis and/or agglutinin synthesis, then normal graft rejection will occur. Second, if the healing of wounds is interfered with by a potentially detrimental chemical, a modified rejection response will occur; it is accelerated in surviving worms. Obviously, clarification is required when interpreting the effects of xenobiotics on mechanisms that include phagocytosis, wound healing and responses to autografts and allografts. Yet to be analyzed are the reactions to xenografts after exposure. At the cell and molecular level, the mechanisms are still obscure.

4.1.2 Where Do Coelomocytes Come From?

After the earthworm's body wall has been injured either by wounding, auto-grafting or xenografting, there is a significant increase in coelomocyte numbers in association with the xenografts. This facilitates their quantitation by measuring the numbers of participating coelomocytes at different times when they appear in direct association with autografts, wounds, and xenografts. When xenografts from *E. foetida* are grafted onto *L. terrestris* hosts, coelomocyte responses to wounds rise most rapidly within 24 h, decline rapidly, and then return to normal by 72 h (Hostetter and Cooper 1973). Autografts elicit a weaker coelomocytic response, rising only slightly above those of ungrafted worms and returning to normal by day 3. In contrast, coelomocytic responses to xenografts are slower, reach peaks in 3–4 days, decline more slowly, and require 7 days for coelomocytes to return to normal levels. This reveals a different, and specific, response to nonself.

Xenograft destruction is complete in a mean time of 17 days at 20 °C, approximately 10 days after the decline in coelomocyte numbers beneath trans-plants. After the transplantation of second-set grafts at this time (17 days posttransplantation of first xenografts) rejection occurs in an accelerated mean time of 6–7 days, indicating a memory component of the rejection response (Hostetter and Cooper 1973). Quantitation of coelomocytes harvested from second grafts or wounds, varied considerably from those associated with first grafts or wounds. After second xenografts, coelomocyte numbers were 20–30% higher than after first grafts, peaking 1–2 days posttransplantation. This response is regulated by temperature (Cooper and Winger 1975).

4.1.3 Adoptive Transfer of Accelerated Responses

Additional evidence for the crucial role of coelomocytes in graft rejection is based upon adoptive transfer experiments in both *L. terrestris* and *E. foetida* (Duprat 1967; Bailey et al. 1971; Valembois 1971a,b). Host *L. terrestris* were first xenografted with *E. foetida*, then coelomocytes were harvested at 5 days posttransplantation and injected into ungrafted *L. terrestris*. This second host was then xenografted with the same *E. foetida* donor used to induce immunity. Because *L. terrestris* shows only negligible allograft response late after transplantation (Cooper 1969a,b), no strong coelomocytic allo-incompatibility was expected prior to the action of primed coelomocytes against the *E. foetida* graft transplanted to a *L. terrestris* host, (Cooper 1968). Thus, the second host showed accelerated rejection of its first transplant, demonstrating short-term memory and confirming that primed coelomocytes can adoptively transfer, since coelomic fluid alone, free of coelomocytes, is ineffective. Moreover, coelomocytes from unprimed earthworms or from earthworms primed with saline, are also unable to transfer the response, and third-party transplants show independent responses. In this adoptive transfer experiment, we must define the cell type(s) and the mechanism by which it effects the response.

5 Proliferative Response of Coelomocytes to Transplantation Antigens

Burnet (1959) postulated, in his generally accepted dogma of vertebrate immunology, that differentiation of immunocytes to effector functions is inseparable from their proliferation. From this point of view, levels of proliferative activity reflect the course of differentiation, one of the basic mechanisms in immune response. Until now, Burnet's view of clonal activity has not been directly confirmed in the immunology of invertebrates; however, some data support this theory. The mesenchymal lining of the coelomic cavity is considered to be the prime and main source of free coelomocytes (Cameron 1932; Liebmann 1942). However, the existence of a leukocytic organ has been found in megascolecid earthworms (Friedman and Weiss 1982), but surprisingly, the leukopoietic function of this organ has not been substantiated.

While the origin of free coelomocytes seems to be well-documented, a disagreement exists as to whether free coelomocytes proliferate. Some authors suggest that free coelomocytes represent a terminal population that cannot undergo mitotic division, (Liebman 1942; Parry 1976). In contrast, there is evidence that free coelomocytes are able to proliferate under certain conditions (Roch et al. 1975; Toupin and Lamoureux 1976; Toupin et al. 1977; Roch 1979a; Valembois et al. 1980a,b; Cooper and Stein 1981; Bilej et al. 1992a). A basal in vitro level of free coelomocyte proliferation has been found to be low, and varied from less than 1% to 8% (Bilej et al. 1992a).

The induction of a proliferative response after grafting or during wound healing has been reviewed by Parry (1976). While eleocytes (i.e., chloragocytes) remained inactive, the proliferative activity of amoebocytes was detected during xenograft rejection, and surprisingly even during autografting. On the other hand, no ^{3}H-thymidine incorporation was demonstrated in free cells after injury to the body wall or after allograft transfer. Similarly, Roch and coworkers compared the proliferative response of *E. foetida* coelomocytes to allo- and xenografts (Roch et al. 1975). DNA synthesis has been detected in nonadherent cells with maximal values observed on day 4 after grafting; however, in the case of xenografts, a short decline in proliferation was found to occur on day 2. The stimulation of DNA synthesis has also been analyzed in an MLR-like reaction

(Valembois et al. 1980b). They also followed ^{3}H-thymidine incorporation in various allogenic mixtures of coelomocytes and found that 8% of the combinations were positive. The stimulation index (S.I.) varied from 1.4 to 8.9, but generally it registered in the range 3 to 4, values comparable to those obtained in chickens (Miggiano et al. 1974), anuran amphibians (DuPasquier and Miggiano 1973), and trout (Etlinger et al. 1977), but lower values than obtained in mammals (Lonai and McDevitt 1977). The responding, stimulated cells were characterized as lymphocyte-like coelomocytes with a diameter 8 to 12 μm.

Lemmi demonstrated the in vivo synthesis of DNA, by ^{3}H-thymidine incorporation, in coelomocytes of xenografted *L. terrestris* at various times postgrafting, and observed a peak incorporation at 4 days (Lemmi et al. 1974; Lemmi and Cooper 1981; Lemmi 1982). Roch investigated in vitro DNA synthesis by *E. foetida* coelomocytes following wounding, allografting and xenografting. Peak incorporation of ^{3}H-thymidine occurred also at 4 days after all three operations, but was greatest after allografting (with an estimated S.I. of approximately 15), less after xenografting (S.I. about 5) and least after wounding (S.I. about 4). Responses following second grafts or wounds were more complex with a peak at day 2 for xenografts (S.I. about 4), at days 2 and 6 for allografts (S.I. about 4), and at day 6 for wounds (S.I. about 3). Using autoradiographic techniques, xenografting (particularly second sets) was found to stimulate DNA synthesis in greater numbers of small coelomocytes ($< 10 \mu$m) than in larger cells ($> 10 \mu$m) (Figs. 2–5).

The stimulation of coelomocytes by both wounding and grafting is greater than that induced by mitogens (Roch 1977; Valembois and Roch 1977). Four days after xenografting, coelomocytes (13%) incorporated ^{3}H-thymidine in

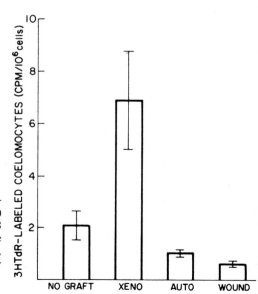

Fig. 2. Histogram representing incorporation of tritiated thymidine (3HTdR) into *Lumbricus terrestris* coelomocytes 48 h after various operative procedures. The values represent counts per minute (cpm), in thousands, per million cells plus or minus standard error

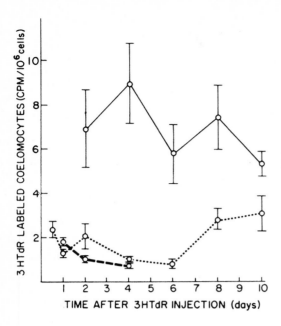

Fig. 3. Concentration of tritiated thymidine (3HTdR) (counts per minute in thousands, per million coelomocytes) at various times after simultaneous grafting and injection of 3HTdR to earthworms. Solid lines represent values obtained from xenografted animals, dashed lines represent values obtained from autografted earthworms, and dotted lines represent values obtained from nongrafted control worms

vitro; after second xenografts fewer (6%) were labeled. After first wounds 6.5% were labeled, and after second wounds, only 4%. Small cells appeared to be stimulated more by grafting than by wounding. Larger coelomocytes were not stimulated by grafts nor by wounds, and were characterized by abundant cytoplasm, a few free ribosomes, polyribosomes, moderate amounts of endoplasmic reticulum, and vesicular elements which incorporated ^{3}H-thymidine more frequently than did small coelomocytes characterized by a high nucleocytoplasmic ratio and many free ribosomes. After second xenografts, however, the situation is reversed, with more of the small cells labeled than large cells. In summary, coelomocytes do respond to mitogens, wounds and grafts with responses which reveal relative specificity. This is a characteristic which strengthens the contention that recognition is an event in the earthworm's primitive immunologic capacity.

According to Lemmi, the proliferative event probably lasts 3–4 days (Lemmi and Cooper 1981). Using thymidine of high specific activity to derange the DNA, in a "thymidine suicide" experiment, revealed no derangement, suggesting that increased uptake of label was due to induced coelomocyte proliferation. Coelomocytes respond in vitro by directional migration to both bacterial and tissue antigens, and the corresponding coelomocytes (Marks et al. 1979) are neutrophils (type-I granulocytes) (Linthicum et al. 1977a).

Using another approach, the effect of antigenic stimulation on proliferative responses has been analyzed by Bilej and co-workers (Bilej et al. 1992a). Coelomocytes collected from earthworms (*E. foetida*) at various time intervals after parenteral administration of a protein antigen, were cultivated in the

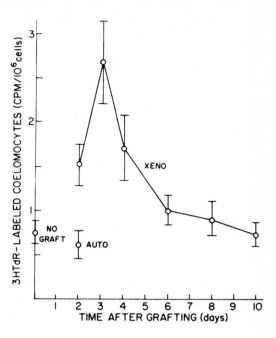

Fig. 4. Duration of the proliferative event as determined by "pulse" labeling of coelomocytes by tritiated thymidine (3HTdR) at various times after grafting. All grafts were performed on day 0. Injection of 3HTdR into the coelomic cavity of earthworms was performed 48 h before harvesting and assay of radioactivity in coelomocytes. Values represent counts per minute, in thousands, per million cells plus or minus standard error

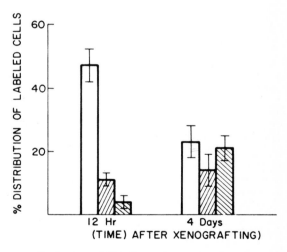

Fig. 5. Histogram representing the percent distribution of "free," nonphagocytic, labeled coelomocytes 12 and 96 h (4 days) after xenografting. Tritiated thymidine (3HTdR) was injected into the coelomic cavity of earthworms at the time of grafting. Clear bars represent the percentage of low-density (<1.031) coelomocytes. Bars with cross-hatched lines slanting upward from left to right represent cells of medium density (1.062), and bars with cross-hatched lines slanting downward from left to right represent the percent distribution of high-density coelomocytes (<1.093). All values are percentage plus or minus standard error

presence of ^{3}H-thymidine. The level of incorporation decreased and was minimal on day 8, i.e., at the time of maximal antigen-binding protein expression. The second stimulation, given on the 16th day, caused a rapid increase in proliferation when coelomocytes from control and antigen-stimulated earthworms were cultivated with or without antigen. The proliferative activity of coelomocytes from stimulated earthworms was significantly lower than the proliferation of the cells observed in control cells. The presence of antigen in the culture medium resulted in decreased levels of ^{3}H-thymidine incorporation by control cells, but augmented proliferation of cells from in vivo prestimulated earthworms. If another (but similar) protein was used for the second in vitro contact, the response reached an intermediate level. In conclusion, results of this study indicate that for successful proliferative response of free coelomocytes, a repeated contact with the same antigen is necessary. In contrast, precursor cells in the mesenchymal lining of the coelomic cavity responded immediately to antigenic stimulation by DNA synthesis. These results suggest that foreign stimuli, in this instance protein antigenic substances, induce proliferation and differentiation of precursor cells which then enter the coelomic cavity to fulfill their defense functions. Only after a second contact with the same antigen can these "primed" or "predetermined" cells undergo further mitotic cycles.

Roch and coworkers have published interesting results that may be relevant, describing the proliferative responses of coelomocytes to concanavalin A (Roch et al. 1975). They found that although all coelomocytes were able to bind Con A on their surfaces, only a small subset (less than 1%) responded to Con A by ^{3}H-thymidine incorporation. Nevertheless, the results obtained by scintillation counting revealed a significant increase in DNA synthesis. Furthermore, as in vertebrates, Con A action was inhibited by carbohydrates, probably due to competition between the carbohydrate and binding sites of coelomocytes for a specific binding site of Con A. Based on the evidence that only 1% of Con A-binding cells were stimulated, it has been postulated that those cell receptors that initially bind to Con A are able to stimulate DNA synthesis only when a second cell receptor interferes with the mitogen or with the first receptor of the mitogen complex. Phytohemagglutinin (PHA) is another mitogen whose effect has been tested in coelomocyte cultures. Augmented ^{3}H-thymidine incorporation after PHA stimulation was detected in nonadherent coelomocytes of *L. terrestris*, (Toupin and Lamoureux 1976; Toupin et al. 1977; Roch 1977) while adherent cells remained inactive. By contrast, in *E. foetida*, large adherent coelomocytes were stimulated more than small nonadherent cells (Toupin et al. 1977).

6 Membrane Components Allied to the Ig Superfamily

All immunological phenomena are no doubt initiated by self/non-self recognition. Although the existence of a histocompatibility-like system in annelids seems to be clear (xeno- and allograft rejection, cell-mediated cytotoxicity, MLR-like reaction etc.), the molecular mechanisms of the recognition are still not understood. Nevertheless, there are data that characterize surface molecules of coelomocytes which can be involved in receptor-mediated recognition. Besides evidence of lectin binding in the formation of T-cell-like rosettes by earthworm coelomocytes (Mohrig et al. 1984), investigations have been focused on finding out whether or not the molecules of the immunoglobulin superfamily are expressed in annelids.

The binding of some vertebrate immunoglobulins to the surface of free coelomocytes is detailed in the paper of Rejnek et al. (1986). These workers observed that sheep and goat IgG molecules and their $F(ab')_2$ fragments reacted with *L. terrestris* and *E. foetida* coelomic fluid. Nevertheless, the antibody binding site is not involved in this reaction and the binding seems to be mediated by some other part of the IgG molecule. Furthermore, approximately 20% of *L. terrestris* and 10% of *E. foetida* coelomocytes were found to bind IgG molecules. The proteins responsible were characterized as 47 and 50 kDa proteins in *L. terrestris* and a 42 kDa protein in *E. foetida*. Immunoblotting analyses of coelomic fluid and coelomocyte lysates revealed that IgG-binding proteins in coelomic fluid and on cell surfaces are similar.

The first putative, analogous molecule followed in annelids was β_2-microglobulin. β_2-microglobulin, present on all mammalian cell types with the exception of mature erythrocytes, is an 11.6 kDa globular protein consisting of 99 amino acid residues in a single polypeptide chain with one intrachain disulfide bond. The exact function of β_2-microglobulin has yet to be determined; however, some results indicate its possible role in the stabilization of class I antigen molecules. Shalev and coworkers demonstrated β_2-microglobulin cross-reactivity in total extracts of several invertebrate species, including earthworms, using radioimmunoassays employing polyclonal antibodies (Shalev et al. 1981, 1983). The authors suggest that the ancestral β_2-microglobulin gene diversified so that the protostomial branch evolved into a "primitive" β_2-microglobulin gene. On the other hand, in deuterostome animals, β_2-microglobulin genes and immunoglobulin genes arose by rapid evolutionary diversification. The relatively slow evolutionary divergence of β_2-microglobulin genes in protostomes, in contrast to rapid diversification in vertebrate species, may be the reason for the considerably high homology of vertebrate β_2-microglobulin and the invertebrate homologue. This homology often exceeds the homology between β_2-microglobulin of mammals and lower vertebrates (Teillaud et al. 1982).

Although the experiments of Shalev's group demonstrated β_2-microglobulin-like molecules in total extracts of earthworms, their analyses did not involve attempts to show the β_2-microglobulin association with coelomocyte membranes. The surface expression of β_2-microglobulin-like molecules has been described by Roch and coworkers (Roch et al. 1983; Roch and Cooper 1983). They found that approximately 50% of coelomocytes, and the whole population of both male and female gonad cells, reacted with the polyclonal antisera tested. The percentage of labeling with monoclonal antibodies was significantly lower, 15% of coelomocytes, and surprisingly, no positivity was shown in gonad cells. The reacting coelomocytes were determined to be weakly adherent acidophils.

Preincubation of cells with β_2-microglobulin prior to the addition of antisera or antibodies, caused a dose-dependent decrease in the level of labeling. Controls with unrelated antibodies were negative, with the exception of normal rabbit serum.

The Thy-1 gene may be closely related in evolutionary history to the ancestral gene for Ig and MHC (Cooper 1986; Williams and Barclay 1988; Stewart 1992) because of its evolutionary significance and structural properties. Finding a Thy-1 homologue in invertebrates may establish whether this molecule is more primitive in evolution than Ig and MHC. Ig and MHC have not as yet been found in species more primitive than vertebrates, except for the primitive Thy-1 homologue of squid, the work on Thy-1 homologue in tunicates, and serological evidence for its presence in earthworms (El Amir et al. 1986; Saad and Cooper 1990).

6.1 Phylogeny of the Immunoglobulin Joining (J) Chain

The J-chain has been detected by a variety of techniques in polymeric Ig of many vertebrate species including representatives of fishes, amphibians, reptiles, birds, and mammals. Furthermore, studies of human cells have revealed that the J-chain may be expressed in lymphocytes from the earliest stage of their differentiation along the B-cell axis. Recently, Takahashi et al. (1992) have examined the phylogeny of J-chain gene expression in various animal species by PCR. RNA was extracted by density centrifugation in cesium trifluoroacetate. The oligonucleotide primers (spanning approximately 300 bp) were chosen from published DNA and amino acid sequences in common to human, mouse, rabbit and bulfrog J-chains. Amplified products were found in earthworm, clam, slug, silkworm, ascidia, lamprey, spotted garpike, African clawed frog, and newt. Southern blot analysis revealed that the amplified 300 bp PCR product from earthworm, hybridized with a human J-chain cDNA probe. Northern blot analysis indicated that the earthworm J-chain gene mRNA also hybridized with the same probe. Thus, their results suggested that the J chain is a phylogenetically primitive component that displays considerable interspecies homologies. Currently, Takahashi and coworkers are analyzing the sequences of earthworm J chain cDNA clones amplified by PCR.

7 Communication Between Immune, Nervous and Endocrine Systems

There has recently been a review of the substances from the nervous system of vertebrates that are also present in the earthworm. According to Kaloustian and Rzasa, earthworms possess Met and Leu-enkephalin-like β-endorphin-like, and gastrin-like immunoreactive peptides (Kaloustian and Rzasa 1986). To test this in an experimental manner in relation to the immune system, Cooper and colleagues have recently found substances that were immunoreactive in an RIA specific for met-enkephalin, and detectable after HPLC fractionation of earthworm coelomic fluid (Cooper et al. 1993). Earthworm coelomocytes and human granulocytes were analyzed for changes in conformation, based on measurements of cellular area and perimeter and expressed mathematically by using the Form Factor (FF). For coelomocytes the form factor decreased following exposure to DAMA, a synthetic enkephalin analog (D-Ala$_2$, Met$_5$-enkephaliamide). DAMA stimulated migration, whereas untreated cells and those exposed to the specific opiate blocker, naloxone, did not move. The enkephalin-like molecule, when exposed to human granulocytes, stimulated an increased number of activated cells. Earthworms have been shown to have demonstrable levels of thymosin alpha$_1$ and thymosin beta$_4$ (Cooper 1991). Clearly these works have a bearing on the concept that the immune, endocrine and nervous systems are interconnected and may have shared a simultaneous evolution (Cooper 1992a,b).

8 Cell-Mediated Cytotoxicity

In 1973, Cooper, using the trypan blue exclusion assay, observed that the viability of cells in xenogeneic culture, decreases rapidly while the cells in allogeneic mixtures are not affected (Cooper 1973a,b). Since coelomocytes must be employed both as effector and target cells, without any possibility of distinguishing them microscopically, the trypan blue exclusion technique does not provide a precise method to study cell-mediated cytotoxicity. A more detailed description of natural and induced cytotoxicity in annelids was reported by Valembois and coworkers (Valembois et al. 1980a). They performed a more sensitive [51]chromium-release assay to compare cytotoxicity in various combinations of allogeneic cultures, and to follow the effect of in vivo prestimulation. A nonadherent fraction of coelomocytes was used both as effector cells and as ^{51}Cr-labeled target cells, and for each culture well both effector and target cells were collected from one earthworm. Although [51]chromium was spontaneously released only slightly in control experiments consisting of a mixture of labeled and nonlabeled cells collected from the same earthworm (6% at 1 h to 19% at 6 h), the release in allogeneic mixtures was significantly higher (15% at 1 h to 39% at 6 h). A spontaneous in vitro cytotoxicity was always detected when cells collected from different subspecies of *Eisenia foetida* (*E. f. foetida* and *E. f. andrei*) were tested, and cytotoxicity was observed even when the cells of *E. f. andrei* obtained from different localities in France were mixed. Other experiments were performed in order to decide if the cytotoxicity can be influenced by in vivo prestimulation. Prestimulation was done by implantation of a small piece of septum into the coelomic cavity of the future donors of effector cells. The results revealed that following a short period after the first dose cell-mediated cytotoxicity increases, and after three implantations it reaches a level comparable with those of untreated earthworms. Cell-mediated cytotoxicity is not caused by the action of proteolytic enzymes, because the values of specific lysis in the presence or absence of inhibitors of proteases are the same. On the other hand, cytotoxicity may be increased, although not significantly, by the addition of a monoclonal antibody against antigen-binding proteins (Tucková et al. 1991a). This is probably due to the binding of Ig molecules to coelomocytes (Rejnek et al. 1986) resulting in a higher level of agglutination and consequently a better contact among cells in culture (Bilej 1993, unpubl.). The cells involved in cytotoxicity are probably small nonadherent coelomocytes, because the large adherent cells and chloragocytes were removed in Valembois's experiments (Valembois et al. 1980a). Nevertheless, the exact cytological observations are still lacking.

9 Mechanisms of Cellular Defense: Adaptive Cellular Response

Obviously the first stage in an immune response is to detect the antigen and to follow its fate after repeated challenge (Bilej et al. 1991b, 1992b). A molecule has been found, in coelomic fluids of *Lumbricus terrestris* and *Eisenia foetida*, which reacts with protein antigens after parenteral stimulation (Tucková et al. 1991a,b; Laulan et al. 1985). This adaptively formed protein has been referred to as antigen-binding protein (ABP) by Tucková. It has been found to be a 56 kDa molecule composed of two disulfide linked polypeptide chains, both of which are believed to participate in the formation of an antigen-binding site. How then is the antigen-binding protein expressed on the surface of free coelomocytes? Superficial binding of an antigen used for in vivo stimulation has been found by direct measurement of radioactivity (Bilej et al. 1990a). After stimulation, there was an increase in both the percentage of antigen-binding cells and the number of antigen-binding sites per single cell. While the percentage of antigen-binding cells reached a level approximately two-times higher on day 8 after in vivo stimulation, the concentration of binding sites was augmented almost three-times. As further evidence, inhibition studies revealed that the coelomocyte's antigenbinding capacity is significantly decreased after preincubating coelomocytes with nonlabeled proteins. To demonstrate the highest inhibiting effect, it was essential to use the same protein as that used for prestimulation. Moreover, comparable data have been obtained by using flow cytometry (Bilej et al. 1990b), and similar results have been observed also with monoclonal antibodies to ABP (Tucková et al. 1991a,b). Indirect fluorescence evaluated by flow cytometry releaved the existence of a small subpopulation of cells that showed an increased density of antigen-binding sites. These highly positive cells did not occur in coelomocyte samples collected from control, non-stimulated earthworms. Interestingly, it disappeared after coelomocytes from stimulated earthworms (*L. terrestris*) were preincubated with nonlabeled proteins. These results have been viewed as excellent, from a quantitative viewpoint; however, there is still the need to identify the cell population involved in antigen recognition.

   In an attempt to solve this question, a hapten-carrier system, labeled with colloidal gold, has been employed for detection by ultrastructural analyses (Bilej et al. 1990c, 1991a,b). The results revealed small numbers of particles in the cytoplasm of clear agranular cells, after the first contact with gold-labeled antigen, when coelomocytes from nonstimulated *E. foetida* earthworms were tested. Coelomocytes were almost always negative; sporadic particles were observed in superficial areas. No differences were found between 1- and 24-h incubation. After a second contract with antigen (when the labeled antigen was added to coelomocytes collected from prestimulated earthworms) significant changes were induced. Gold particles were observed, either free in the cell matrix or associated with intracellular channels. Although the labeling was well-expressed in agranular coelomocytes, its density varied among single cells and it was somewhat higher after 24-h than after 1-h incubation. However, during both incubation times, the levels of labeling significantly surpassed those of non-stimulated cells. In control experiments, the active process of internalization of gold labeled antigen was confirmed since non conjugated gold particles were detected only rarely in coelomocytes. As further support for an active process, there was an increased uptake of gold-labeled antigen by cells from prestimulated earthworms compared with the uptake of nonstimulated earthworms. According to these results, earthworms most probably do possess a cell population (agranular coelomocytes) that is capable of recognizing and internalizing antigen, although the mechanism of antigen processing remains to be solved.

These results were encouraging in the light of additional experiments that had analyzed the fate of administered antigen, both in vivo (Laulan et al. 1985) and in vitro (Bilej et al. 1990). Briefly, protein antigen is rapidly digested by a powerful proteolytic system, and their enzymes are present in the coelomic fluid; lower levels of intracellular proteases were also demonstrated. Labeled antigen administered into the coelomic cavity disappeared from coelomic fluid after 5 days, (Laulan et al. 1985). As early as 2 h after antigen injection, 40–50% of the antigen present in the coelomic fluid was cleaved into the form of nonprecipitable fragments, and after 24 h this proportion was 70–80%. In contrast, antigen present in free coelomocytes was more intact. However, the decline in the amount of antigen and the number of its fragments was higher than in coelomic fluid.

According to these results, one can consider the rapid release of cleaved antigen from coelomocytes into the coelomic fluid. In vitro experiments confirmed the existence of intracellular and extracellular proteolytic systems (Bilej et al. 1990a). The rate of proteolysis in collected cell-free coelomic fluid was lower in comparison with in vivo evaluations. However, within 48 h, 75–80% of antigen was detected in nonprecipitable form and was also digested in cell cultures of free coelomocytes (in the absence of the coelomic fluid), both intracellularly and by proteolytic enzymes synthesized by coelomocytes and released into culture medium. Moreover, proteolysis has also been detected in supernatants of nonstimulated and antigen-stimulated coelomocyte cultures. Evidence for an inducible mechanism of proteolytic enzymes production is derived from the observation of higher levels of proteolytic activity demonstrated in supernatants from antigen-stimulated coelomocytes.

Two questions provoked the question of how the response is induced. First, there was the rapid digestion of protein antigens introduced into the coelomic cavity (within 1 or 2 days) combined with, second, the relatively delayed antigenic response (maximal levels of antigen-binding protein, ABP, were detected on day 8). As an approach, autoradiograms of transverse sections of earthworms revealed the presence of ^{125}I-labeled antigen within the first 24 h after stimulation in all tissues except the epidermis and cuticle. After 4 days the radioactivity was detected only in chloragogen tissue around the gut and dorsal vein, and in the typhlosole. Moreover, the character of antigen deposition was changed. Instead of the homogeneous dispersion that had been observed on the first day, the antigen was found to be concentrated in interstitial spaces among chloragogen cells where it was often found to be bound apically. These results are in agreement with the description of antigen-induced proliferative response (Bilej et al. 1992a). While the proliferation of free coelomocytes was not increased after first contact with an antigen, precursor cells in the mesenchymal lining responded immediately, as revealed by ^{3}H-thymidine incorporation. Therefore, after antigenic stimulation, precursor cells may recognize and bind antigen that triggers a proliferative response. Consequently, differentiation of these cells is followed in turn by their liberation into the coelomic cavity.

To search for the origins of cells, experiments were performed in which small pieces of chloragogenic tissue were cultivated with antigenic material (Tucková et al. 1995; Bilej unpubl. 1993). For the method of cultivating tissue explants, they modified the one first described by Janda and Bohuslav (1934). The observed release of cells from tissue into culture medium is in agreement with the concept of the precursor role of mesenchymal tissue. Furthermore, the presence of antigen-binding protein in the culture medium was evidenced in antigen-stimulated cultures. Surprisingly, the antigen-binding protein response occurred even if small proteolytic fragments of antigen (<10 kDa) predigested by the coelomic fluid were tested. In contrast, larger fragments (>10 kDA) did not provide a sufficient stimulating signal. The addition of the serine protease inhibitor Pefabloc to intact antigen, resulted in the absence of antigen-binding protein formation. Thus, it was assumed that antigen must undergo processing before the antigenic response is triggered.

10 Lysozyme-Like Substances

Vertebrate lysozyme is a light globular polypeptide which splits mucopeptides in bacterial cell walls. N-acetylmuramidases, which functionally resemble vertebrate lysozyme but structurally are quite different, have been observed in invertebrates. The lysozyme-like activity in the coelomic fluid and coelomocytes of the earthworm *Eisenia foetida*, has been detected by Cotuk and Dales (1984a,b) and Lassalle et al. (1988). The activity was connected with a 15-kDa protein, reacting against both Gram-positive and Gram-negative bacteria, having pH optimum at pH 6.2. Vaccination with fixed bacterial suspension resulted in enhanced levels of lysozyme-like activity in the coelomic fluid.

Hirigoyenberry et al. (1990, 1992) performed a kinetic study observing levels of antibacterial substances in *E. foetida*. From the results, they suggested that antibacterial defense is effected by two different humoral systems (lysozyme-like and heat-sensitive antibacterial system) supported by cellular mechanisms (phagocytosis). After bacterial infection, lysozyme-like and nonlysozymal antibacterial activities are increased reaching the maximal levels at 4 h and 3 days, respectively. Both humoral responses were accompanied by RNA and protein synthesis, though basal levels of stable RNAs could be detected in control noninfected earthworms. In summary, at least two bacteriolytic factors are involved in humoral defense of annelids: one corresponds to *"Eisenia foetida andrei* factor" (EFAF), (Roch et al. 1981) and the second resembles vertebrate lysozyme. Both these factors, combined with phagocytic cellular defense can easily prevent the coelomic flora from excessive proliferation.

10.1 Hemolytic and Hemagglutinating Systems of the Coelomic Fluid

Since 1968, when the hemolytic thermo-labile lipoprotein produced by *Eisenia foetida andrei* chloragocytes was first described by Du Pasquier and Duprat (1968), there have been several reports on hemolytic systems in earthworms (DuPasquier 1971; Chateaureynaud-Duprat and Izoard 1973; Lassègues et al. 1984; Valembois et al. 1984, 1985). Agglutinins have been isolated from lysins in earthworm coelomic fluid by gel filtration followed by chromatography (Roch et al. 1984). Lytic activity has also been demonstrated in the coelomic fluid of *Lumbricus* (Tucková et al. 1986). Nevertheless, the most complex investigation has been performed in *Eisenia foetida*. The *E. f. andrei* coelomic fluid has been subjected to polyacrylamide gel electrophoresis (PAGE) under native conditions, and hemolytic activity has been estimated directly on the gel (Roch et al. 1981). Two clear zones of hemolysis of sheep erythrocytes were detected corresponding to molecular weights of 40 and 45 kDa. Using analytical isoelectric focusing combined with erythrocyte overlay, Roch succeeded in characterization of four different proteins of pI ranging from 5.9 to 6.3. All worms possess either two or three isoforms, with one isoform invariably present. In the European population of earthworms, three proteins of pI 5.9, 5.95, and 6.3 are coded by the same gene with three allelic forms. The combinations of three alleles (a, b, and c) can give rise to six genotypes (three homozygous aa, bb, cc, and three heterozygous, ab, ac, bc) that correspond to six phenotypes (hemolysin families) (Roch 1979b; Vaillier et al. 1985; Roch et al. 1986, 1987, 1991a). The significant bacteriostatic activity (against Gram-positive *Bacillus megaterium* and Gram-negative *Aeromonas hydrophila*) was found only in phenotypes B (bb genotype) and K (nonelucidated genotype). While the B family is the most frequent, the frequency of the K-phenotype is intermediate. Furthermore, the evidence of a relatively low bacteriostatic activity in some frequent phenotypes indicates that alternative mechanisms may be involved in antibacterial defense (Roch et al. 1991a,b). Rather different results have been observed in a population that originated from Californian ancestors. One third of earthworms expressed a fourth allele (d) that has never been found in the European population. The occurrence of the allele d in the Californian population is explained as a very rate allele in the ancestral European worms. After migration in a new biotope during the fifteenth century, it acquired more favorable conditions for its expression (Valembois et al. 1986).

10.1.1 Cytolytic Substances

Coelomic fluid of some annelid species not only exhibits the hemolytic and bacteriostatic activities, but cytolytic substances have also been observed (Mohrig et al. 1989). The toxic effect of *Eisenia foetida* coelomic fluid on a variety of cell types, like chicken fibroblasts, guinea-pig polymorphonuclear leukocytes, and insect hemocytes, has been described by Kauschke and Mohrig (1987a,b). In contrast, it has been found that the coelomic fluid does not affect the viability of coelomocytes of other lumbricids nor the cells of some mollusks, nematodes, and protozoans. The cytolytic activity has been characterized as having properties similar to those of the hemolytic activity. For example, both hemolytic and cytolytic substances are thermo-labile, absorbable by sheep erythrocytes, and lyse the cytoplasmic membrane of the target cell. Based on these similarities, the authors suggest that some compounds of the hemolytic and cytolytic systems may be identical. Their view can be supported by the fact that three out of seven isolated hemolytic fractions possess cytolytic activity.

As an extension of previous work (Valembois et al. 1982b; Lassègues et al. 1989) Lassègues and Valembois (1995) have recently launched a structural study of a glycoprotein system involved in immune defense. Their previous studies supported the assumption of a main role played by two glycoproteins of respectively 40 and 45 kDa in the immune defense of *Eisenia fetida andrei*. These two multifunctional molecules, having close structural and functional relationships, are known to mediate hemolysis, bacteriolysis, agglutination, and peroxidation. Their synthesis by chloragocytes, which are free cells in the coelom, is stimulated in response to an injection of bacteria into the coelomic cavity. Lassègues and Valembois have produced a cDNA library in order to elucidate the structural particularities of the two molecules, as well as to form an understanding of their mechanism of action at the molecular level. A strand of 1 440 nucleotides was cloned and sequenced. The molecular weight of the peptide coded by the 1 440 nucleotide sequence was calculated to be 34 kDa, a value equivalent to that of the peptide component obtained after deglycosylation of the two glycoproteins of 40 and 45 kDa. The peptide chain, thus determined, contains, in position 50–60, a sequence very similar to the proximal heme-binding ligand of several peroxidases. An attempt to obtain fusion proteins using different expression systems is now in progress in order to determine the respective roles of the peptidic chain and the carbohydrate and acyl components in hemolytic and bacteriolytic activities.

10.1.2 Adaptively Formed Substances

Since the sixties, when the results of transplantation experiments in earthworms were published, the interest of invertebrate immunologists has extended to determine whether earthworms and other invertebrates are able to adaptively form specific substances in response to an antigenic stimulation. For example, the inducible 48 kDa protein, called hemolin, that has been detected in the giant silk moth, *Hyalophora cecropia*, has been subjected to molecular analysis, revealing that this protein belongs to the immunoglobulin superfamily (Sun et al. 1990). This makes it the first invertebrate molecule which is related to the immune system to be classified as such. Returning to earthworms, the inducible character of their agglutinins has also been described (Wojdani et al. 1982; Stein and Cooper 1981, 1982, 1983, 1988; Stein et al. 1980, 1982, 1986, 1990; Cooper 1981, 1982a). The antibacterial molecules in *Lumbricus terrestris* are less well documented. Stein and coworkers detected substances agglutinating both Gram-positive and Gram-negative bacteria. Each of the bacterial strains reacted with a different agglutinin or combination of agglutinins. Furthermore, after vaccinating earthworms, there was a significant increase in agglutinin levels within 24 h. Inducible characteristics of antibacterial molecules have been confirmed by Anderson (1988) and Kauschke and Mohrig (1987a,b). These can occur even after nonspecific stimulation with rabbit erythrocytes. There is one dilemma; it is difficult to conclude whether the increase in agglutinin levels was caused by a specific stimulus or by a nonspecific increase in coelomic fluid protein levels (Tucková et al. 1988).

When analyzing an immune response, it is clearly essential to know something about the nature of the antigen and its fate (Valembois et al. 1973; Rejnek et al. 1991, 1993; Richman et al. 1982). The first attempt at a detailed analysis of adaptively synthesized substances was performed by Laulan et al. (1985). They used two synthetic haptens (intensain-398 daltons and AXAB152-416 daltons; both are synthetic drugs in mammals) that most probably had never been in contact with experimental earthworms, and two carrier proteins, bovine serum albumin (BSA) and keyhole limpet hemocyanin (KLH). In response to a parenteral immunization, earthworms (*L. terrestris*) formed specific molecules that bound a hapten-carrier complex used previously for immunization. The significance of the response depended on the amount of introduced antigen and on the carrier protein (the amount of antihapten substances that appeared after in vivo immunization with intensain-KLH complex was twice that which appeared with intensain-BSA). The kinetics of antihapten substance formation revealed that the response reached a maximal level between the 5th and 8th day after the first immunization. The second administration of the same hapten-carrier complex (on day 24 after the first immunization, when the response had decreased to control levels) resulted in a faster and more intense response. The reciprocal tests yielded a good estimation of the specificity of antihapten substances. When the coelomic fluid collected from earthworms immunized with one hapten was tested for the binding of the second one, the values reached those detected in non-immunized earthworms.

Similar results have been published by Tucková and her co-workers (Tucková et al. 1988, 1991a,b). They used arsanilic acid as a hapten coupled with human serum albumin (ARS-HSA). The injection of ARS-HSA into the coelomic cavity of *L. terrestris* and *E. foetida* led to a marked increase in coelomic fluid protein concentration and to the formation of a protein which bound the stimulating antigen. As mentioned earlier, antigen-binding protein (ABP) was isolated by affinity chromatography and retained its original binding activity. The molecular weight of both the native ABP in coelomic fluid and the isolated ABP was 56 kDa, when analyzed by SDS-PAGE. Under reducing conditions, two bands with molecular weights of 31 and 33 kDa were detected. After further biochemical analyses, the 56-kDa protein was found to consist of two disulfide-linked polypeptide chains, both of which are assumed to form the antigen-binding site. The kinetic studies of ABP formation confirmed the results obtained by Laulan et al. (1985). The highest response after the primary stimulation occurred between the 4th and 8th day, and the secondary response was faster and more pronounced. However, the specificity of antigen binding was lower than that reported by Laulan's group.

10.1.3 Tumorostatic Activity of Coelomic Fluid

In addition to synthesizing humoral substances that serve to protect earthworms themselves, earthworms are also the source of other substances which, upon purification, possess certain potential therapeutic actions, notably their capacity to inhibit growth. Nagasawa et al. (1991) have extracted and purified an active compound called lombricine from earthworms (*Lumbricus terrestris*) which inhibits the growth of spontaneous mammary tumors in SHN mice. Lombricine was extracted from dry earthworm skin using 60% methanol, the supernatant was concentrated, subjected to adsorption on an Amberlite IR 120 column, solubilized in NaOH, and concentrated again. Cool extract was then crystallized and recrystallized after washing with methanol. The efficiency of this procedure is relatively low (from 5 kg of dry earthworm skin 10 mg of lombricine can be isolated), but due to the high rate of multiplication of the earthworm biomass, this method represents a sufficient tool for isolating lombricine or other active compounds.

To test its anticancer effects, lombricine was administered either subcutaneously (0.3 mg/0.05 ml of live oil daily) or as a diet (120 mg/kg). As early as day 6 after the first subcutaneous injection a significant retardation of mammary tumor growth was observed, but not in the control group of nontreated mice. The rate of tumor growth during the entire experimental period decreased significantly in the lombricine-injected mice, so that on day 21 tumor size was increased by only 1.88 times, whereas in non-treated mice it was three times larger. Furthermore, a slower growth of preneoplastic mammary hyperplastic alveolar nodules was also observed. Perorally administered lombricine did not cause inhibitory effects to that extent, and significant differences were observed only after 21 days. The ^{1}HNMR spectral analysis revealed that lombricine-treated mice had lower serum levels of lactic acid and glucose, while the urine levels of allantoin creatinine, and creatine were elevated in comparison with those of non-treated mice. Although the mechanism of antitumor activity of lombricine is not clear, the effect is due to an imbalance in homeostasis of the internal milieu. Lower levels of glucose and lactic acid in treated mice indicate that lombricine inhibits an excessive uptake of glucose as an energy source, and that indirectly leads to the improvement of immunosurveillance.

Apparently, there are several antitumor substances present in earthworms. Hrzenjak et al. (1992) have described another substance with potentially anti-tumor activity. They isolated a glycoprotein fraction called GP-90 that displays tumorostatic activity from body homogenates of *Eisenia foetida* and *Lumbricus rubelus*, using chloroform-methanol extraction. For controls, they first injected melanoma B_{16} cells into the hind limb of mice causing a rapid growth of tumors that were palpable 5 days later. Then, treating mice by intraperitoneal injections of 0.28 or 2.8 mg/kg of GP-90 daily, starting on the 5th day a significant decrease in the tumor's growth rate was observed. At 8 days, the relative volume of tumors observed in the experimental mice was almost half that observed in a sham-treated group. In vitro experiments revealed that GP-90 has a stimulatory effect on ^{3}H-thymidine incorporation in melanoma B_{16} and fibrosarcoma CMC_1 cell lines, when cultured in serum-free medium. By contrast, in the presence of 10% of fetal calf serum, GP-90 decreased the proliferation of tumor cells two to four times, it also affected the incorporation of nucleic acids and amino acids and the proliferation of normal splenocytes. Subjected to SDS PAGE, GP-90 was separated into 17 fractions with molecular weights from 24 to 97.4 kDa, and by isoelectric focusing, ten bands were analyzed with pIs from 2.5 to 5.2 and one separate band at pI 8.5. Tests for mutagenicity showed neither mutagenic nor carcinogenic effects of GP-90.

11 Evolution of Immune Responses and Where Earthworms Fit In

This chapter has focused on the immunodefense reactions and other peculiar characteristics among two well-known earthworm species, *Lumbricus terrestris* and *Eisenia foetida* (Cooper 1974, 1979a,b; Hostetter and Cooper 1974). In the context of evolution, this seems highly appropriate since earthworms were one subject among the many of Darwin's prodigious works (Darwin 1881; Cooper 1982b). According to Šíma and Větvička (1990), annelids may be considered a key group of coelomate metazoans, most probably direct descendants of the first segmented coelomates, that evolved through their own evolutionary pathways away from the mainstream toward the higher protostomes and deuterostomes. For this reason, the annelids, and especially the oligochaetes, deserve attention since, from the phylogenetic point of view, they probably occupy an intermediate position. They possess a relatively simple body organization, which has been conserved, that in turn allowed the retention of their present immune system since we assume that fundamentally new defense mechanisms did not evolve (Šíma and Větvička 1990). When thinking about the evolution of immunity, it is imperative that we continuously keep in mind that all types of immunodefense reactions of extant forms, presumably the descendants of extinct species, are optimal for them as long as they continue to survive. Whatever the level that they occupy in the phylogenetic sequence, their various immune systems accommodate and indeed ensure their survival.

Despite a certain progress (albeit sporadic), there is clear evidence for the existence of homology between invertebrate and vertebrate immunity (Marchalonis and Schluter 1990; Cooper et al. 1992). In all probability, the components of immunity, i.e., recognition, memory, and effectors, evolved in these two groups by means of different mechanisms (Šíma and Větvička 1990; Větvička et al. 1994). However, for now, it may be too early in our various approaches to declare definitively that homology does not exist, rather that molecular approaches may provide certain undiscovered relationships. Undoubtedly, the basis of all immune systems known so far is the self/nonself recognition. Yet when we consider this concept and its broadest ramifications, it need not lie at the core of the immune system since recognition is crucial to the basic workings of any system. Transplantation of foreign tissues have been performed and have served to produce a major breakthrough in demonstrating that allopolymorphism is universal. Moreover, it is often found that the histocompatibility reactions are under the control of a single or major histocompatibility locus, which suggest strong analogy, if not homology, with the MHC (Du Pasquier 1974, 1992). Although the reasons are unknown, situations in nature have revealed that competition for space may be one explanation. In other words, territories are maintained by this kind of recognition followed by rejection or killing which in many respects resembles what happens when invertebrate effector cells are faced with tumor targets, as has just been recently accomplished using earthworms (Suzuki and Cooper 1995a,b,c; Suzuki et al. 1995). Moreover, it has also been presumed that the primordial histocompatibility system might avoid parasitism by germ cells of another animal (Buss 1982).

Turning now to the molecular level, the origin of members of the immunoglobulin superfamily presents another problem (Cooper et al. 1992). The variety of molecules used in invertebrates in immunodefense is extremely broad, and these substances are not a simplified version of the ones known in endothermic vertebrates, i.e., only a few have been conserved in vertebrates and structural data suggest that these peptides strongly differ from their mammalian counterparts. Several members of the immunoglobulin superfamily, for whom the reasons for their existence are as yet unknown, have been observed in invertebrates. A nonintegral *Drosophila* membrane protein amalgam has three immunoglobulin-like domains assembled in a single exon, and has a significant homology with the NCAM and V_H domains of the immunoglobulin molecule (Seeger et al. 1988). As mentioned earlier, Sun et al. (1990) analyzed the cDNA of an insect immune protein, hemolin, and showed that it contains four internal repeats which are characteristic of immunoglobulin-like domains.

The most commonly found member of the immunoglobulin superfamily is the Thy-1 antigen. It was demonstrated in tunicates (Cooper and Mansour 1989), earthworms (Saad and Cooper 1990), and locusts (Shalev et al. 1985). The significance of these findings is unknown, as the function of the Thy-1 antigen (a ubiquitous antigen found on brain cells, thymocytes and other cells) is not known. Studies using a strain of mouse referred to as knockout, showed that no major abnormalities of the immune system exist in Thy-1 deficient mice, suggesting the minor, if any, function of Thy-1 antigen, at least in certain analyses of immune function (Haas and Kuhn 1992). Those who would like more in the way of speculation, as well as the history of the Ig superfamily, should consult early work by Williams and Barclay (1988).

Stewart (1992) in his interesting paper on variable region molecules (VRM), approached the problem from a different angle and suggested that immunoglobulins did not arise in evolution to fight infection. This created controversy when compared to the recent proposal of Janeway (1992) modified from an earlier one (Janeway 1989) and which is the exact opposite; the immune system did evolve to prevent infection. Thinking strictly about immunoglobulins and not about other members of the immunoglobulin superfamily, the question as to why circulating immunoglobulins can be found in all living vertebrates (despite sometimes with limited number of classes), but never in any invertebrates could, but need not, be perplexing. Stewart presumed that these VRM began as cell adhesion molecules with a primary function to maintain homeostasis (interacting in cell differentiation and tissue organization, they would play a role in cell-versus-organism conflicts). This view has been challenged (Cohn and Langman 1990; Langman 1992). Although all efforts to find a primitive form of immunoglobulin have been unsuccessful, this does not negate early rootlets that may have been present in invertebrates (Langman 1989).

Acknowledgments. The author expresses appreciation to the following persons for providing access to unpublished material: Professor Pierre Valembois and Dr. Patrice Ville who kindly provided Fig. 1. done in collaboration with ELC.

References

Anderson RS (1988) Bacteriostatic factor(s) in the coelomic fluid of *Lumbricus terrestris.* Dev Comp Immunol 12: 189–194

Bailey S, Miller BJ, Cooper EL (1971) Transplantation immunity in annelids. II. Adoptive transfer of the xenograft reaction. Immunology 21: 81–86

Bang FB (1973) A survey of phagocytosis as a protective mechanism against disease among invertebrates. In: Braun W, Unger J (eds) Non-specific factors influencing host resistance. Karger, Basel, pp 2–10

Beck G, O'Brien RF, Habicht GS, Stillman DL, Cooper EL, Raftos DA (1993) Invertebrate cytokines. III. Invertebrate interleukin-1-like molecules stimulate phagocytosis by tunicate and echinoderm cells. Cell Immunol 146: 284–299

Bilej M, Tuckova L, Rejnek J, Větvička V (1990a) In vitro antigen-binding properties of coelomocytes of *Eisenia foetida* (Annelida). Immunol Lett 26: 183–188

Bilej M, Scheerlinck J-P, VandenDriessche T, De Baetselier P, Větvička V (1990b) The flow cytometric analysis of in vitro phagocytic activity of earthworm coelomocytes *Eisenia foetida* (Annelida). Cell Biol Int Rep 14: 831–837

Bilej M, Větvička V, Tucková L, Trebichavský I, Koukal M, Šíma P (1990c) Phagocytosis of synthetic particles in earthworms. Effect of antigenic stimulation and opsonization. Folia Biol (Prague) 36: 273–280

Bilej M, De Baetselier P, Trebichavsky I, Větvička V (1991a) Phagocytosis of synthetic particles in earthworms: Absence of oxidative burst and possible role of lytic enzymes. Folia Biol 37: 227–233

Bilej M, Rossmann P, VandenDriessche T, Scheerlinck J-P, De Baetselier P, Tuckova L, Větvička V, Rejnek J (1991b) Detection of antigen in the coelomocytes of the earthworm, *Eisenia foetida* (Annelida). Immunol Lett 29: 241–246

Bilej M, Rejnek J, Tucková L (1992a) The interaction of *Staphylococcal* protein A with free coelomocytes of annelids. Cell Biol Int Rep 16: 481–485

Bilej M, Sima P, Slipka J (1992b) Repeated antigenic challenge induces earthworm coelomocyte proliferation. Immunol Lett 32: 181–184

Burke JM (1974a) Wound healing in *Eisenia foetida* (Oligochaeta) I: Histology and ^{3}H-thymidine radiography of the epidermis. J Exp Zool 188: 49–63

Burke JM (1974b) Wound healing in *Eisenia foetida* (Oligochaeta). II. A fine structural study of the role of the epidermis. Cell Tissue Res 154: 61–82

Burke JM (1974c) Wound healing in *Eisenia foetida* (Oligochaeta). III. A fine structural study of the role of non-epidermal tissues. Cell Tissue Res 154: 83–102

Burnet FM (1959) The clonal selection theory of acquired immunity. Cambridge Univ Press, London

Buss LW (1982) Somatic cell parasitism and the evolution of somatic tissue compatibility. Proc Natl Acad Sci USA 79: 5337–5344

Cameron GR (1932) Inflammation in earthworms. J Pathol Bacteriol 35: 933–972

Châteaureynaud-Duprat P, Izoard F (1973) Etude des mécanismes de défense chez *Lumbricus terrestris*. C R Acad Sci Paris 276: 2859

Chen SC, Fitzpatrick LC, Goven AJ, Venables BJ, Cooper EL (1991) Nitroblue tetrazolium dye reduction by earthworm *Lumbricus terrestris* coelomocytes: an enzyme assay for non-specific immunotoxicity of xenobiotics. Environ Toxicol Chem 10: 1037–1043

Cohn M, Langman R (1990) The protecton: the unit of humoral immunity selected by evolution. Immunol Rev 115: 1–131

Cooper EL (1968) Transplantation immunity in annelids. I. Rejection of xenografts exchanged between *Lumbricus terrestris* and *Eisenia foetida*. Transplantation 6: 322–337

Cooper EL (1969a) Chronic allograft rejection in *Lumbricus terrestris*. J Exp Zool 171: 69–73

Cooper EL (1969b) Specific tissue graft rejection in earthworms. Science 166: 1414–1415

Cooper EL (1971) Phylogeny of transplantation immunity. Graft rejection in earthworms. Transplant Proc 3: 214–216

Cooper EL (1973a) Evolution of cellular immunity In: Braun W, Ungar J (eds) Non-specific factors influencing host resistance. Karger, Basel, pp 11–23

Cooper EL (1973b) Earthworm coelomocytes: role in understanding the evolution of cellular immunity. I. Formation of monolayers and cytotoxicity. In: Rehácek J, Blaškovic D, Hink WF (eds) Proc 3rd Int Coll Invertebr Tissue Cult. Publ House Slovak Acad Sci, Bratislava, pp 381–404

Cooper EL (1974) Phylogeny of leukocytes: Earthworm coelomocytes in vitro and in vivo. In: Lindahl-Kiessling K, Osaba D (eds) Lymphocyte recognition and effector mechanisms. Academic Press, New York, pp 155–162

Cooper EL (1979a) The earthworm coelomocyte: A mediator of cellular immunity. In: Wright RK, Cooper EL (eds) Phylogeny of thymus and bone marrow-bursa cells. Elsevier, Amsterdam, pp 9–18

Cooper EL (1979b) Earthworms and immunology. TIBS 4: 295–296

Cooper EL (1981) Immunity in invertebrates. CRC Crit Rev Immunol 2: 1–32

Cooper EL (1982a) Invertebrate defense systems: an overview. In: Cohen N, Sigel MM (eds) The reticuloendothelial system. A comprehensive treatise 3. Plenum Press, New York, pp 1–35

Cooper EL (1982b) Did Darwinism help comparative immunology? Am Zool 22: 890

Cooper EL (1986) Evolution of histocompatibility. In: Brehelin M (ed) Immunity in invertebrates. Springer, Berlin Heidelberg New York, pp 139–150

Cooper EL (1991) Evolutionary development of neuroendocrineimmune system. Adv Neuroimmunol 1: 83–96

Cooper EL (1992a) Perspectives in neuroimmunomodulation: lessons from the comparative approach. An Biol 1: 169–180

Cooper EL (1992b) Overview of Immunoevolution. Boll Zool 59: 119–128

Cooper EL, Mansour MH (1989) Distribution of Thy-1 in invertebrates and ectothermic vertebrates. In: Reif AE, Schlesinger M (eds) Cell surface antigen Thy-1. Immunology, neurology, and therapeutic applications. Marcel Dekker, New York, pp 197–219

Cooper EL, Roch P (1984) Earthworm leukocyte interactions during early stages of graft rejection. J Exp Zool 232: 67–72

Cooper EL, Roch P (1986) Second-set allograft responses in the earthworm *Lumbricus terrestris*. Transplantation 41: 514–520

Cooper EL, Roch P (1992) The capacities of earthworms to heal wounds and to destroy allografts are modified by polychlorinated biphenyls (PCB). J Invertebr Pathol 60: 59–63

Cooper EL, Stein EA (1981) Oligochaetes. In: Ratcliffe NA, Rowley AF (eds) Invertebrate blood cells. Academic Press, New York, pp 75–140

Cooper EL, Winger LA (1975) Transplantation immunity in annelids. III. Effects of temperature on xenograft rejection in earthworms. Am Zool 15: 7–11

Cooper EL, MacDonald HR, Sordat B (1979) Separation of earthworm coelomocytes by velocity sedimentation. In: Ruchholtz WM, Muller-Hermelink HK (eds) Function and structure of the immune system. Plenum Press, New York, pp 101–106

Cooper EL, Rinkevich B, Uhlenbruck G, Valembois P (1992) Invertebrate immunity: another viewpoint. Scand J Immunol 35: 247–266

Cooper EL, Leung MK, Suzuki MM, Vick K, Cadet P, Stefano GB (1993) An enkephalin-like molecule in earthworm coelomic fluid modifies leukocyte behavior. Dev Comp Immunol 17: 201–209

Çotuk A, Dales RP (1984a) The effect of the coelomic fluid of the earthworm *Eisenia foetida* Sav. on certain bacteria and the role of the coelomocytes in internal defense. Comp Biochem Physiol 78A: 271–275

Çotuk A, Dales RP (1984b) Lysozyme activity in the coelomic fluid and coelomocytes of the earthworm *Eisenia foetida* Sav. in relation to bacterial infection. Comp Biochem Physiol 78A: 469–474

Dales RP, Kalac Y (1992) Phagocytic defense by the earthworm *Eisenia foetida* against certain pathogenic bacteria. Comp Biochem Physiol 101A: 487–490

Darwin CR (1881) The formation of vegetable mould through the action of worms with observations on their habits. Murray, London

Du Pasquier L (1971) Etude comparée d'un facteur cytolytique humoral chez une larve d'amphibien et chez un Oligochète. Arch Zool Exp Gen 112: 81–84

Du Pasquier L (1974) The genetic control of histocompatibility reactions: phylogenetic aspects. Acad Biol Brux 85: 91–103

Du Pasquier L (1992) Origin and evolution of the vertebrate immune system. APMIS 100: 383–392

Du Pasquier L, Duprat P (1968) Aspects humoraux et cellularies d'une immunité naturelle non-spécifique chez l'Oligochète *Eisenia foetida* (Oligochaeta). C R Acad Sci Paris 266: 538–546

Du Pasquier L, Miggiano VC (1973) The mixed leukocyte reaction in the toad *Xenopus laevis*. A family study. Transplant Proc 3: 1457–1461

Duprat P (1967) Etude de la prise et du maintien d'un greffon de paroi du corps chez le lombricien *Eisenia fetida*. Ann Inst Pastèur 113: 867–881

El Amir A, Saad AH, El Deeb S, Wahby AF, Soliman AW, Cooper EL (1986) Serological evidence for a Thy-1 homolog in earthworms. Proc Zool Soc AR Egypt 12: 287–302

Etilinger HM, Hodgins HO, Chiller JM (1977) Evolution of the lymphoid system. II. Evidence for immunoglobulin determinants on all rainbow trout lymphocytes and demonstration of mixed lymphocyte reaction. Eur J Immunol 7: 881–887

Eyambe GS, Goven AJ, Fitzpatrick LC, Venables BJ, Cooper EL (1991) Extrusion protocol for use in chronic immunotoxicity studies with earthworm *Lumbricus terrestris* coelomic leukocytes. Lab Anim 25: 61–67

Fitzpatrick LC, Goven AJ, Venables BJ, Rodriguez J, Cooper EL (1990) Earthworm immunoassay for evaluating biological effects of exposure to hazardous materials. In: Sandh S, Lower WR, de Serres FJ, Suk WA, Tice RR (eds) In situ. evaluation of biological hazards of environmental pollutants. Plenum Press, New York, pp 119–129

Friedman MM, Weiss L (1982) The leukocytic organ of the megascolecid earthworm *Amynthas diffringens* (Annelida, Oligochaeta). J Morphol 174: 251–268

Goven AJ, Kennedy J (1995) Environmental pollution and toxicity in invertebrates: an earthworm model for immunotoxicology. In: Cooper EL (ed) Invertebrate immune responses. ACEP 24. Springer, Berlin Heidelberg New York (in press)

Goven AJ, Eyambe GS, Fitzpatrick LC, Venables BJ, Cooper EL (1993) Cellular biomarkers for measuring toxicity of xenobiotics: effects of polychlorinated biphenyls on earthworm *Lumbricus terrestris* coelomocytes. Environ Toxicol Chem 12: 863–870

Goven AJ, Chen SC, Fitzpatrick LC, Venables BJ (1994) Lysozyme activity in earthworm (*Lumbricus terrestris*) coelomic fluid and coelomocytes: an enzyme assay for immunotoxicity of xenobiotics. Environ Toxicol Chem 13: 607–613

Haas W, Kuhn R (1992) Knock out mice models for immunodeficiency diseases. In: Gergeley J, Benczur M, Erdei A, Falus A, Fust G, Medayesi G, Petranyis G, Rajnavolgyi E (eds) Progress in immunology, vol II. Springer, Budapest, p 561

Hirigoyenberry F, Lassalle F, Lassegues M (1990) Antibacterial activity of *Eisenia fetida andrei* coelomic fluid: Transcription and translation regulation of lysozyme and proteins evidenced after bacterial infestation. Comp Biochem Physiol 95B: 71–75

Hirigoyenberry F, Lassègues M, Roch P (1992) Antibacterial activity of *Eisenia fetida andrei* coelomic fluid: immunological study of the two major antibacterial proteins. J Invertebr Pathol 59: 69–74

Hostetter RK, Cooper EL (1972) Coelomocytes as effector cells in earthworm immunity. Immnol Commun 1: 155–183

Hostetter RK, Cooper EL (1973) Cellular anamnesis in earthworms. Cell Immunol 9: 384–392

Hostetter RK, Cooper EL (1974) Earthworm coelomocyte immunity. In: Cooper EL (ed) Contemporary topics in immunoglobiology 4. Plenum Press, New York, pp 91–107

Hrzenjak T, Hrzenjak M, Kasuba V, Efenberger-Marinculic P, Levanat S (1992) A new source of biologically active compounds: earthworm tissue (*Eisenia foetida, Lumbricus rubelus*). Comp Biochem Physiol 102A: 441–447

Janda V, Bohuslav P (1934) Sur l'explantation du tissue de la paroie intestinale et des amebocytes de *Lumbricus terrestris* L. et des cellules d'epithelium intestinal d'*Anodonta cygnaee* L. Publ Fact Sci Univ Charles 133: 1–23 (in Czech with French Summary)

Janeway CA Jr (1989) Approaching the asymptote? Evolution and revolution in immunology. Cold Spring Harbor Symp Quant Biol 54: 1–13

Janeway CA Jr (1992) The immune system evolved to discriminate infectious nonself from non infectious self. Immunol Today 13: 11–16

Kaloustian KV, Rzasa PJ (1986) Immunochemical evidence on the occurrence of opioid and gastrin-like peptides in tissues of the earthworm *Lumbricus terrestris*. In: Stefano GB (ed) CRC handbook of comparative opioid and related neuropeptide mechanisms, vol 1. CRC Press, Boca Raton, pp 73–85

Kauschke E, Mohrig W (1987a) Comparative analysis of hemolytic and hemagglutinating activities in the coelomic fluid of *Eisenia foetida* and *Lumbricus terrestris* (Annelida, Lubricidae). Dev Comp Immunol 11: 331–341

Kauschke E, Mohrig W (1987b) Cytotoxic activity in the coelomic fluid of the annelid *Eisenia foetida* Sav. J Comp Physiol B 157: 77–83

Keilin ND (1925) Parasitic autotomy of the host as a mode of liberation of coelomic parasites from the body of the earthworm. Parasitology 17: 170–172

Klein J (1989) Are invertebrates capable of anticipatory immune response? Scand J Immunol 29: 499–505

Langman RE (1989) The immune system. Evolutionary principles guide our understanding of this complex biological defense system. Academic Press, New York

Langman RE (1992) Comment. Immunol Today 13: 399–400

Lassalle F, Lassègues M, Roch P (1988) Protein analysis of earthworm coelomic fluid IV. Evidence, activity, induction and purification of *Eisenia fetida andrei* lysozyme (Annelida). Comp Biochem Physiol 91B: 187–192

Lassègues M, Valembois P (1994) Structural study of a glycoprotein system involved in immune defense in an earthworm. Proc 6th ISDCI Congr, Wageningen. Pergamon Press, New York, 18 Suppl 1: S122

Lassègues M, Roch P, Cadoret MA, Valembois P (1984) Mise en évidence de protéines hémolytiques et hémagglutinantes spécifiques de l'albumen des cocons du Lombricien *Eisenia fetida andrei*. C R Acad Sci Paris Ser III 299: 691–696

Lassègues M, Roch P, Valembois P (1989) Antibacterial activity of *Eisenia fetida andrei* coelomic fluid: Evidence, induction and animal protection. J Invertebr Pathol 53: 1–6

Laulan A, Lestage J, Bouc AM, Chateaureynaud-Duprat P, Fontaine M (1983) Mise en evidence de substances contenues dans le liquide coelomique de *Lumbricus terrestris* possédant des fonctions communes avec celles de certains composants du complement humain. Ann Immunol Inst Pasteur 1340: 223–232

Laulan A, Morel A, Lestage J, Delaage M, Chateaureynaud P (1985) Evidence of synthesis by *Lumbricus terrestris* of specific substances in response to an immunization with a synthetic hapten. Immunology 56: 751–758

Laulan A, Lestage J, Bouc AM, Chateaureynaud-Duprat (1988) The phagocytic activity of *Lumbricus terrestris* coelomocytes is enhanced by the vertebrate opsonins: IgG and complement C3b fragment. Dev Comp Immunol 12: 269–278

Lemmi CAE (1982) Characteristics of primitive leukocytes equipped with receptors for xenogeneic grafts. In: Cooper EL, Brazier MAB (eds) Developmental immunology: clinical problems and Aging. Academic Press, New York

Lemmi CAE, Cooper EL (1981) Induction of coelomocyte proliferation by xenografts in the earthworm *Lumbricus terrestris*. Dev Comp Immunol 5: 73–80

Lemmi CA, Cooper EL, Moore TC (1974) An approach to studying evolution of cellular immunity. In: Cooper EL (ed) Contemporary topics in immunobiology. Plenum Press, New York, pp 4, 109–119

Liebmann E (1942) The coelomocytes of Lumbricidae. J Morphol 71: 221–245

Linthicum DS, Stein EA, Marks DH, Cooper EL (1977a) Electron microscopic observations of normal coelomocytes from the earthworm *Lumbricus terrestris*. Cell Tissue Res 185: 315–330

Linthicum DS, Marks DH, Stein EA, Cooper EL (1977b) Graft rejection in earthworms: an electron-microscopic study. Eur J Immunol 7: 871–876

Lonai P, McDevitt HO (1977) The expression of I-region gene products on lymphocytes. I. Demonstration of MLR determinants on T cells. Immunogenetics 4: 17–31

Marchalonis JJ, Schluter SF (1990) On the relevance of invertebrate recognition and defense mechanisms to the emergence of the immune response of vertebrates. Scand J Immunol 32: 13–20

Marks DH, Stein EA, Cooper EL (1979) Chemotactic attraction of *Lumbricus terrestris* coelomocytes to foreign tissue. Dev Comp Immunol 3: 277–285

Miggiano VC, Birgen I, Pink JRK (1974) The mixed leukocyte reaction in chickens. Evidence for control by the major histocompatibility complex. Eur J Immunol 4: 397–401

Mohrig W, Kauschke E, Ehlers M (1984) Rosette formation of the coelomocytes of the earthworm *Lumbricus terrestris* L. with sheep erythrocytes. Dev Comp Immunol 8: 471–476

Mohrig W, Eue I, Kauschke E (1989) Proteolytic activities in the coelomic fluid of earthworms (Annelida, *Lumbricidae*). Zool Jahrb Physiol 93: 303–317

Nagasawa H, Sawaki K, Fujii Y, Kobayashi M, Segawa T, Suzuki R, Inatomi H (1991) Inhibition by lombricine from earthworm (*Lumbricus terrestris*) of the growth of spontaneous mammary tumours in SHN mice. Anticancer Res 11: 1061–1064

Parry MJ (1976) Evidence of mitotic division of coelomocytes in the normal, wounded and grafted earthworm *Eisenia foetida*. Experientia 32: 449–451

Raftos DA, Cooper EL, Habicht GS, Beck G (1991a) Invertebrate cytokines: tunicate cell proliferation stimulated by an interleukin 1-like molecule. Proc Natl Acad Sci 88: 9518–9522

Raftos DA, Stillman DL, Cooper EL (1991b) Interleukin-2 and phytohaemagglutinin stimulate the proliferation of tunicate cells. Immunol Cell Biol 69: 225–234

Raftos DA, Cooper EL, Stillman DL, Habicht GS, Beck G (1992) Invertebrate cytokines II: Release of interleukin-1-like molecules from tunicate hemocytes stimulated with zymosan. Lymphokine Cytokine Res 11: 235–240

Ratcliffe NA, Rowley AF, Fitzgerald SW, Rhodes CP (1985) Invertebrate immunity: basic concepts and recent advances. Int Rev Cytol 97: 183–384

Rejnek J, TuckováL, Šíma P, Kostka J (1986) The proteins in *Lumbricus terrestris* and *Eisenia foetida* coelomic fluids and on coelomocytes reacting with sheep and goat IgG molecules. Dev Comp Immunol 10: 467–475

Rejnek J, Tucková L, Zikán J, Tomana M (1991) The interaction of a protein from the coelomic fluid of earthworms with staphylococcal protein A. Dev Comp Immunol 15: 269–277

Rejnek J, Tucková L, Síma P, Bilej M (1993) The fate of protein antigen in earthworms: study in vivo. Immunol Lett 36: 131–136

Richman DD, Cleveland PH, Oxman MN, Johnson KM (1982) The binding of *Staphylococcal* protein A by the sera of different animal species. J Immunol 128: 2300–2305

Roch P (1977) Réactivité in vitro des leukocytes du lombricien *Eisenia foetida* Sav. a quelques substances mitogéniques. CR Acad Sci Sér D 284: 705–708

Roch P (1979a) Leukocyte DNA synthesis in grafted lumbricids: an approach to study histo-compatibility in invertebrates. Dev Comp Immunol 3: 417–428

Roch P (1979b) Protein analysis of earthworm coelomic fluid. 1. Polymorphic system of the natural hemolysin of *Eisenia fetida andrei*. Dev Comp Immunol 3: 599–608

Roch P (1995) A definition of cytolytic responses in invertebrates In: Cooper EL (ed) Invertebrate immune responses. ACEP 24: Springer, Berlin Heidelberg New York, (in press)

Roch PG, Cooper EL (1983) A β_2-microglobulin-like molecule on earthworm (*L. terrestris*) leukocyte membranes. Dev Comp Immunol 7: 633–636

Roch P, Valembois P, Du Pasquier L (1975) Response of earthworm leukocytes to concanavalin A and transplantation antigens. In: Hildemann WH, Benedict AA (eds) Immunologic phylogeny. Plenum, New York, pp 45–54

Roch P, Valembois P, Davant N, Lassègues M (1981) Protein analysis of earthworm coelomic fluid II. Isolation and biochemical characterization of the *Eisenia fetida andrei* factor (EFAF). Comp Biochem Physiol 69B: 829–836

Roch P, Cooper EL, Eskinazi DP (1983) Serological evidence for a membrane structure related to human β_2-microglobulin expressed by certain earthworm leukocytes. Eur J Immunol 13: 1037–1042

Roch P, Davant N, Lassègues M (1984) Isolation of agglutinins from lysins in earthworm coelomic fluid by gel filtration followed by chromatofocusing. J Chromatogr 290: 231–235

Roch P, Valembois P, Vaillier J (1986) Amino acid composition and relationships of 5 earthworm defense proteins. Comp Biochem Physiol 85B: 747–751

Roch P, Valembois P, Lassègues M (1987) Genetic and biochemical polymorphism of earthworm humoral defenses. In Developmental and comparative immunology. Cooper EL, Langlet C, Bierne J (eds) Alan R Liss, New York, pp 91–102

Roch P, Lassègues M, Valembois P (1991a) Antibacterial activity of *Eisenia fetida andrei* coelomic fluid: III Relationship within the polymorphic hemolysins. Dev Comp Immunol 15: 27–32

Roch P, Stabili L, Pagliara P (1991b) Purification of three serine proteases from the coelomic cells of earthworms (*Eisenia foetida*). Comp Biochem Physiol 98B: 597–602

Rodriguez-Grau JB, Venables BJ, Fitzpatrick LC, Goven AJ, Cooper EL (1989) Suppression of secretory rosette formation by PCBs in *Lumbricus terrestris*: an earthworm immunoassay for humoral immunotoxicity of xenobiotics. Environ Toxicol Chem 8: 1201–1207

Saad A-H, Cooper E (1990) Evidence for a Thy-1 like molecule expressed on earthworm leukocytes. Zool Sci 7: 217–222

Seeger MA, Haffley L, Kaufman TC (1988) Characterization of amalgam: a member of the immuno-globulin superfamily from *Drosophila*. Cell 55: 589–600

Seymour J, Nappi A, Valembois P (1995) Characterization of a phenoloxidase of the coelomic fluid of the eartworm *Eisenia fetida andrei* by electrochemical detection and electrophoresis. Anim Biol (in press)

Shalev A, Greenberg AH, Lögdberg L, Björck L (1981) β_2-microglobulin-like molecules in low vertebrates and invertebrates. J Immunol 127: 1186–1191

Shalev A, Pla M, Ginsburger-Vogel T, Echalier G, Lögdberg L, Björck L, Colombani J, Segal S (1983) Evidence for β_2-microglobulin-like and H-2-like antigenic determinants in *Drosophila*. J Immunol 130: 297–302

Shalev A, Segal S, Eli MB (1985) Evolutionary conservation of brain Thy-1 glycoprotein in vertebrates and invertebrates. Dev Comp Immunol 9: 494–506

Šíma P, Větvička V (1990) Evolution of Immune Reactions. CRC Press, Boca Raton

Stein EA, Cooper EL (1978) Cytochemical observations of coelomocytes from the earthworm *Lumbricus terrestris*. Histochem J 10: 657–678

Stein EA, Cooper EL (1981) The role of opsonins in phagocytosis by coelomocytes of the earthworm *Lumbricus terrestris*. Dev Comp Immunol 5: 415–425

Stein EA, Cooper EL (1982) Agglutinins as receptor molecules: a phylogenetic approach. In: Cooper EL, Brazier MAB (eds) Developmental immunology: clinical problems and aging. Academic Press, New York, pp 85–98

Stein EA, Cooper EL (1983) Carbohydrate and glycoprotein inhibitors of naturally occurring and induced agglutinins in the earthworm *Lumbricus terrestris*. Comp Biochem Physiol 76B: 197–206

Stein EA, Cooper EL (1988) In vitro agglutinin production by earthworm leukocytes. Dev Comp Immunol 12: 531–547

Stein EA, Avtalion RR, Cooper EL (1977) The coelomocytes of the earthworm *Lumbricus terrestris*: morphology and phagocytic properties. J Morphol 153: 467–476

Stein EA, Moravati A, Rahimian P, Cooper EL (1980) Lipid agglutinins from coelomic fluid of the earthworm, *Lumbricus terrestris*. Comp Biochem Physiol 94B: 703–707

Stein EA, Wojdani A, Cooper EL (1982) Agglutinins in the earthworm *Lumbricus terrestris*: naturally occurring and induced. Dev Comp Immunol 6: 407–421

Stein EA, Younai S, Cooper EL (1986) Bacterial agglutinins of the earthworm, *Lumbricus terrestris*. Comp Biochem Physiol 84B: 409–415

Stein EA, Younai S, Cooper EL (1990) Separation and partial purification of agglutinins from coelomic fluid of the earthworm, *Lumbricus terrestris*. Comp Biochem Physiol 97B: 701–705

Stewart J (1992) Immunoglobulins did not arise in evolution to fight infection. Immunol Today 13: 396–395

Sun SC, Lindström I, Boman HG, Faye I, Schmidt O (1990) Hemolin: an insect-immune protein belonging to the immunoglobulin superfamily. Science 250: 1729–1732

Suzuki MM, Cooper EL (1995a) Allogeneic killing by earthworm effector cells. Nat Immun (in press)

Suzuki MM, Cooper EL (1995b) Killing of intrafamilial and xenogeneic targets by earthworm effector cells. Immunol Lett 44: 45–49

Suzuki MM, Cooper EL (1995c) Characteristics of effector cells – kinetics and electron microscopy using mammalian targets. Zool Sci (in press)

Suzuki MM, Cooper EL, Eyambe GS, Goven AJ, Fitzpatrick LC, Venables BJ (1995) Effects of exposure to polychlorinated biphenyls PCBs) on natural cytotoxicity of earthworm coelomocytes, vol 14 (No. 10) Environ Toxicol Chem

Takahashi T, Iwase T, Kobayashi K, Rejnek J, Mestecky J, Moro I (1992) Phylogeny of the immunoglobulin joining (J) chain. 7th Int Cogr Mucosal Immunol, Prague, Czech Republic 16–20 August 1992, 234 pp

Teillaud JL, Crevat D, Chardon P, Kalil J, Goujet-Zalc C, Mahouy G, Vaiman M, Fellous M, Pious D (1982) Monoclonal antibodies as a tool for phylogenetic studies of major histocompatibility antigens and β_2-microglobulin. Immunogenetics 15: 377–384

Toupin J, Lamoureux G (1976) Coelomocytes of earthworms: phytohemagglutinin (PHA) responsiveness. In: Wright RK, Cooper EL (eds) Phylogeny of thymus and bone marrow-bursa cells. Elsevier, Amsterdam, pp 19–27

Toupin J, Leyva F, Lamoureux G (1977) Transformation blastique par la PHA des coelomocytes de *Lumbricus terrestris*. Ann Immunol Inst Pasteur 128C: 29–32

Tucková L, Bilej M (1995) Mechanisms of antigen processing in invertebrates: are there receptors? In: Cooper EL (ed) Intertebrate immune responses. ACEP 24. Springer, Berlin Heidelberg New York (in press)

Tucková L, Rejnek J, Šíma P, Ondejová R (1986) Lytic activities in coelomic fluid of *Eisenia foetida* and *Lumbricus terrestris*. Dev Comp Immunol 10: 181–189

Tucková L, Rejnek J, Síma P (1988) Response to parenteral stimulation in earthworms *L. terrestris* and *E. foetida*. Dev Comp Immunol 12: 287–296

Tucková L, Rejnek J, Bilej M, Pospíšil R (1991a) Characterization of antigen-binding protein in earthworms *Lumbricus terrestris* and *Eisenia foetida*. Dev Comp Immunol 15: 263–268

Tucková L, Rejnek J, Bilej M, Hajkova H, Romanovsky A (1991b) Monoclonal antibodies to antigen-binding protein of annelids *Lumbricus terrestris*. Comp Biochem Physiol 100B: 19–23

Tucková L, Bilej M, Rejnek J (1995) The fate of protein antigen in annelids in vivo and in vitro study. Adv Exp Med Biol (in press)

Vaillier J, Cadoret M-A, Roch P, Valembois P (1985) Protein analysis of earthworm coelomic fluid. III. Isolation and characterization of several bacteriostatic molecules from *Eisenia fetida andrei*. Dev Comp Immunol 9: 11–20

Valembois P (1971a) Role des leucocytes dans l'acquisition d'une immunite antigreffe specifique chez les lombriciens. Arch Zool Exp Gen 112: 97–104

Valembois P (1971b) Etude ultrastructurale des coelomocytes du lombricien *Eisenia foetida* Sav. Bull Soc Zool Fr 96: 59–72

Valembois P (1974) Cellular aspects of graft rejection in earthworms and some other metazoa. In: Cooper EL (ed) Contemporary topics in immunology. Plenum Press, New York, pp 75–90

Valembois P, Roch P (1977) Identification par autoradiographie des leucocytes stimules à la suite de plaies ou de greffes chez un ver de terre. Biol Cell 28: 81–82

Valembois P, Roch P, DuPasquier L (1973) Dégradation in vitro de protéine étrangère par les macrophages du Lombricien *Eisenia foetida* Sav. CR Acad Sci Paris Sér III 277: 57–60

Valembois P, Roch P, Boiledieu D (1980a) Natural and induced cytotoxicities in sipunculids and annelids. In: Manning MJ (ed) Phylogeny of immunological memory. Elsevier, Amsterdam, pp 47–55

Valembois P, Roch P, Du Pasquier L (1980b) Evidence of MLR-like reaction in an invertebrate, the earthworm *Eisenia foetida*. In: Solomon JB (ed) Aspects of developmental and comparative immunology. Pergamon Press, Oxford, pp 23–30

Valembois P, Roch P, Boiledieu D (1982a) Cellular denfense system of the Platyhelminths, Nemertinea, Sipunculidea and Annelida. In: Cohen N, Sigel M (eds) The reticuloendothelial system: a comprehensive treatise, vol 3. Plenum Press, New York, pp 89–139

Valembois P, Roch P, Lassègues M, Cassand P (1982b) Antibacterial activity of the hemolytic system from the earthworm *Eisenia fetida andrei*. J Invertebr Pathol 40: 21–27

Valembois P, Roch P, Lassègues M (1984) Simultaneous existence of hemolysins and hemagglutinins in the coelomic fluid and in the cocoon albumen of the earthworm *Eisenia fetida andrei*. Comp Biochem Physiol 78A: 141–145

Valembois P, Lassègues M, Roch P, Vaillier J (1985) Scanning electron microscopic study of the involvement of coelomic cells in earthworm antibacterial defense. Cell Tissue Res 240: 479–484

Valembois P, Roch P, Lassègues M (1986) Antibacterial molecules in annelids. In: Brehélin M (ed) Immunity in invertebrates. Springer, Berlin Heidelberg New York, pp 74–93

Valembois P, Seymour J, Roch P (1991) Evidence and cellular localization of an oxidative activity in the coelomic fluid of the earthworm *Eisenia foetida andrei*. J Invertebr Pathol 57: 177–183

Valembois P, Lassègues M, Roch P (1992) Formation of brown bodies in the coelomic cavity of the earthworm *Eisenia fetida andrei* and attendant changes in shape and adhesive capacity of constitutive cells. Dev Comp Immunol 16: 95–101

Valembois P, Seymour J, Lasségues M (1994) Evidence of lipofuscin and melanin in the brown body of the earthworm *Eisenia fetida andrei*. Cell Tissue Res 277: 183–188

Venables BJ, Fitzpatrick LC, Goven AJ (1992) Earthworms as indicators of ecotoxicity. In: Greig-Smith PW, Becker H, Edwards PJ, Heimbach F (eds) Ecotoxicology of earthworms. Intercept, Andover, pp 197–206

Větvička V (1994) Concluding remarks. In: Větvička V, Šíma P, Cooper EL, Bilej M, Roch P (eds) The immunology of annelids. CRC Press, Boca Raton, pp 281–286

Větvička V, Šíma P, Cooper EL, Bilej M, Roch P (1994) The immunology of annelids. CRC Press, Boca Raton

Ville P, Roch P, Cooper EL, Masson P, Narbonne J-F (1995) PCBs increase molecular-related activities (lysozyme, pathogenic antibacterial, hemolysis, preteases) but inhibit macrophage-related functions (phagocytosis, wound healing) in earthworms. J Invertebr Pathol (in press)

Williams AF, Barclay AN (1988) The immunoglobulin superfamily domains for cell surface recognition. Annu Rev Immunol 6: 381–405

Wojdani A, Stein EA, Lemmi CA, Cooper EL (1982) Agglutinins and proteins in the earthworm, *Lumbricus terrestris*, before and after injection of erythrocytes, carbohydrates, and other materials. Dev Comp Immunol 6: 613–624

The Prophenoloxidase Activating System and Associated Proteins in Invertebrates

M.W. JOHANSSON and K. SÖDERHÄLL

1 Introduction

Invertebrates lack antibodies, lymphocytes or other features of the vertebrate adaptive immune system, but they have innate defence reactions (Ratcliffe et al. 1985). Many of them have an open circulatory system and therefore they need immediate and constitutive mechanisms for the recognition and immobilization of microorganisms and parasites, and for clotting to prevent blood loss upon wounding. Some invertebrates also have inducible defence reactions which are dependent on new or increased synthesis of, for example, antimicrobial proteins, but these will not be dealt with in this chapter.

One candidate for an immediate non-inducible system in invertebrates is the prophenoloxidase activating system, and evidence is accumulating that this system has a role in recognition and defence (for previous reviews see Ashida et al. 1982; Söderhäll 1982; Ratcliffe et al. 1985; Söderhäll and Smith 1986; Johansson and Söderhäll 1989a; Ashida and Yamazaki 1990; Söderhäll et al. 1990; Söderhäll 1992; Söderhäll and Aspán 1993; Söderhäll et al. 1994a). It has long been recognized that defence reactions in many invertebrates are often accompanied by melanization. The key enzyme in melanin formation is phenoloxidase (PO). PO activity has been detected in the haemolymph or coelom of many invertebrate groups, both protostomes and deuterostomes (Table 1). It catalyzes the oxidation of phenols to quinones, which then polymerize non-enzymatically to melanin. The enzyme has, wherever carefully studied, been found to exist in the blood in an inactive pro-form, prophenoloxidase (proPO), that is activated in a stepwise process involving serine protease by microbial cell-wall constituents (Table 1). This specific elicitation of the so-called proPO activating system by certain polysaccharides, e.g. by β-1,3-glucans but not by other glucans (Unestam and Söderhäll 1977; Söderhäll and Unestam 1979), is an argument in favour of this system being a recognition system. The ability of a host to recognize certain conserved structures on microorganisms has been termed pattern recognition (Janeway 1989), so, accordingly, the proPO system may be considered a pattern recognition system.

Department of Physiological Botany, University of Uppsala, Villavägen 6, 75236 Uppsala, Sweden

Table 1. Phenoloxidase activities and prophenoloxidase activation in invertebrate blood

Animal group	Phenoloxidase activity	Activation of prophenoloxidase by		
		β-1,3-glucan	LPS	Peptidoglycan
Annelids:				
Polychaetes	No[a]			
Oligochaetes	Yes[b]			
Molluscs				
Bivalves	No[a]			
Gastropods	Yes[c]			
Brachiopods	Yes[a]			
Arthropods				
Crustaceans	Yes[d]	Yes[e]	Yes[f]	
Insects	Yes[d,g]	Yes[h]	Yes[i]	Yes[j]
Onycophorans	Yes[k]			
Chelicerates	No[l]			
Echinoderms				
Crinoids	?[m]			
Echinoids	Yes[a]			
Asteroids	Yes[a]			
Holothurians	Yes[n]	Yes[o]		
Urochordates				
Ascidians	Yes[a]	Yes[p]	Yes[p]	

[a]Smith and Söderhäll (1991); [b]Valembois et al. (1988); [c]De Aragao and Bacila (1976); [d]for review, see Söderhäll (1982); [e]Unestam and Söderhäll (1977); [f]Söderhäll and Häll (1984); [g]for review, see Ashida et al. (1982); [h]Ashida et al. (1983); [i]Saul and Sugumaran (1987); [j]Yoshida and Ashida (1986); [k]Krishnan and Ravindranath (1973); [l]Söderhäll et al. (1985); [m]Smith (1991); [n]Canicatti and Seymour (1991); [o]Canicatti and Götz (1991); [p]Jackson et al. (1993).

Several observations have indicated that the proPO system also is a defence system. First, melanin, or reactive compounds that are intermediates in its formation, have been shown to be fungistatic (Söderhäll and Ajaxon 1982; St Leger et al. 1988; Rowley et al. 1990). Recently, it has also been claimed that PO generates a factor that immobilizes bacteria (Marmaras et al. 1993) and even that it can generate an antiviral activity (Ourth and Renis 1993). Second, many studies indicate a correlation between elicitation by the microbial polysaccharides and stimulation of cellular defence reactions or cellular responses in vitro or in vivo, including phagocytosis (Smith and Söderhäll 1983a; Goldenberg et al. 1984; Ratcliffe et al. 1984; Leonard et al. 1985a; Söderhäll et al. 1986; Brookman et al. 1988; Anggraeni and Ratcliffe 1991; Smith and Peddie 1992; Thörnqvist et al. 1994), nodule formation (Smith et al. 1984; Gunnarsson and Lackie 1985; Brookman et al. 1989a), encapsulation (Kobayashi et al. 1990), blood-cell adhesion (Johansson and Söderhäll 1988, 1992), degranulation (Smith and Söderhäll 1983b; Söderhäll et al. 1986; Johansson and Söderhäll 1989b) and locomotion (Takle and Lackie 1986; Huxham and Lackie 1988). These observations are compatible with the interpretation that when the proPO system is triggered, a factor (or factors) is activated which in turn binds to blood cells and mediates these cellular activities.

Consistent with the idea of the proPO system as a defence system, there are also reports that certain successful parasites avoid eliciting the system or inhibit its activation or activities. Thus, some entomopathogenic fungi produce wall-less cells (protoplasts) without the β-1,3-glucan elicitor when entering the host blood (Söderhäll 1982). The protoplasts are not encapsulated and do not degranulate host blood cells (Beauvais et al. 1989). Other fungi synthesize toxins that can inhibit β-1,3-glucan-stimulated blood-cell motility (Huxham et al. 1989). Some parasitic wasps, at the time of oviposition, inject viruses into the host that inhibit the activation of proPO or suppress PO activity (Stoltz and Cook 1983; Beckage et al. 1990). Microfilarial infection can reduce melanization in mosquito (Christensen and LaFond 1986), gregarious endoparasitoids suppress PO in their host (Kitano et al. 1990) and entomopathogenic nematodes can inhibit the β-1,3-glucan-triggered activation of the proPO system (Yokoo et al. 1992). Other examples of a correlation between the proPO system and defence are the melanization that occurs in a *Drosophila* tumour mutant, which is regarded as an "autoimmune" reaction (Rizki and Rizki 1980), and the melanization of the malarial parasite in a refractory strain of *Anopheles* (Collins et al. 1986).

In this chapter we will mainly review information on the components of the proPO system and on proteins apparently associated with it, considering especially those cases in which components have been purified and characterized.

2 Components of the Prophenoloxidase Activating System

2.1 Prophenoloxidase

Although phenoloxidase activity has been detected in the blood of many invertebrate groups (Table 1), the enzyme or its inactive pro-form proPO has been isolated only from arthropod blood. The literature contains a confusing array of molecular masses for proPO and PO. However, considering only purified proteins analyzed by sodium dodecyl-sulphate-polyarylamide electrophoresis (SDS-PAGE), blood proPO from the different species seems to have a molecular mass between 70 and 90 kDa and the active PO a molecular mass between 60 and 70 kDa (Table 2). ProPO from the crayfish *Pacifastacus leniusculus* has an isoelectric point of 5.4 and contains two Cu^{2+} ions (Aspán and Söderhäll 1991). Its cDNA was recently cloned and sequenced (Aspán et al. 1995). PO is a "sticky", hydrophobic protein that non-specifically attaches to various surfaces in vitro (Söderhäll et al. 1979; Ashida and Dohke 1980; Leonard et al. 1985b). PO activity has also been claimed to lead to covalent linking of proteins (Ashida and Yoshida 1988). In conclusion, the higher reported weights of proPO and PO are due to aggregation that may or may not be relevant in vivo.

ProPO can be activated through limited proteolysis by various proteases in vitro. Crayfish PO is the only one studied after cleavage by a purified endogenous

Table 2. Molecular masses (kDa)[a] of purified prophenoloxidase, phenoloxidase and prophenoloxidase activating enzyme from invertebrate blood

Species	Prophenoloxidase	Phenoloxidase	Prophenoloxidase activating enzyme
Crustaceans:			
Pacifastacus leniusculus	76[b]	60, 62[b]	36[c]
Insects:			
Blaberus discoidalis	76[d]		
Bombyx mori	80[e,f]	70[f,g]	32 (from cuticle)[f,h]
Calliphora erythrocephala	87[i]		
Drosophila melanogaster	77, 78[j]		
Hyalophora cecropia	76[k]		43, 53[k,l]
Manduca sexta	71, 77[m]		
Musca domestica		60[n]	

[a]Determined by SDS-PAGE; [b]Aspán and Söderhäll (1991), Aspán et al. (1995); [c]Aspán et al. (1990b); [d]Durrant et al. (1993); [e]Ashida (1971); [f]Ashida and Yamazaki (1990); [g]Ashida et al. (1974); [h]Dohke (1973); [i]Naqvi and Karlson (1979); [j]Fujimoto et al. (1993); [k]Andersson et al. (1989); [l]Activates prophenoloxidase only in the presence of another blood protein, C1; [m]Aso et al. (1985); [n]Hara et al. (1993).

activating enzyme from the blood (Aspán and Söderhäll 1991; Table 2). The site of cleavage in crayfish proPO by the endogenous enzyme was recently determined (Aspán et al. 1995). In the case of the silkworm *Bombyx mori*, an enzyme from cuticle was used to activate blood proPO (Dohke 1973; Table 2). PO from the housefly *Musca domestica* was only isolated in an already active form (Table 2). ProPO from the mosquito *Aedes aegypti* was not isolated but studied with an antibody against silkworm proPO (Ashida et al. 1990). Mosquito proPO had a molecular mass of 74 kDa and PO 63 kDa.

At least in vitro, proPO can also be activated without proteolysis by different treatments such as heat or detergents (Ashida and Söderhäll 1984; Leonard et al. 1985b; Söderhäll and Smith 1986; Ashida and Yamazaki 1990), presumably through a conformational change. The biological significance of this observation is unclear. Importantly, these treatments do not activate the whole proPO system or associated proteins, neither serine proteases nor the cell-adhesion protein are activated (see below).

The localization of proPO in different species has been controversial. In various crustaceans it seems clear that proPO is localized in the semigranular and granular blood cells (Söderhäll and Smith 1983; Smith and Söderhäll 1983b; Johansson and Söderhäll 1985). During degranulation, proPO is secreted from these cells in its inactive form and the released proPO system can then be activated outside the cells by the microbial elicitors (Johansson and Söderhäll 1985). Thus, the secretion and the activation are two distinct processes. Also, in some insects, proPO has been found in blood cells (Leonard et al. 1985b) that contain granules (Huxham and Lackie 1988), whereas in others it has been claimed to mainly exist in the plasma (Ashida 1971; Saul and Sugumaran 1987), even though the same authors have also described its synthesis and localization in certain blood cells (Iwama and Ashida 1986; Ashida et al. 1988). More recently, proPO was detected in the morula cells in ascidians (Smith and Söderhäll 1991). Generally, it appears that proPO is stored in the granules of some blood-cell types from which it is released, and that the activation of the proPO system takes place in the plasma.

2.2 Prophenoloxidase Activating Enzyme

As ProPO can be activated by serine proteases, an enzyme responsible for this in vivo should be localized in the blood. Only in two cases does the relevant prophenoloxidase-activating enzyme (ppA) seem to have been isolated (Table 2). From crayfish blood cells an active ppA with a molecular mass of 36 kDa under non-reducing conditions was purified (Aspán et al. 1990b). After reduction, a 28 kDa active fragment was seen as well as smaller fragments. This enzyme probably has a pro-form of unknown size, since in a blood cell homogenate, protease activity of the 36 kDa protein was not seen until after addition of a β-1,3-glucan (Aspán and Söderhäll 1991). In the crude homogenate, a minor band of 38 kDa was also activated in the presence of β-1,3-glucan; in a crude system proPO could also be activated by this band (Aspán and Söderhäll 1991). A fraction with serine protease activity was isolated from blood of the moth *Hyalophora cecropia* (Andersson et al. 1989). It contained two bands of 43 and 53 kDa. However, this fraction could only activate proPO in the presence of another blood protein, called C1, with a molecular mass of 112 kDa.

The cuticle has been used as a source for proPO activating protease from several insects (Dohke 1973; Aso et al. 1985; Saul and Sugumaran 1986, Table 2), but the function of this enzyme in vivo is unclear. However, in the same species, there are also (not yet isolated) serine proteases in the blood that are activated in the presence of the microbial elicitors (Yoshida and Ashida 1986). One of these enzymes should be the relevant ppA. Recently, activation of serine protease from ascidian blood cells by β-1,3-glucan or lipopolysaccharide (LPS) was described (Jackson and Smith 1993) and this enzyme may be the ppA in that system.

It has been reported from several species that in a crude system, protease and proPO are activated at a low Ca^{2+} concentration in vitro in the absence of microbial compounds (Söderhäll 1981; Söderhäll and Häll 1984; Dularay and Lackie 1985; Leonard et al. 1985b). This has been suggested to represent a response to altered ionic composition during wounding.

2.3 Prophenoloxidase Activation Inhibitors

Obviously, there needs to be some means of regulating the proPO system after its activation in order to prevent massive melanization throughout the entire circulatory system of a host. A high molecular mass trypsin inhibitor, which is a quite strong inhibitor of ppA, was purified from plasma of the crayfish *P. leniusculus* and characterized (Hergenhahn et al. 1987; Aspán et al. 1990a; Table 3). Another protease inhibitor, which is a disulfide-linked dimer of 190 kDa subunits and a homologue of vertebrate α_2-macroglobulin, has been purified and partially sequenced from some invertebrates (Table 4; see also Armstrong and Quigley, this Vol.). It has been found that α_2-macroglobulin from crayfish plasma (Hergenhahn et al. 1988) decreased the activity of ppA to some degree, but less efficiently than did the 155 kDa trypsin inhibitor (Aspán et al. 1990a). A fraction which inhibited proPO activation by the 43/53 kDa protease and the

Table 3. Prophenoloxidase activation inhibitors (prophenoloxidase activating enzyme inhibitors) from invertebrate blood

Species	Molecular mass (kDa)	Reference
Crustaceans:		
Pacifastacus leniusculus	155	Hergenhahn et al. (1987); Aspán et al. (1990a)
Insects:		
Locusta migratoria	14[a]	Brehélin et al. (1991)
Manduca sexta	8, 14[b]	Sugumaran et al. (1985); Saul and Sugumaran (1986)
Sarcophaga bullata	5[c]	Sugumaran et al. (1985)

[a] Inhibits trypsin-triggered indirect activation of prophenoloxidase, not β-1,3-glucan-triggered.
[b] Inhibits activation of prophenoloxidase by cuticular protease.
[c] Inhibits chymotrypsin-triggered activation of prophenoloxidase.

Table 4. Invertebrate proteins with a structure similar to those of vertebrate complement proteins

Protein	Structure
	thiolester
α_2-Macroglobulin[a,b,c]	
Factor C[d]	

▦ Complement-like "Sushi" domain
▤ C-type lectin domain
▨ serine protease domain

[a] Spycher et al. (1987) [*Homarus americanus*].
[b] Hall et al. (1989) [*Pacifastacus leniusculus*].
[c] Sottrup-Jensen et al. (1990) [*Limulus polyphemus*] (these proteins have not been completely sequenced, the position of the thiolester is assumed according to the human sequence; Sottrup-Jensen et al. (1984).
[d] Muta et al. (1992) [*Tachypleus tridentatus*].

protein C1 was isolated from *H. cecropia* blood (Andersson et al. 1989); however, this inhibitor was not characterized.

Smaller inhibitors have been purified from the blood of several insects (Table 3). These inhibited proPO activation by various means, including the cuticular protease in vitro, but it is unclear whether they would also be inhibitors of ppA from blood (Table 3).

In addition, Tsukamoto et al. (1992) recently claimed to have found peptide inhibitors of PO activity (not of proPO activation) in the housefly *M. domestica*.

2.4 Elicitors and Elicitor-Binding Proteins

As mentioned above, the proPO system is specifically activated by some microbial cell-wall components. These include fungal β-1,3-glucan, lipopolysaccharide (LPS) of Gram-negative bacteria and peptidoglycan of Gram-positive bacteria. The recognition of glucan is specific for those with β-1,3 linkages; glucose polymers with, for example, α- or β-1,4 bonds do not cause activation (Unestam and Söderhäll 1977; Söderhäll and Unestam 1979). The minimal requirement for this elicitor activity is a β-1,3-glucan pentasaccharide (Söderhäll and Unestam 1979).

Activation of the proPO system by the microbial elicitors has been demonstrated in greatest detail in arthropods, but it has also been shown quite recently in holothurians and ascidians (see Table 1). To our knowledge, microbial activators have not been identified for PO in annelids, molluscs, brachiopods, echinoids or asteroids. In addition to β-1,3-glucan (Table 1; Ashida et al. 1983; Dularay and Lackie 1985; Leonard et al. 1985b), peptidoglycan appears to function as an elicitor in insects (Table 1: Yoshida and Ashida 1986; Brookman et al. 1989b). LPS has been reported as an elicitor in several insects (Table 1; Saul and Sugumaran 1987; Brehélin et al. 1989); however, investigators working with other species failed to detect activation by LPS (Leonard et al. 1985b; Ashida and Yamazaki 1990). It is unclear at present whether this discrepancy really reflects a difference in vivo or whether it is due to the assay conditions in vitro.

One would expect factors to exist which are part of the proPO system and which bind the elicitors. Such factors would be pattern-recognition molecules according to Janeway's (1989) nomenclature. Proteins able to bind β-1,3-glucan were first purified from the plasma of two insects (Table 5). One has a molecular mass of 62 kDa and comes from the silkworm *B. mori* (Ochiai and Ashida 1988), the other, from the cockroach *Blaberus craniifer* has a molecular mass of 90 kDa (Söderhäll et al. 1988). More recently, similar proteins of molecular masses 95–110 kDa, which cross-react immunologically with each other, have been purified from plasma of several crustaceans (Tables 5 and 6 for review; Söderhäll et al. 1994b). After removal of the β-1,3-glucan-binding protein (βGBP), the proPO system in silkworm blood could no longer be activated by β-1,3-glucan (Yoshida et al. 1986). *B. craniifer* or *P. leniusculus* βGBP enhances protease and phenoloxidase activity in the presence of β-1,3-glucan (Söderhäll et al. 1988;

Table 5. Proteins from invertebrate blood that bind elicitors of prophenoloxidase system activation

Protein and species	Structure
β-1,3-glucan-binding proteins:	
Crustaceans: *Astacus astacus*[a]	
Carcinus maenas[b]	RGD
Pacifastacus leniusculus[c]	
Procambarus clarkii[a]	
Insects: *Blaberus craniifer*[d]	
Bombyx mori[e]	
Chelicerates: *Tachypleus tridentatus*: factor G[f]	
LPS-binding proteins:	
Crustaceans: *Pacifastacus leniusculus*[g,h]	
Penaeus californiensis[h,i]	
Insects: *Periplaneta americana*: 28 kDa[h,j]	
30 kDa lectin[h,k]	
Chelicerates: *Tachypleus tridentatus*: factor C[l]	
anti-LPS factor[m]	
tachyplesins I, II[n]	
L-6[o]	
Ascidians: *Halocynthia roretzi*[h,p]	
Peptidoglycan-binding protein:	
Insects: *Bombyx mori*[q]	

$\boxtimes$ β-1,3-Glucanase-like domain; $\blacksquare$ xylanase-like domain; $\blacksquare$ serine protease domain; $\blacksquare$ C-type lectin domain; $\square$ complement-like "Sushi" domain.

[a]Duvic and Söderhäll (1993); [b]Thörnqvist et al. (1994); [c]Duvic and Söderhäll (1990); Cerenius et al. (1994); [d]Söderhäll et al. (1988); [e]Ochiai and Ashida (1988); [f]Seki et al. (1994); (mediates clotting system activation); [g]Kopácek et al. (1993a); [h]not yet demonstrated to mediate prophenoloxidase system activation; [i]Vargas-Albores et al. (1993); [j]Jomori et al. (1990); Jomori and Natori (1991); [k]Kawasaki et al. (1993); [l]Muta et al. (1991) (mediates clotting system activation); [m]Aketagawa et al. (1986) (inhibits LPS-triggered clotting system activation); [n]Shigenaga et al. (1990); [o]Iwanaga (1993); [p]Azumi et al. (1991); [q]Ashida and Yamazaki (1990).

Duvic and Söderhäll 1990). It seems that βGBPs from insects and crustaceans are not enzymes; no protease or β-1,3-glucanase activity has been detected (Ashida and Yamazaki 1990; Söderhäll et al. 1994b).

Importantly, βGBP (at least from crustaceans) also has other functions apart from apparently being involved in proPO system activation. After reacting with β-1,3-glucan it specifically binds a blood-cell membrane receptor and mediates several blood-cell activities (Table 6). The partial degranulation triggered by β-1,3-glucan-treated βGBP (Barracco et al. 1991) probably represents the way the proPO system is initially released into the plasma in vivo during a fungal infection. Notably, crustacean βGBP also stimulates the phagocytosis of fungal

Table 6. The β-1,3-glucan-binding protein from *Pacifastacus leniusculus*

Molecular mass:	100 kDa[a]
Isoelectric point:	5.0[a]
Other properties:	Specific binding of β-1,3-glucan
	(not of β-1,4-glucan or α-glucans)[a]
	Has an RGD putative cell binding sequence[b]
Site of synthesis:	Hepatopancreas[b]
Localization:	Plasma[a]
Generation of activity:	After reacting with β-1,3-glucan[c,d]
Activities:	Binding to blood cell membrane receptor[d]
	Blood cell spreading[c]
	Partial blood cell degranulation[c]
	Stimulation of phagocytosis[e]

[a]Duvic and Söderhäll (1990).
[b]Cerenius et al. (1994).
[c]Barracco et al. (1991).
[d]Duvic and Söderhäll (1992).
[e]Thörnqvist et al. (1994).

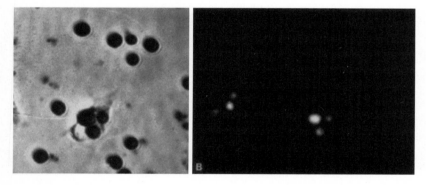

Fig. 1. Phagocytosis of β-1,3-glucan binding protein-opsonized fluorescent-labelled yeast cells by hyaline blood cells of the shore crab *Carcinus maenas*. **A** Light microscopy. **B** Fluorescence microscopy. Yeast cells outside the blood cells have been quenched, yeast cells that have been phagocytosed by the blood cells are fluorescent. (Photo courtesy of P.-O. Thörnqvist)

cells by the blood cells (Thörnqvist et al. 1994; Fig. 1), indicating a link between proPO system components and cellular defence.

Recently, cDNA for βGBP from the crayfish *P. leniusculus* was cloned and sequenced (Cerenius et al. 1994; Tables 5 and 6). The sequence is not significantly similar to any known protein. However, it has an arginine-glycine-aspartic acid (RGD) triplet, a motif that is responsible for the binding of fibronectin and several other vertebrate extracellular proteins to their receptors of the integrin family (Ruoslahti 1991). Thus, it may be that βGBP, after binding β-1,3-glucan exposes its RGD sequence and uses it for cell binding. An RGD (arginine-glycine-glutamic acid) containing peptide, but not a control (RGE) peptide, can degranulate crayfish blood cells (Johansson and Söderhäll 1989c), showing that the cells have the ability to recognize such a sequence.

A β-1,3-glucan-binding protein that is different from crustacean and insect βGBP has been purified and cloned from the horseshoe crab of the arthropod group chelicerates (Table 5; see also Muta and Iwanaga, this Vol.). This protein, factor G, consists of two subunits, one with a serine protease domain and the other similar to bacterial β-1,3-glucanase. Factor G is autocatalytically cleaved in the presence of β-1,3-glucan, and can then mediate activation of the clotting system in this animal (for review, see Iwanaga 1993). Horseshoe crabs do not have blood PO (Table 1), but their clotting system is a cascade which consists of several serine proteases and which like the proPO system, is activated by LPS or β-1,3-glucans (Table 5; for reviews, Iwanaga et al. 1992; Iwanaga 1993).

Several LPS-binding proteins have been purified from arthropod blood and, recently, from an ascidian (see Table 5). At least some of these proteins may be involved in proPO system activation, but this has not yet been shown. The crustacean LPS-binding proteins are agglutinins. The LPS-binding proteins from plasma of the cockroach *Periplaneta americana* seem to be opsonins that facilitate uptake or clearance of bacteria (Jomori et al. 1990; Kawasaki et al. 1993). Horseshoe crab factor C is, like factor G, a serine protease which is autocatalytically activated in the presence of the microbial compound and which then mediates clotting system activation (Iwanaga et al. 1992; Iwanaga 1993). The other small LPS-binding proteins from horseshoe crab are antimicrobial and/or inhibitors of LPS-elicited clotting system activation.

Finally, a peptidoglycan-binding factor that is necessary for proPO system activation by peptidoglycan was identified in silkworm plasma (Yoshida et al. 1986). This factor was later reported to be a 19 kDa protein (Ashida and Yamazaki 1990). As for insect and crustacean βGBP, no enzyme activity of this protein has been detected.

2.5 Associated Proteins

2.5.1 The Cell-Adhesion Protein

As mentioned in the introduction, many observations suggest an association between proPO system activation and cellular responses. A 76 kDa cell-adhesion protein which is apparently connected to the proPO system and which mediates several cellular defence activities was first identified in, and purified from, blood cells of the crayfish *P. leniusculus* (Tables 7 and 8; Johansson and Söderhäll 1988, 1989b, 1992). This protein seems to be responsible for the reported activities. As mentioned above, the PO enzyme is itself "sticky" and it could therefore be a candidate for an opsonin. However, it has not been possible to substantiate this (Söderhäll et al. 1986; Kobayashi et al. 1990), and cellular encapsulation can occur in PO-deficient *Drosophila* mutant (Rizki and Rizki 1990).

Immunologically cross-reactive proteins of similar sizes and with the same activities as the crayfish 76 kDa cell-adhesion protein, have been found in the

Table 7. Cell adhesion proteins from invertebrate blood cells

Species	Molecular mass (kDa)[a]	Isoelectric point
Crustaceans:		
Carcinus maenas	80[b]	
Pacifastacus leniusculus	76[c]	7.2[d]
Insects:		
Blaberus craniifer	90[e]	
Galleria mellonella	90–100[f]	

[a] Determined by SDS-PAGE under reducing conditions.
[b] Thörnqvist et al. (1994).
[c] Johansson and Söderhäll (1988).
[d] Johansson and Söderhäll (1989b).
[e] Rantamäki et al. (1991).
[f] Mullett et al. (1993a,b).

Table 8. The *Pacifastacus leniusculus* cell adhesion protein and related proteins

Molecular masses and isoelectric point:	See Table 7
Other biochemical and immunochemical properties:	Heparin binding after urea treatment[a]
	Reaction with and inhibition by anti-vertebrate vitronectin antibodies[a,b]
Site of synthesis:	Granular and semigranular haemocytes[c-e]
Localization:	Blood cell granules[c]
Generation of activity:	Concomitant with prophenoloxidase system activation[f]
Activities:	Binding to blood cell membrane receptor[g]
	Activation of protein kinase C and protein tyrosine phosphorylation[h]
	Blood cell attachment and spreading[f,i]
	Blood cell degranulation[j]
	Stimulation of phagocytosis[e]
	Stimulation of encapsulation[k]
	Bacterial attachment to blood cells[d,i]
	Nodule formation in vivo[i]

[a] P.-O. Thörnqvist, M.W. Johansson and K. Söderhäll, unpubl.; [b] Johansson and Söderhäll (1992); [c] Liang et al.(1992); [d] Mullett et al. (1993a); [e] Thörnqvist et al. (1994); [f] Johansson and Söderhäll (1988); [g] M. Cammarata, M.W. Johansson, M. Kobayashi, N. Parrinello and K. Söderhäll, unpubl.; [h] Johansson and Söderhäll (1993); [i] Mullett et al. (1993b); [j] Johansson and Söderhäll (1989b); [k] Kobayashi et al. (1990).

blood cells of other crustaceans and in the cockroach *B. craniifer* (Table 7). The insect protein can degranulate and attach crayfish blood cells and vice versa (Rantamäki et al. 1991). Recently, monoclonal antibodies were raised against blood cells from the wax moth *Galleria mellonella* (Mullett et al. 1993a); several of these antibodies recognized a 90–100 kDa granulocyte protein (Mullett et al. 1993a,b). The antibodies decreased bacterial attachment to blood cells in vitro (Mullett et al. 1993a) and nodule formation in vivo (Mullett et al. 1993b; Table

8). Thus, the *Galleria* protein also seems to be a cell-adhesion protein, and is most probably a homologue of the crustacean and the *B. craniifer* protein (Table 7). Furthermore, the arthropod cell-adhesion protein cross-reacts with the 75 kDa vertebrate cell-adhesion protein vitronectin (Table 8, Johansson and Söderhäll 1992; P.-O. Thörnqvist, M.W. Johansson and K. Söderhäll, unpubl.), which is involved in the coagulation and complement systems (for reviews, Tomasini and Mosher 1990; Preissner 1991). In addition, anti-vitronectin antiserum inhibits adhesion of crayfish blood cells to the crayfish 76 kDa protein (P.-O. Thörnqvist et al., unpubl.).

In crustaceans, the cell-adhesion protein is synthesized by the semigranular and granular bloods cells, stored in their secretory granules (Table 8) and released from there during degranulation (Johansson and Söderhäll 1988). It gains its biological activities under the same conditions as when the proPO system is triggered, i.e. in the presence of microbial elicitors, or low Ca^{2+} concentration (Johansson and Söderhäll 1988). As for proPO activation, the generation of the activity of the cell-adhesion protein is inhibited by serine protease inhibitors (M.W. Johansson and K. Söderhäll, unpubl.), but no difference in molecular mass has been detected between the inactive and the active form (Johansson and Söderhäll 1988). It is not activated when proPO is activated non-proteolytically by heat or detergent (Johansson and Söderhäll 1988). In conclusion, there seems to be an association between activation of the whole proPO system and the cell-adhesion protein, but the detailed mechanism for this is unknown. One could speculate that, like mammalian vitronectin, it may undergo a conformational change in the presence of a serine protease–protease inhibitor complex (Tomasini and Mosher 1990). Moreover, the cell-adhesion protein may be regulated not only by protease inhibitors, but also by degradation. In a crude system, it can be degraded in the presence of the microbial cell-wall compounds to a 30 kDa fragment with lower biological activity (Kobayashi et al. 1990).

Finally, an opsonic activity that also appears to be correlated with activation of the proPO system was recently found in ascidian blood cells (Smith and Peddie 1992). In an annelid, cellular communication between different blood cell types has been described (Porchet-Honnere 1990), reminiscent of the model involving the cell-adhesion protein proposed earlier for arthropods (Söderhäll et al. 1986; Johansson and Söderhäll 1989a; Anggraeni and Ratcliffe 1991). Thus, it could be suggested that a homologous cell-adhesion protein may exist in several invertebrate groups to mediate cell communication and cellular defence.

2.5.2 The Receptor for the β-1,3-Glucan Binding Protein and the Cell-Adhesion Protein

As both βGBP and the cell-adhesion protein mediate or stimulate cellular activities like spreading, degranulation and phagocytosis (Tables 6 and 8), one would expect the blood cells to have cell-surface receptors for these proteins. Both βGBP (after reacting with β-1,3-glucan) and the cell-adhesion protein have been shown to bind to the blood cell surface (Barracco et al. 1991; M. Cammarata,

Table 9. The haemocyte membrane receptor for the β-1,3-glucan-binding protein and the 76 kDa cell adhesion protein from *Pacifastacus leniusculus*

Molecular mass:	90 kDa (SDS-PAGE, strong reduction)[a]
	90 + 230 kDa (SDS-PAGE, no or weak reduction)[b]
	350 kDa (native PAGE)[b]
	320 kDa (gel filtration)[b]
Isoelectric point:	4.4[a]
Affinity:	K_d for β-1,3-glucan-binding protein = 0.35 μM[b]
	(apparently higher affinity for the 76 kDa protein[a])

[a]M. Cammarata, M.W. Johansson, M. Kobayashi, N. Parrinello and K. Söderhäll, (unpubl.).
[b]Duvic and Söderhäll (1992).

M.W. Johansson, M. Kobayashi, N. Parrinello and K. Söderhäll, unpubl.). Preliminary experiments indicated that one ligand inhibited the binding of the other one (Cammarata et al., unpubl.). A blood-cell membrane protein which binds βGBP has been identified and purified (Duvic and Söderhäll 1992; Table 9). It is a multimer of 90 kDa subunits. This protein was recently also shown to bind the cell-adhesion protein and this binding was inhibited by β-1,3-glucan-treated βGBP (Cammarata et al., unpubl. Thus, this protein seems to be a blood-cell membrane receptor for both βGBP and the cell-adhesion protein. The latter bound with an apparently higher affinity, which is in concordance with the fact that the cell-adhesion protein degranulates cells at μg/ml levels (Johansson and Söderhäll 1989b), whereas much higher concentrations (two orders of magnitude) of βGBP are required (Barracco et al. 1991).

Regarding the consequences of ligand binding to cells, it was recently shown that degranulation triggered by the cell-adhesion protein of β-1,3-glucan-reacted βGBP, was inhibited by a protein kinase C inhibitor or by a tyrosine kinase inhibitor (Johansson and Söderhäll 1993; Table 8). Moreover, protein kinase C-activating phorbol esters could degranulate the cells, whereas a control non-activating isomer did not. Thus, it seems that ligation of the receptor leads to an intracellular signalling pathway involving protein kinase C activation and tyrosine phosphorylation of cellular protein, and then secretion of the proPO system.

Presumably, related receptors should exist in several invertebrate groups. It would also be interesting to investigate whether LPS-, peptidoglycan-binding proteins or other ligands would be able to bind such receptors and degranulate cells.

3 Complement-Like Proteins

It has been suggested that the proPO system is a complement-like system (Söderhäll 1982). Functional similarities include activation by microbial cell-wall components or by proteins that bind these compounds, and the generation of opsonic factors. Until recently, the only invertebrate proteins shown to be structurally similar to proteins of the vertebrate complement system were the α_2-macroglobulin-like protease inhibitor from various invertebrates, and horseshoe factor C (Table 4). However, recently, proPO from the crayfish *Pacifastacus leniusculus* was found to contain a complement-like motif (similar to the thiolester-containing region of e.g. α_2-macroglobulin) (Aspán et al. 1995), indicating that more invertebrate proteins with complement domains may be found in the future.

4 Clotting Proteins

The horseshoe crab clotting system is, as mentioned in Section 2.4, apparently similar to the proPO system; a cascade of proteases triggered by microbial elicitors which is stored in blood-cell granules and released during degranulation. Activation ends in the cleavage of a 20 kDa coagulogen or clotting protein, and subsequent protein polymerization (Iwanaga et al. 1992; Iwanaga 1993; Table 10). Similarly, in crustaceans, gelation of a blood cell lysate, not including any plasma component, can be triggered by microbial elicitors in a process involving serine proteases, reminiscent of proPO system activation (Söderhäll 1981; Durliat 1985). Whether this is biologically significant is unclear. It now seems that a firm clot is instead formed by polymers of a large plasma clotting protein, apparently unrelated to horseshoe crab coagulogen (Table 10). The clotting protein is crosslinked by a transglutaminase from the blood cells in a manner seemingly independent of the proPO system (Kopácek et al. 1993b). It is not known how the transglutaminase is released and it is also unclear whether it is already present in an active form in the cells or if its activity is affected by, for example, microbial cell-well constituents.

Table 10. Clotting proteins from invertebrate blood

Species	Molecular mass (kDa)[a]	Amino terminal sequence
Crustaceans:		
Pacifastacus leniusculus[b]	210	LHSNLEYQYR...
Panulirus interruptus[c]	220	LQPKLEYQYK...
Chelicerates:		
Carcinoscorpius rotundicauda[d]	20	ADTNAPLCLC...
Limulus polyphemus[e]	20	GDPNVPTCLC...
Tachypleus gigas[f]	20	DDTNAPLCLC...
Tachypleus tridentatus[g]	20	ADTNAPICLC...

[a]Determined by SDS-PAGE under reducing conditions.
[b]Kopácek et al. (1993b).
[c]Doolittle and Riley (1990).
[d]Srimal et al. (1985).
[e]Miyata et al. (1984a).
[f]Miyata et al. (1984b).
[g]Miyata et al. (1986).

Thus, even in different arthropod groups, the mechanism of blood clotting seems to differ. At present it is not clearly understood how clotting occurs in insects and other invertebrate groups.

5 Summary

In this review, we present arguments indicating that the prophenoloxidase (proPO) activating system acts as a pattern recognition and defence system in invertebrate blood. Phenoloxidase (PO) activity has been found in the blood of many invertebrates. At least in arthropods, echinoderms and urochordates, the inactive pro-form, proPO has been found to be elicited by the microbial cell-wall components β-1,3-glucans, lipopolysaccharide and/or peptidoglycan. This activation seems to involve elicitor-binding proteins and serine protease(s). ProPO, the proPO-activating enzyme (ppA) and plasma elicitor-binding proteins, have been purified from some arthropods, and proPO and the β-1,3-glucan binding protein (βGBP) have been cloned and sequenced from crayfish. Arthropod proPO has a molecular mass of 70–90 kDa and PO has a molecular mass of 60–70 kDa.

The βGBP also stimulates phagocytosis of fungal cells and, after reacting with β-1,3-glucan, blood-cell degranulation (and release of the proPO system). In addition, a cell-adhesion protein (of 70–100 kDa), apparently associated with the proPO system, has been purified from arthropods. This mediates blood-cell adhesion, degranulation, phagocytosis and encapsulation. The cell-adhesion protein and βGBP bind to a common blood-cell membrane receptor.

It would be interesting to see the sequences of more proPO system components and investigate whether the scheme for cellular communication and defence, involving the cell-adhesion protein, elicitor-binding proteins and the membrane receptor described in arthropods, applies to invertebrates in general.

Acknolwedgment. This review was supported mainly by grants from the Swedish Council for Forestry and Agricultural Research and the Swedish Natural Science Research Council.

References

Aketagawa J, Miyata T, Ohtsubo S, Nakamura T, Hayashida H, Miyata T, Iwanaga S (1986) Primary structure of *Limulus* anticoagulant anti-lipopolysaccharide factor. J Biol Chem 261: 7357–7365
Andersson K, Sun S-C, Boman HG, Steiner H (1989) Purification of Cecropia prophenoloxidase and four proteins involved in its activation. Insect Biochem 19: 629–638
Anggraeni T, Ratcliffe NA (1991) Studies on cell–cell co-operation during phagocytosis by purified haemocyte populations of the wax moth, *Galleria mellonella*. J Insect Physiol 37: 453–460
Ashida M (1971) Purification and characterization of prephenoloxidase from hemolymph of the silkworm, *Bombyx mori*. Arch Biochem Biophys 144: 749–762
Ashida M, Dohke K (1980) Activation of prophenoloxidase by the activating enzyme of the silkworm, *Bombyx mori*. Insect Biochem 10: 37–47
Ashida M, Söderhäll K (1984) The prophenoloxidase activating system in crayfish. Comp Biochem Physiol 77B: 21–26

Ashida M, Yamazaki HI (1990) Biochemistry of the phenoloxidase system in insects: with special reference to its activation. In: Ohnishi E, Ishizaki H (eds) Molting and metamorphosis. Jpn Sci Soc Press/Springer, Berlin, pp 237–263

Ashida M, Yoshida H (1988) Limited proteolysis of prophenoloxidase during activiation by microbial products in insect plasma and effect of phenoloxidase on electrophoretic mobilities of plasma proteins. Insect Biochem 18: 11–19

Ashida M, Dohke K, Ohnishi E (1974) Activation of prephenoloxidase. III. Release of a peptide from prephenoloxidase by the activating enzyme. Biochem Biophys Res Commun 57: 1089–1095

Ashida M, Iwama R, Iwahana H, Yoshida H (1982) Control and function of the prophenoloxidase activating system. In: Payne CC, Burges HD (eds) Proceedings of the 3rd international colloquium on invertebrate pathology. Univ Sussex, Brighton, pp 81–86

Ashida M, Ishizaki Y, Iwahana H (1983) Activation of pro-phenoloxidase by bacterial cell walls or β-1,3-glucan in plasma of the silkworm, Bombyx mori. Biochem Biophys Res Commun 113: 562–568

Ashida M, Ochiai O, Nlki T (1988) Immunolocalization of prophenoloxidase among hemocytes of the silkworm, Bombyx mori. Tissue Cell 20: 599–610

Ashida M, Kinoshita K, Brey PT (1990) Studies on prophenoloxidase activation in the mosquito Aedes aegypti. Eur J Biochem 188: 507–515

Aso Y, Kramer KJ, Hopkins TL, Lookhart GL (1985) Characterization of haemolymph protyrosinase and a cuticular activator from Manduca sexta (L). Insect Biochem 15: 9–17

Aspán A, Söderhäll K (1991) Purification of prophenoloxidase from crayfish blood cells and its activation by an endogenous serine proteinase. Insect Biochem 21: 363–373

Aspán A, Hall M, Söderhäll K (1990a) The effect of endogenous proteinase inhibitors on the prophenoloxidase activating enzyme, a serine proteinase from crayfish haemocytes. Insect Biochem 20: 485–492

Aspán A, Sturtevant J, Smith VJ, Söderhäll K (1990b) Purification and characterization of a prophenoloxidase activating enzyme from crayfish blood cells. Insect Biochem 20: 709–718

Aspán A, Huang T-s, Cerenius L, Söderhäll K (1995) cDNA cloning of prophenoloxidase from the freshwater crayfish Pacifastacus leniusculus and its activation. Proc Natl Acad Sci USA 92: 939–943

Azumi K, Ozeki S, Yokosawa H, Ishii S-i (1991) A novel lipopolysaccharide-binding hemagglutinin isolated from hemocytes of the solitary ascidian, Halocynthia roretzi: it can agglutinate bacteria. Dev Comp Immunol 15: 9–16

Barracco MA, Duvic B, Söderhäll K (1991) The β-1,3-glucan-binding protein from the crayfish Pacifastacus leniusculus, when reacted with β-1,3-glucan, induces spreading and degranulation of crayfish granular cells. Cell Tissue Res 266: 491–497

Beauvais A, Latgé J-P, Vey A, Prévost M-C (1989) The role of surface components of the entomopathogenic fungus Entomophaga aulicae in the cellular immune response of Galleria mellonella (Lepidoptera). J Gen Microbiol 135: 489–498

Beckage NE, Metcalf JS, Nesbit DJ, Schleifer KW, Zetlan SR, De Buron I (1990) Host hemolymph monophenoloxidase activity in parasitized Manduca sexta larvae and evidence for inhibition by wasp polydnavirus. Insect Biochem 20: 285–294

Brehélin M, Drif I, Baud I, Boemare N (1989) Activation of pro-phenoloxidase in insect haemolymph: cooperation between humoral and cellular factors in Locusta migratoria. Insect Biochem 19: 301–307

Brehélin M, Boigegrain RA, Drif L, Coletti-Previero MA (1991) Purification of a protease inhibitor which controls prophenoloxidase activation in hemolymph of Locusta migratoria (Insecta). Biochem Biophys Res Commun 179: 841–846

Brookman JA, Ratcliffe NA, Rowley AF (1988) Optimization of a monolayer phagocytosis assay and its application for studying the role of the prophenoloxidase system in the wax moth Galleria mellonella. J Insect Physiol 34: 337–345

Brookman JA, Ratcliffe NA, Rowley AF (1989a) Studies on nodule formation in locusts following injection of microbial products. J Invertebr Pathol 53: 315–323

Brookman JA, Ratciffe NA, Rowley AF (1989b) Studies on the activation of the prophenoloxidase system in insects by bacterial cell wall components. Insect Biochem 19: 47–57

Canicatti C, Götz P (1991) DOPA oxidation by *Holothuria polii* coelomocyte lysate. J Invertebr Pathol 58: 305–310

Canicatti C, Seymour J (1991) Evidence for phenoloxidase activity in *Holothuria tubulosa* (Echinodermata) brown bodies and cells. Parasitol Res 77: 50–53

Cerenius L, Liang Z, Duvic B, Keyser P, Hellman U, Palva ET, Iwanaga S, Söderhäll K (1994) A (1,3)-β-D-glucan binding protein in crustacean blood. Structure and biological activity of a fungal recognition protein. J Biol Chem 269: 29462–29467

Christensen BM, LaFond MM (1986) Parasite-induced suppression of the immune response in *Aedes aegypti* by *Brugia pahangi*. J Parasitol 72: 216–219

Collins FH, Sakai RK, Vernick KD, Paskewitz S, Seeley DC, Miller LH, Collins WE, Campbell CC, Gwadz RW (1986) Genetic selection of a *Plasmodium*-refractory strain of the malaria vector *Anopheles gambiae*. Science (Wash DC) 234: 607–610

De Aragao GA, Bacila M (1976) Purification and properties of a polyphenoloxidase from the freshwater snail, *Biomphalaria glabrata*. Comp Biochem Physiol 54B: 179–182

Dohke K (1973) Studies on prephenoloxidase-activating enzyme from cuticle of the silkworm *Bombyx mori*. II. Purification and characterization of the enzyme. Arch Biochem Biophys 157: 210–221

Doolittle RF, Riley M (1990) The amino-terminal sequence of lobster fibrinogen reveals common ancestry with vitellogenins. Biochem Biophys Res Commun 167: 16–19

Dularay B, Lackie AM (1985) Haemocytic encapsulation and the prophenoloxidase-activiating pathway in the locust *Schistocerca gregaria* Forsk. Insect Biochem 15: 827–834

Durliat M (1985) Clotting processes in Crustacea Decapoda. Biol Rev 60: 473–498

Durrant HJ, Ratcliffe NA, Hipkin CR, Aspán A, Söderhäll K (1993) Purification of the prophenol oxidase enzyme from haemocytes of the cockroach *Blaberus discoidalis*. Biochem J 289: 87–91

Duvic B, Söderhäll K (1990) Purification and characterization of a β-1,3-glucan binding protein from plasma of the crayfish *Pacifastacus leniusculus*. J Biol Chem 265: 9327–9332

Duvic B, Söderhäll K (1992) Purification and characterization of a β-1,3-glucan binding protein membrane receptor from blood cells of the crayfish *Pacifastacus leniusculus*. Eur J Biochem 207: 223–228

Duvic B, Söderhäll K (1993) β-1,3-glucan-binding proteins from plasma of the freshwater crayfish *Astacus astacus* and *Procambarus clarkii*. J Crust Biol 13: 403–408

Fujimoto K, Masuda K-i, Asada N, Ohnishi E (1993) Purification and characterization of prophenoloxidase from pupae of *Drosophila melanogaster*. J Biochem (Tokyo) 113: 285–291

Goldenberg PZ, Huebner E, Greenberg AH (1984) Activation of lobster hemocytes for phagocytosis. J Invertebr Pathol 43: 77–88

Gunnarsson S, Lackie AM (1995) Haemocyte aggregation in the locust, *Schistocerca gregaria*, and the cockroach, *Periplaneta americana*, in response to injected molecules of microbial orgin. J Invertebr Pathol 46: 312–319

Hall M, Söderhäll K, Sottrup-Jensen L (1989) Amino acid sequence around the thiolester of α_2-macroglobulin from plasma of the crayfish *Pacifastacus leniusculus*. FEBS Lett 254: 111–114

Hara T, Miyoshi T, Tsukamoto T (1993) Comparative studies on larval and pupal phenoloxidase of the housefly, *Musca domestica* L. Comp Biochem Physiol 106B: 287–292

Hergenhahn H-G, Aspán A, Söderhäll K (1987) Purification and characterization of a high-M_r proteinase inhibitor of pro-phenoloxidase activation from crayfish plasma. Biochem J 248: 223–228

Hergenhahn H-G, Hall M, Söderhäll K (1988) Purification and characterization of an α_2-macroglobulin-like proteinase inhibitor from plasma of the crayfish *Pacifastacus leniusculus*. Biochem J 255: 801–806

Huxham IM, Lackie AM (1988) Behaviour in vitro of separated haemocytes from the locust, *Schistocerca gregaria*. Cell Tissue Res 251: 677–684

Huxham IM, Lackie AM, McCorkindale NJ (1989) Inhibitory effects of cyclidepsipeptides, destruxins, from the fungus *Metarhizium anisopliae*, on cellular immunity in insects. J Insect Physiol 35: 97–105

Iwama R, Ashida M (1986) Biosynthesis of prophenoloxidase in hemocytes of larval hemolymph of the silkworm, *Bombyx mori*. Insect Biochem 16: 547–555

Iwanaga S (1993) Primitive coagulation systems and their message to modern biology. Thromb Haemostasis 70: 48–55

Iwanaga S, Miyata T, Tokunaga F, Muta T (1992) Molecular mechanism of hemolymph clotting in *Limulus*. Thromb Res 68: 1–32

Jackson AD, Smith VJ (1993) LPS-sensitive protease activity in the blood cells of the solitary ascidian *Ciona intestinalis* (L.) Comp Biochem Physiol 106B: 505–512

Jackson AD, Smith VJ, Peddie CM (1993) In vitro prophenoloxidase activity in the blood of *Ciona intestinalis* and other ascidians. Dev Comp Immunol 17: 97–108

Janeway Jr CA (1989) Approaching the asymptote? Evolution and revolution in immunology. Cold Spring Harbor Symp Quant Biol 54: 1–13

Johansson MW, Söderhäll K (1985) Exocytosis of the prophenoloxidase activating system from crayfish haemocytes. J Comp Physiol B 156: 175–181

Johansson MW, Söderhäll K (1988) Isolation and purification of a cell adhesion factor from crayfish blood cells. J Cell Biol 106: 1795–1803

Johansson MW, Söderhäll K (1989a) Cellular immunity in crustaceans and the proPO system. Parasitol Today 5: 171–176

Johansson MW, Söderhäll K (1989b) A cell adhesion factor from crayfish haemocytes has degranulating activity towards crayfish granular cells. Insect Biochem 19: 183–190

Johansson MW, Söderhäll K (1989c) A peptide containing the cell adhesion sequence RGD can mediate degranulation and cell adhesion of crayfish granular haemocytes in vitro. Insect Biochem 19: 573–579

Johansson MW, Söderhäll K (1992) Cellular defence and cell adhesion in crustaceans. Anim Biol 1: 97–107

Johansson MW, Söderhäll K (1993) Intracellular signaling in arthropod blood cells: Involvement of protein kinase C and protein tyrosine phosphorylation in the response to the 76 kDa protein or the β-1,3-glucan-binding protein in crayfish. Dev Comp Immunol 17: 495–500

Jomori T, Natori S (1991) Molecular cloning of cDNA for lipopolysaccharide-binding protein from the haemolymph of the American cockroach *Periplaneta americana*. J Biol Chem 266: 13318–13323

Jomori T, Kubo T, Natori S (1990) Purification and characterization of lipopolysaccharide-binding protein from haemolymph of American cockroach *Periplaneta americana*. Eur J Biochem 190: 201–206

Kawasaki K, Kubo T, Natori S (1993) A novel role of *Periplaneta* lectin as an opsonin to recognize 2-keto-3 deoxy octonate residues of bacterial lipopolysaccharide. Comp Biochem Physiol 106B: 675–680

Kitano H, Wago H, Arakawa T (1990) Possible role of teratocytes of the gregarious parasitoid, *Cotesia* (=*Apanteles*) *glomerata* in the suppression of phenoloxidase activity in the larval host, *Pieris rapae crucivora*. Arch Insect Biochem Physiol 13: 177–185

Kobayashi M, Johansson MW, Söderhäll K (1990) The 76 kDa cell-adhesion factor from crayfish haemocytes promotes encapsulation in vitro. Cell Tissue Res 260: 113–118

Kopácek P, Grubhoffer L, Söderhäll K (1993a) Isolation and characterization of a hemagglutinin with affinity for lipopolysaccharide from plasma of the crayfish *Pacifastacus leniusculus*. Dev Comp Immunol 17: 407–418

Kopácek P, Hall M, Söderhäll K (1993b) Characterization of a clotting protein, isolated from plasma of the freshwater crayfish *Pacifastacus leniusculus*. Eur J Biochem 21: 591–597

Krishnan G, Ravindranath MH (1973) Blood cell phenoloxidase of millipeds. J Insect Physiol 19: 647–653

Leonard CM, Ratcliffe NA, Rowley AF (1985a) The role of prophenoloxidase acitivation in non-self recognition and phagocytosis by insect blood cells. J Insect Physiol 31: 789–799

Leonard CM, Söderhäll K, Ratcliffe NA (1985b) Studies on prophenoloxidase and protease activity of *Blaberus craniifer* haemocytes. Insect Biochem 15: 803–810

Liang Z, Lindblad P, Beauvais A, Johansson MW, Latgé J-P, Hall M, Cerenius L, Söderhäll K (1992) Crayfish α-macroglobulin and 76 kD protein; their biosynthesis and subcellular localization of the 76 kD protein. J Insect Physiol 38: 987–995

Marmaras VJ, Bournazos SN, Katsoris PG, Lambropoulou M (1993) Defense mechanisms in insects: Certain integumental proteins and tyrosinase are responsible for non-self recognition and immobilization of *Escherichia coli* in the cuticle of developing *Ceratitis capitata*. Arch Insect Biochem Physiol 23: 169–180

Miyata T, Hiranaga M, Umezu M, Iwanaga S (1984a) Amino acid sequence of the coagulogen from *Limulus polyphemus* hemocytes. J Biol Chem 259: 8924–8933

Miyata T, Usui K, Iwanaga S (1984b) The amino acid sequence of coagulogen isolated from Southeast Asian horseshoe crab *Tachypleus gigas*. J Biochem (Tokyo) 95: 1793–1801

Miyata T, Matsumoto H, Hattori M, Sasaki Y, Iwanaga S (1986) Two types of coagulogen mRNAs found in horseshoe (*Tachypleus tridentatus*) hemocytes: molecular cloning and nucleotide sequences. J Biochem (Tokyo) 100: 213–220

Mullett H, Ratcliffe NA, Rowley AF (1993a) The generation and characterisation of anti-insect blood cell monoclonal antibodies. J Cell Sci 105: 93–100

Mullett H, Ratcliffe NA, Rowley AF (1993b) Analysis of immune defences of the wax moth, *Galleria mellonella*, with anti-haemocytic antibodies. J Insect Physiol 39: 897–902

Muta T, Miyata T, Misumi Y, Tokunaga F, Nakamura T, Toh Y, Ikehara Y, Iwanaga S (1991) Limulus factor C: An endotoxin sensitive serine protease zymogen with a mosaic structure of complement-like, epidermal growth factor-like, and lectin-like domains. J Biol Chem 266: 6554–6561

Naqvi SNH, Karlson P (1979) Purification of prophenoloxidase in the haemolymph of *Calliphora vicina* (R. & D.). Arch Int Physiol Biochem 87: 687–695

Ochiai M, Ashida M (1988) Purification of a β-1,3-glucan recognition protein in the prophenoloxidase activating system from hemolymph of silkworm, *Bombyx mori*. J Biol Chem 263: 12056–12062

Ourth DD, Renis HE (1993) Antiviral melanization reaction of *Heliothis virescens* hemolymph against DNA and RNA viruses in vitro. Comp Biochem Physiol 105B: 719–723

Porchet-Honnere E (1990) Cooperation between different coelomocyte populations during the encapsulation response of *Nereis diversicolor* demonstrated by using monoclonal antibodies. J Invertebr Pathol 56: 353–361

Preissner KT (1991) Structure and biological role of vitronectin. Annu Rev Cell Biol 7: 275–310

Rantamäki J, Durrant H, Liang Z, Ratcliffe NA, Duvic B, Söderhäll K (1991) Isolation of a 90 kDa protein from haemocytes of *Blaberus craniifer* which has similar properties to the 76 kDa protein from crayfish haemocytes. J Insect Physiol 37: 627–634

Ratcliffe NA, Leonard CM, Rowley AF (1984) Prophenoloxidase activation: nonself recognition and cell cooperation in insect immunity. Science (Wash DC) 226: 557–559

Ratcliffe NA, Rowley AF, Fitzgerald SW, Rhodes CP (1985) Invertebrate immunity: basic concepts and recent advances. Int Rev Cytol 97: 183–350

Rizki RM, Rizki TM (1980) Developmental analysis of a temperature-sensitive mutant in *Drosophila melanogaster*. Roux's Arch Dev Biol 189: 197–206

Rizki RM, Rizki TM (1990) Encapsulation of parasitoid eggs in phenoloxidase-deficient mutants of *Drosophila melanogaster*. J Insect Physiol 36: 523–529

Rowley AF, Brookman JL, Ratcliffe NA (1990) Possible involvement of the prophenoloxidase system of the locust, *Locusta migratoria*, in antimicrobial activity. J Invertebr Pathol 56: 31–38

Ruoslahti E (1991) Integrins. J Clin Invest 87: 1–5

Saul S, Sugumaran M (1986) Protease inhibitor controls prophenoloxidase activation in *Manduca sexta*. FEBS Lett 208: 113–116

Saul S, Sugumaran M (1987) Protease mediated prophenoloxidase activation in the hemolymph of the tobacco hornworm, *Manduca sexta*. Arch Insect Biochem Physiol 5: 1–11

Seki N, Muta T, Oda T, Iwaki D, Kuma K, Miyata T, Iwanaga S (1994) Horseshoe crab (1,3)-β-D-glucan-sensitive coagulation factor G. A serine protease zymogen heterodimer with similarities of β-glucan-binding proteins. J Biol Chem 269: 1370–1374

Shigenaga T, Muta T, Toh Y, Tokunaga F, Iwanaga S (1990) Antimicrobial tachyplesin peptide precursor: cDNA cloning and cellular localization in horseshoe crab (*Tachypleus tridentatus*). J Biol Chem 265: 21350–21354

Smith VJ (1991) Invertebrate immunology: phylogenetic, ecotoxicological and biomedical applications. Comp Haematol Int 1: 61–76

Smith VJ, Peddie CM (1992) Cell cooperation during host defense in the solitary tunicate *Ciona intestinalis* (L.). Biol Bull (Woods Hole) 183: 211–219

Smith VJ, Söderhäll K (1983a) β-1,3-glucan activation of crustacean haemocytes in vitro and in vivo. Biol Bull (Woods Hole) 164: 299–314

Smith VJ, Söderhäll K (1983b) Induction of degranulation and lysis of haemocytes in the freshwater crayfish, *Astacus astacus* by components of the prophenoloxidase activating system in vitro. Cell Tissue Res 233: 295–303

Smith VJ, Söderhäll K (1991) A comparison of phenoloxidase activity in the blood of marine invertebrates. Dev Comp Immunol 15: 251–261

Smith VJ, Söderhäll K, Hamilton M (1984) β-1,3-glucan induced cellular defence reactions in the shore crab, *Carcinus maenas*. Comp Biochem Physiol 77A: 635–639

Söderhäll K (1981) Fungal cell wall β-1,3-glucans induce clotting and phenoloxidase attachment to foreign surfaces of crayfish hemocyte lysate. Dev Comp Immunol 5: 565–573

Söderhäll K (1982) Prophenoloxidase activating system and melanization – a recognition mechanism of arthropods? A review. Dev Comp Immunol 6: 601–611

Söderhäll K, Ajaxon R (1982) Effect of quinones and melanin on mycelial growth of *Aphanomyces* spp. and extracellular protease of *Aphanomyces astaci*, a parasite on crayfish. J Invertebr Pathol 39: 105–109

Söderhäll K, Aspán A (1993) Prophenoloxidase activating system and its role in cellular communication. In: Pathak JPN (ed) Insect immunity. Oxford and IBH, New Delhi, pp 113–129

Söderhäll K, Häll L (1984) Lipopolysaccharide-induced activation of prophenoloxidase activating system in crayfish haemocyte lysate. Biochim Biophys Acta 797: 99–104

Söderhäll K, Smith VJ (1983) Separation of the haemocyte populations of *Carcinus maenas* and other marine decapods, and prophenoloxidase distribution. Dev Comp Immunol 7: 229–239

Söderhäll K, Smith VJ (1986) The prophenoloxidase activating system: the biochemistry of its activation and role in arthropod cellular immunity, with special reference to crustaceans. In: Brehélin M (ed) Immunity in invertebrates. Springer, Berlin Heidelberg New York, pp 208–223

Söderhäll K, Unestam T (1979) Activation of serum prophenoloxidase in arthropod immunity. The specificity of cell wall glucan activation and activation by purified fungal glycoproteins of crayfish phenoloxidase. Can J Microbiol 25: 406–414

Söderhäll K, Häll L, Unestam T, Nyhlén L (1979) Attachment of serum prophenoloxidase to fungal cell walls in arthropod immunity. J Invertebr Pathol 34: 285–294

Söderhäll K, Levin J, Armstrong PB (1985) The effects of β-1,3-glucans on blood coagulation and amoebocyte release in the horseshoe crab, *Limulus polyphemus*. Biol Bull (Woods Hole) 169: 661–674

Söderhäll K, Smith VJ, Johansson MW (1986) Exocytosis and uptake of bacteria by isolated haemocyte populations of two crustaceans: evidence for cellular co-operation in the defence reactions of arthropods. Cell Tissue Res 245: 43–49

Söderhäll K, Rögener W, Söderhäll I, Newton RP, Ratcliffe NA (1988) The properties and purification of a *Blaberus craniifer* plasma protein which enhances the activation of haemocyte prophenoloxidase by a β-1,3-glucan. Insect Biochem 18: 323–330

Söderhäll K, Aspán A, Duvic B (1990) The proPO system and associated proteins; role in cellular communication in arthropods. Res Immunol 141: 896–904

Söderhäll K, Cerenius L, Johansson MW (1994a) The prophenoloxidase activating system and its role in invertebrate defence. In: Beck G, Cooper EL, Habicht GS, Marchalonis JJ (eds) Primordial immunity: foundations of the vertebrate immune system, vol 712. Ann NY Acad Sci, New York, pp 155–161

Söderhäll K, Johansson MW, Cerenius L (1994b) Pattern recognition in invertebrates: The β-1,3-glucan binding proteins. In: Hoffmann J, Janeway Jr C, Natori S (eds) Phylogenetic perspectives in immunity: the insect host defense. Molecular Biology Intelligence Unit. Landes, Austin, pp 97–104

Sottrup-Jensen L, Stepanik TM, Kristensen T, Wierzbicki DM, Jones CM, Lønblad PB, Magnusson S, Petersen TE (1984) Primary structure of human α_2-macroglobulin. V. The complete structure. J Biol Chem 259: 8318–8327

Sottrup-Jensen L, Borth W, Hall M, Quigley P, Armstrong PB (1990) Sequence similarity between α_2-macroglobulin from the horseshoe crab, *Limulus polyphemus*, and proteins of the α_2-macroglobulin family from mammals. Comp Biochem Physiol 96B: 621–625

Spycher SE, Arya S, Isenman DE, Painter RH (1987) A functional, thiolester-containing α_2-macroglobulin homologue isolated from the hemolymph of the American lobster (*Homarus americanus*). J Biol Chem 262: 14606–14611

Srimal S, Miyata T, Kawabata S, Iwanaga S (1985) The complete amino acid sequence of coagulogen isolated from Southeast Asian horseshoe crab. *Carcinoscorpius rotundicauda*. J Biochem (Tokyo) 98: 305–318

St Leger RJ, Cooper RM, Charnely AK (1988) The effect of melanization of *Manduca sexta* cuticle on growth and infection by *Metarhizium anisopliae*. J Invertebr Pathol 52: 459–470

Stoltz DB, Cook DI (1983) Inhibition of host phenoloxidase activity by parasitoid *Hymenoptera*. Experientia 39: 1022–1024

Sugumaran M, Saul SJ, Ramesh N (1985) Endogenous protease inhibitors prevent undesired activation of prophenoloxidase in insect haemolymph. Biochem Biophys Res Commun 132: 1124–1129

Takle GB, Lackie AM (1986) Chemokinetic behaviour of insect haemocytes in vitro. J Cell Sci 85: 85–94

Thörnqvist P-O, Johansson MW, Söderhäll K (1994) Opsonic activity of cell adhesion proteins and β-1,3-glucan-binding proteins from two crustaceans. Dev Comp Immunol 18: 3–12

Tomasini BR, Mosher DF (1990) Vitronectin. In: Coller BS (ed) Progress in hemostasis and thrombosis, vol 10. Saunders, Philadelphia, pp 269–305

Tsukamoto, T, Ichimaru Y, Kanegae N, Watanabe K, Yamaura I, Katsura Y, Funatsu M (1992) Identification and isolation of endogenous insect phenoloxidase inhibitors. Biochem Biophys Res Commun 184: 86–92

Unestam T, Söderhäll K (1977) Soluble fragments from fungal cell walls elicit defence reactions in crayfish. Nature (London) 267: 45–46

Valembois P, Roch P, Götz P (1988) Phenoloxidase activity in the coelomic fluid of earthworms. Dev Comp Immunol 13: 429–430

Vargas-Albores F, Guzmán M-A, Ochoa J-L (1993) A lipopolysaccharide-binding agglutinin isolated from brown shrimp (*Penaeus californiensis* Holmes). Comp Biochem Physiol 104B: 407–413

Yokoo S, Tojo S, Ishibashi N (1992) Suppression of the prophenoloxidase cascade in the larval haemolymph of the turnip moth, *Agrotis segetum* by an entomopathogenic nematode, *Steinernema carpocapsae* and its symbiotic bacterium. J Insect Physiol 38: 915–924

Yoshida H, Ashida M (1986) Microbial activation of two serine enzymes in the plasma fraction of the silkworm, *Bombyx mori*. Insect Biochem 16: 539–545

Yoshida H, Ochiai M, Ashida (1986) β-1,3-glucan receptor and peptidoglycan receptor are present as separate entities within insect prophenoloxidase activating system. Biochem Biophys Res Commun 141: 1177–1184

Inducible Humoral Immune Defense Responses in Insects

R.D. KARP

1 Introduction

There is a revolution taking place in the field of immunobiology that is challenging the widely held dogmatic belief that invertebrates are incapable of mounting true adaptive immune responses when exposed to foreign substances. If the definition of adaptive immunity is predicated on the possession of a thymus or lymph nodes, then invertebrates will never satisfy these requirements. However, a growing body of evidence from representatives of various invertebrate phyla, clearly indicate that the two basic criteria of an adaptive immune response, specificity and immunologic memory, have evolved in some of these animals. The data from these studies will hopefully force us to rethink, and redefine, what we mean by immunity. This will require that we no longer use anatomical complexity to gauge what is physiologically possible, but rather measure an animal's potential by the degree of molecular complexity that is engendered in its genome. This will surely open up our minds and lead to a better perspective concerning an understanding of how the sophisticated mechanisms of adaptive immunity evolved.

Insects have become one of the leading models for the study of invertebrate immune defense reactions for several reasons. First, insects of themselves are important to us. There are an estimated 850000 to 1 million known species of insects, comprising 85% of the animal life on Earth. As such, they represent our major competitor in the ecosystem, since they eat our food and vector many diseases serious for man. Obviously, any knowledge we can gain concerning their biology would be of immense importance in effecting their control. Second, one of the difficulties in studying invertebrates has been that it usually necessitates working with wild animals, which means that the investigator has no knowledge of the animal's background viz a viz prior exposure to environmental stresses or even age. Insects represent one group of invertebrates that can be raised relatively easily in captivity under controlled conditions, and thus they provide an excellent model for genetic and molecular analysis.

Much of the early work in invertebrate immunology was focused on determining if any of the members of this group had defense responses that were

Department of Biological Sciences, University of Cincinnati, Cincinnati, Ohio, USA analogous to those that had been defined in mammals. Along the way, many other types of inducible defense responses were discovered, which although nonspecific and lacking memory, were nevertheless very significant to the survival of these animals. Insects, for instance, have evolved an impressive array of mechanisms that put out a broad net to protect against microbial and other noxious agents in their environment. The American cockroach, an example of a longer-lived species, has been found to have developed effector mechanisms more akin to higher animals. Whether or not the cells and/or molecules mediating these reactions prove to be homologous to mammalian immune mediators remains to be seen. However, we have already learned that the concept of differentiating between self vs. nonself is basic to the survival of all animals. This review will discuss how insects act on the recognition of nonself utilizing inducible humoral responses. This discussion is not meant to be exhaustive, and will not cover defense mechanisms found elsewhere in this volume.

2 Antibacterial Responses

As can be seen in Table 1, the great majority of known immune mediators in insects have been discovered in animals which have been injected with bacteria. Although there is quite a variety of immune proteins induced in animals immunized with bacteria, many of these molecules have been found to exist in a number of genera, either in identical form or as analogs carrying out the same functions. There has obviously been a very strong selective pressure during the evolution of insects to generate mechanisms that can effectively deal with microbial pathogens.

Table 1. Inducible protein mediators of insect immunity

Name	M_r (kDa)	Activity
Lysozyme	14–16	Bactericidal
Lectins	72–2000	Opsonin, recognition?
Cecropins	3.5–4	Bactericidal
Sarcotoxin I	5	Bactericidal
Attacins	20–23	Bactericidal
Sarcotoxin II	24	Bactericidal
Sarcotoxin III	7	Bactericidal
Diptericins	9	Bactericidal
Defensins	4	Bactericidal
Apidaecin	2	Bactericidal
Abaecin	4	Bactericidal
Roach Antibacterial Factor	?	Bactericidal
Hemolin	48	Opsonin, recognition?
Roach IHF	400–700	Binding, recognition?

2.1 Lysozyme

The lysozymes represent a highly conserved group of antibacterial agents that are ubiquitous throughout the animal kingdom. There are basically two forms of lysozymes: The so-called goose type and the chicken type. Apparently all of the insect lysozymes have shared characteristics with the chicken type (Jollès et al. 1979; Dunn 1986; Sun et al. 1991a). The presence of lysozyme is widespread throughout the Lepidoptera. Various studies have demonstrated the induction of lysozyme in *Galleria* (Chadwick 1970; Powning and Davidson 1973), *Bombyx* (Powning and Davidson 1973), *Spodoptera* (Anderson and Cook 1979), *Hyalophora* (Hultmark et al. 1980) and *Manduca* (Dunn et al. 1985). Lysozyme has also been reported in the locust (Zachary and Hoffmann 1984) and the American cockroach (Powning and Irzykiewicz 1967). There is some suspicion that lysozyme plays a large role in antibacterial defense in the Lepidoptera since it is easily induced in these animals, whereas it has been very difficult to demonstrate induction in other insects such as roaches. Roaches seem to store the enzyme in hemocytes or have lysozyme-like activity present in the form of chitinases (Powning and Irzykiewicz 1967). Lysozyme can be induced not only by bacteria, but also by their cell wall fragments (Kanost et al. 1988). In addition, many studies have claimed that lysozyme can be induced either by trauma or the injection of saline. However, Dunn and his collaborators (Dunn et al. 1985; Kanost et al. 1988) have observed that when *Manduca* is raised in a sterile environment, fed sterile food and injected with pyrogen-free filtered saline, there is no appreciable induction of lysozyme in control animals.

The lysozymes are relatively small molecules with a reported M_r ranging from 14–16 kDa. Lysozyme genes have now been cloned and sequenced in *Hyalophora* (Sun et al. 1991a) *Drosophila* (Kylsten et al. 1992), and *Manduca* (Mulnix and Dunn 1994). Furthermore, it has now been established in *Manduca* that the regulation of lysozyme gene expression is at the transcriptional level (Mulnix and Dunn 1994). The results from the studies in *Manduca* reveal a logical sequence of events by which the degradative action of lysozyme on the cell walls of invading Gram-positive bacteria brings about the release of peptidoglycan fragments which then act as very potent elicitors of antibacterial peptides such as the cecropins and attacins (Kanost et al. 1988). Thus, it is hypothesized that an integrative cooperation can occur amongst the panel of antibacterial proteins that are induced by the exposure to bacteria, which stimulates the induction of a highly effective protective response.

2.2 Lectins

Lectins are a group of substances which can be loosely defined as molecules which bind to cells of various sorts through sugar–sugar interactions, bringing about either the agglutination of such cells or the enhanced phagocytosis of those cells.

The M_r range of known lectin molecules in insects is enormous, ranging from the 72 kDa M13 lectin of *Manduca* (Minnick et al. 1986) all the way up to the 2000 kDa lectin found in *Periplaneta* (Kubo et al. 1990). Lectin molecules are quite well represented in the Lepidoptera, with such examples as a hemagglutinating lectin in *Spodoptera exigua* (Pendland et al. 1988) and *Anticarsia gemmatalis* (Pendland and Boucias 1985), the 72 kDa molecule in *Manduca sexta* (Minnick et al. 1986), a 160 kDa lectin in *Hyalophora cecropia* (Castro et al. 1987), a 190 kDa lectin in *Sarcophaga peregrina* (Komano et al. 1980), and a 260 kDa lectin in *Bombyx mori* (Suzuki and Natori 1983). In addition, lectins have been identified in *Glossina* (Ibrahim et al. 1984), the honeybee (Gilliam and Jeter 1970), the grasshopper (Jurenka et al. 1982) and locust (Drif and Brehelin 1989), as well as a 1000 kDa lectin that has been observed in the cricket (Hapner and Jermyn 1981). Various lectins have been identified in cockroaches, for example the 140 and 390 kDa lectins in *Blaberus discoides* (Chen et al. 1993) and the 450, 1500 and 2000 kDa lectins in *Periplaneta americana* (Jomori et al. 1990; Kubo and Natori 1987; Kubo et al. 1990). All of the insect lectins that have been studied at the biochemical level have shown subunit structure.

One of the controversial areas in invertebrate immunology has been determining if lectins plays some significant role in defense. One would logically assume that a surface binding protein would serve as a recognition molecule that would lead to enhanced phagocytosis of whatever had been bound by the lectin. However, this has not been so easy to prove in many invertebrate taxa. There have been reports from studies on various insect genera that lectins may indeed participate in defense and perhaps even during development. Natori and his coworkers have isolated, purified and cloned the gene for a lectin found in *Sarcophaga peregrina* (Komano et al. 1980; Kubo et al. 1984; Takahashi et al. 1985). The lectin is induced in *Sarcophaga* by merely injuring the body wall with a hypodermic syringe needle. Natori and coworkers (Komano and Natori 1985; Takahashi et al. 1986) have also demonstrated that the induction of the lectin leads to enhanced lysis of sheep red blood cells injected into the body cavities of *Sarcophaga* larvae. In addition, Komano et al. (1980) observed that the lectin was induced during pupation, and so have concluded that this molecule may play a role in development of the insect by facilitating the clearance of disintegrated larval fragments. The inducible lipopolysaccharide-binding lectin discovered in the American cockroach has now been purified (Jomori et al. 1990), and the gene cloned and sequenced (Jomori and Natori 1991). The investigators have concluded that this molecule may very well be active in helping to bring about elimination of bacterial substances, again possibly during molting periods in the development of the roach. Finally, Pendland et al. (1988) have demonstrated that the galactose-binding lectin in *Spodoptera exigua* will act as an opsonin and facilitate the phagocytosis of fungal blastospores. So it would appear that there is some evidence to believe that lectins play a significant role in the life history of an insect, be it during development or for purposes of warding off infectious challenges from the environment.

2.3 Cecropins and Allied Factors

The cecropins represent a well-studied group of bactericidal peptides that were so named because much of the original work by Boman and his coworkers was carried out using the giant silk moth *Hyalophora cecropia* (Boman et al. 1974; Faye et al. 1975; Hultmark et al. 1980). These basic peptides have a wide range of activity, and will kill both Gram-negative and Gram-positive microorganisms (Hultmark et al. 1982). Three major and four minor cecropins have now been identified in *Hyalophora* (Hultmark et al. 1980; Steiner et al. 1981; Hultmark et al. 1982), with a M_r = 4 kDa. Since their original discovery, cecropins and their homologues have been found to be widespread among the Lepidopterans (Hoffmann et al. 1981; Okada and Natori 1983; Dunn et al. 1985; Morishima et al. 1990) as well as other insects such as *Glossina* (Kaaya et al. 1987) and *Drosophila* (Robertson and Postlethwait 1986; Samakovlis et al. 1990). Lee et al. (1989) have even reported finding a representative of the cecropins in pig intestine. Time will tell if cecropin-type molecules turn out to be as ubiquitous as the lysozymes.

It is now known that the mode of action of the cecropins is to attack the bacterial cell membrane (Okada and Natori 1984; Christensen et al. 1988; Samakovlis et al. 1990). The most likely scenario for this activity is that when the cecropins insert themselves into the bacterial membrane, they form voltage-dependent ion channels causing leakage, and eventually cell death (Okada and Natori 1985; Christensen et al. 1988). Steiner et al. (1988) believes that the target cell must be saturated with cecropin molecules before membrane disruption will occur. This activity is nonspecific and so cecropins have a fairly broad range of activity, with no evidence that the various species of cecropins exhibit selective killing (Faye and Hultmark 1993). The cecropins also seem to be produced in various tissues besides the fat body (Dickinson et al. 1988), which would indicate that they also have a very wide range of biological utility. Gudmundsson et al. (1991) have reported that cecropins start out as preproteins that go through four stages of processing before becoming the active molecule. In addition, there is a sequential expression of cecropin genes following induction in which transcripts for cecropins A and B appear within hours of the stimulatory event, whereas, cecropin D may be delayed for as long as 96 h post stimulus (Gudmundsson et al. 1991). The cecropins have been characterized very well, with amino acid sequences available from six species of insects (as reviewed in Faye and Hultmark 1993), and the genes cloned and sequenced in *Hyalophora* (Gudmundsson et al. 1991) and *Sarcophaga* (Kanai and Natori 1989). Many aspects of the biological significance of the cecropins wait to be studied.

A cecropin-like molecule that was discovered in *Sarcophaga peregrina* (Okada and Natori 1983) has also been well-studied. Sarcotoxin I has a similar molecular weight (5 kDa) to that of cecropins (Okada and Natori 1983), and also seems to have the same mode of action of disrupting the membranes of bacteria (Okada and Natori 1984, 1985).

Other factors that may be related to the cecropins are: Andropin, which is an antibacterial peptide found in the male reproductive tract of *Drosophila* (Samakovlis et al. 1991); and a 5 kDa antibacterial factor discovered in the migratory locust, *Locusta migratoria* (Hoffmann 1980).

2.4 Attacins and Allied Factors

A second major group of antibacterial proteins that were originally discovered in *Hyalophora* were named the attacins and have a M_r = 20–23 kDa (Hultmark et al. 1983). The genomic cloning carried out by Lee et al. (1983) indicated that there were two different attacins encoded in the genome. It was then found that the attacins, like the cecropins, are actually a family of proteins comprising four species that are basic, and two others that are acidic (Hultmark et al. 1983; Sun et al. 1991b). Unlike the cecropins, the range of activity for attacins against bacteria is quite narrow, and actually confined to a limited number of Gram-negative organisms (Faye and Hultmark 1993). The mode of action of the attacins is to alter the permeability of the outer membrane of the bacterium by inhibiting the synthesis of outer membrane proteins (Carlsson et al. 1991). It is thought that the result of this action by attacins allows improved access to the cell wall for lysozyme and to the cytoplasmic membrane for the cecropins (Carlsson et al. 1991).

A group of antibacterial peptides discovered in *Sarcophaga,* called sarcotoxin II, were found to have attacin-like activity (Ando et al. 1987). Sarcotoxin II is also polymorphic, encoded by a gene cluster (Kanai and Natori 1990) that yields three closely related forms (Ando et al. 1987).

Another attacin-like group of proteins has been discovered in the dipteran *Phormia terranovae* (Dimarcq et al. 1988). Although the M_r (9 kDa) is well below that of the *Hyalophora* attacins, the so-called diptericins are nevertheless considered, on the basis of gene sequencing studies (Sun et al. 1991b), to be related. Like the attacins, the diptericins show activity against Gram-negative organisms. Interestingly, diptericin activity can be induced by a sterile injury in axenically raised animals, which underscores the nonspecific nature of this response (Dimarcq et al. 1990). In addition, a secondary immunization in primed animals did not induce a secondary response, so immunologic memory is also not characteristic of this reactivity (Dimarcq et al. 1990). A diptericin homologue has been reported to exist in *Sarcophaga peregrina* (Ishikawa et al. 1992) together with another protein called sarcotoxin III (M_r = 7 kDa) which may also be related (Baba et al. 1987).

Other examples of attacin-like activity have now been reported in *Manduca sexta* (Hurlbert et al. 1985) and the tsetse fly, *Glossina morsitans morsitans* (Kaaya et al. 1987). A third possibility is a protein called coleoptericin, which has been found in the beetle *Zophobas atratus* (Bulet et al. 1991), which although only 8 kDa in M_r shares an activity spectrum with the attacins.

2.5 Defensins

The growing body of information concerning the group of antibacterial proteins known as defensins is indicating that this may represent yet another group of ubiquitous defense molecules that can be found throughout the evolutionary tree in a similar manner to the lysozymes. For instance, Lehrer et al. (1991) have reported their abundance in mammalian phagocytic cells. Insect defensins are small molecules with a $M_r = 4$ kDa. Defensins have been reported in the flesh fly *Sarcophaga peregrina* (they are known as sapecin; Matsuyama and Natori 1988a), in *Phormia terranovae* (Lambert et al. 1989), in the beetle *Zophobas artratus* (Bulet et al. 1991) and are known as royalisin in the honeybee (Fujiwara et al. 1990). The defensins constitute another antibacterial factor that is active against Gram-positive organisms. Although the defensins show a predilection for different bacterial targets from that of the cecropins, their mode of action appears to be similar in that they have been found to also attack the bacterial cell membrane (Matsuyama and Natori 1990). Matsuyama and Natori (1988b), during their cloning and sequencing studies of sapecin, found that this molecule was not only inducible, but was spontaneously expressed in embryonic and pupal stages during the development of *Sarcophaga*. These findings led the two authors to conclude that sapecin not only plays a role as a defense molecule, but also may play a role in the development of the animal.

2.6 The Antibacterial Response in the American Cockroach

The animal that we became interested in studying was the American cockroach, which not only represents a type of insect that does not go through the dramatic metamorphic changes experienced during the development of insects such as moths, but it is also found to have a longer life span. Since there had been so much reported data establishing the existence of very formidable antibacterial factors in holometabolous insects, we naturally became curious as to how the roach dealt with bacterial threats.

Groups of adult male roaches were injected with either sterile, pyrogen-free Burns-Tracey saline (BTS; Ludwig et al. 1957) or 10^6 glutaraldehyde-killed *Pseudomonas aeuruginosa* cells. Following various periods of rest, all animals were challenged with an LD_{100} dose of viable *Pseudomonas,* and the percent survival in each group was calculated 48-h post challenge. The results indicated that the injection of killed *Pseudomonas* had generated a protective response against the lethal dose of living bacteria. The surprise was that the response rather than lasting only a few days, as might be expected from the studies in Lepidopterans, still showed protection 3 weeks after the challenge (Faulhaber and Karp 1992). The longevity of the roach response to *Pseudomonas* thus could not be explained either by acute cellular reactivity or the presence of short-lived cecropin-type molecules.

It is known that cecropins exhibit a broad spectrum of bactericidal activity, which is to say that they are not specific. Since the kinetics of the roach antibacterial response was so unlike the Lepidopteran response, we thought perhaps other characteristics such as specificity might also differ between the roach and holometabolous insects. For this purpose, a panel of unrelated bacteria were used for immunization of adult male roaches prior to challenge with *Pseudomonas*. Two Gram-positive organisms, *Streptococcus lactis* and *Micrococcus lysodeikticus*, and two Gram-negative organisms, *Enterobacter cloacae* and *Serratia marcescens* were each killed by treatment with glutaraldehyde and used for immunization in doses comparable to that used for *Pseudomonas*. Immunized animals, as well as controls injected with sterile, pyrogen-free BTS, were rested for various periods of time and then challenged with an LD_{100} dose of *Pseudomonas*. The results (Faulhaber and Karp 1992) proved to be quite interesting since during the initial 3 days of the response, any of the injected unrelated bacteria generated protection that was statistically equal to animals being immunized with *Pseudomonas*. However, by day 4 post immunization, there was a dramatic change in the kinetics, since now there was a significant difference in the number of survivors in groups in which animals were immunized with killed *Pseudomonas* as opposed to the unrelated bacteria. The responses induced by Gram-negative organisms were still higher than controls, but significantly below that of groups immunized with *Pseudomonas;* whereas the Gram-positive organisms were no better at induction than BTS injection. A picture began to emerge indicating that the response to bacteria in the American cockroach was biphasic: the first phase being acute and nonspecific, and the second phase being more long term and showing specificity. Our initial thoughts were that the acute first phase may be dominated by nonspecific cellular events (i.e. phagocytosis, encapsulation), whereas the second phase may be mediated by some induced factor showing selective activity.

We developed a simple in vitro assay in order to determine if a humoral factor was being induced in animals immunized with bacteria. Cell-free hemolymph was obtained from immune as well as BTS-injected control animals of days 2 and 7 post immunization, and diluted to various protein concentrations. To aliquots of each sample, 2×10^6 viable *P. aeruginosa* were added and incubated at 27 °C for 24 h in a shaking water bath. A 10 μl drop of each sample was then placed on a Tripticase Soy agar plate, and incubated overnight. The results indicated that the injection of killed bacteria into roaches had induced a humoral bactericidal factor during both phases of the response (Faulhaber and Karp 1995). Continuing studies are presently attempting to determine the nature of the factor(s), and whether the same factor(s) is acting in both components of the antibacterial response. There are many scenarios that can be considered, and it may well be that a whole panoply of defense mechanisms come into play during the acute phase, such as phagocytic cells, lysozyme, perhaps even cecropin-like molecules, which mask the specific nature of the antibacterial humoral factor. As time goes on in the response, these acute factors fall by the wayside, revealing the specific activity of the induced antibacterial factor. Further analysis will hopefully sort all this out.

An additional piece of important information was revealed to us by treating both day 2 and 7 post immunization hemolymph with a proteolytic enzyme. The highest concentration of immune and control hemolymph samples were incubated with proteinase K for 1.5 h at 37 °C, and then tested in the in vitro assay for bactericidal activity. The enzyme treatment totally destroyed the killing activity of the immune hemolymph, indicating that the induced antibacterial factor was protein in nature. We are just beginning studies aimed at characterizing this factor to determine if it is at all related to the inducible humoral factor (IHF) generated in the roach against foreign soluble proteins or whether it represents a separate family of inducible immune proteins in this animal.

2.7 Hemolin

Among the nine antibacterial proteins originally induced in *Hyalophora* (Faye et al. 1975), the protein designated P4 seemed to be the major protein produced following bacterial stimulation. However, P4 did not seem to demonstrate any antibacterial activity like that of the cecropins, attacins and lysozymes that were also induced. It has now been reported that P4, which has been renamed hemolin, does in fact share sequence structure with molecules that are members of the immunoglobulin superfamily (Sun et al. 1990). The M_r for hemolin was found to be 48 kDa. A homologous protein has also been found to be induced in *Manduca sexta* (Ladendorff and Kanost 1990). The *Manduca* hemolin has also been found to be the major hemolymph protein induced by bacterial immunization (Ladendorff and Kanost 1991), and is generated in both larval and adult stages. Sun et al. (1990) found that hemolin binds to the surface of bacteria. Ladendorff and Kanost (1991) observed that hemolin also associates with the surface of hemocytes and inhibits hemocyte aggregation responses, suggesting a possible role for hemolin in regulating hemocyte adhesion during the actual recognition and subsequent response to bacteria. The molecule which hemolin is most similar to is *Drosophila* neuroglian, an adhesion molecule (Sun et al. 1990). Wang et al. (1995) have observed that hemolin mRNA not only accumulates in the fat body following stimulation, but also in granular hemocytes. This may be the first clue to the biologic activity of hemolin, since Wang et al. (1995) surmise that following stimulation, granular hemocytes begin to synthesize hemolin and store it in cytoplasmic granules. Degranulation of these cells during phagocytic events could then be visualized as delivering a very strong stimulus to other hemocytes in the microenvironment, resulting in a more effective response to the bacterial invasion. However, it must be kept in mind that just because a molecule shares a structure with members of the immunoglobulin superfamily, it does not necessarily follow that it plays a direct role in a defense response. In fact there are two other molecules in insects, fasciclin II (Harrelson and Goodman 1988) and amalgam (Seeger et al. 1988), which are known to be members of the immunoglobulin superfamily, but participate in the development of the insect rather than as mediators of immunity. Perhaps hemolin will turn out to participate in both defense and development as has been proposed for *Sarcophaga* lectin and sapecin.

2.8 Unclassified Antibacterial Factors

There are several other types of hemolymph proteins that have been induced by the injection of bacteria which do not fall into any of the above categories. Casteels et al. (1989) observed the induction of a group of antibacterial peptides in the honeybee *Apis mellifera* that have a M_r of only 2.1 kDa. These so-called apidaecins are quite non specific, and can be induced just as well by the injection of latex particles or india ink (although the authors did not check these materials for bacterial or pyrogen contamination). The apidaecins are basic peptides that are bactericidal and will inhibit the growth of most Gram-negative organisms tested, but showed only weak activity against Gram-positive bacteria. Thus, unlike cecropins, the apidaecins were not found to disturb the cell membrane of bacteria. A possibly related molecule in the honeybee, abaecin, is of higher M_r (4 kDa) and shows moderate killing activity against both Gram-negative and Gram-positive bacteria (Casteels et al. 1990).

Postlethwait et al. (1988) have reported finding an antibacterial response in the medfly *Ceratitis capitata*. The response is evident after only 3 h following induction and lasts for about 8 days. Apparently, there are four factors produced following stimulation, but at this point, we do not know if they are related to other known antibacterial factors, such as the cecropins.

3 The Response to Soluble Proteins

The antibacterial proteins were by and large discovered through studies with either pupal or larval stages of holometabolous insects. Holometabolous insects are generally short-lived, with very finite stages, in that the adults serve as egg laying machines that may only exist a few weeks depending on the species. Since most of the work on insect immunity at that point involved the use of bacteria or other particulate antigens, we were interested in trying a different approach by studying the insect response to soluble proteins. In addition, we chose to study a paurometabolous insect, which does not pass through the complete metamorphic stages typical of the holometabolous insect, and in the case of the American cockroach, one which has a longer life span of up to 4 years. We speculated that an animal such as the long-lived roach might have had reason to evolve mechanisms more closely attuned to classical adaptive immunity in order to deal with repeated exposures to environmental threats.

Our first experiments were based on immunizing adult, male American cockroaches with an inactivated invertebrate toxin (honey bee venom, HBT), and after various periods of rest, determining if these animals could then survive a challenge with active HBT. By enumerating the number of survivors at each time period, we found that a single injection of HBT toxoid stimulated a protective response that was already apparent by 3 days post immunization. The intensity of the response continued to increase with time until peak levels of protection were reached at two weeks post immunization (Rheins et al. 1980). At this time, 80% of the immunized roaches were able to survive a lethal challenge of HBT, whereas less than 10% of the control animals that received injections of sterile BTS survived the lethal challenge. Since the response did not begin to decline until 3 to 4 weeks following immunization, it was obvious that the primary response to this soluble protein toxin in the American cockroach was of much longer duration than the established antibacterial responses that had been reported to exist in holometabolous insects.

The next issue to address was whether the response in the roach displayed antigenic specificity. For this purpose, we chose Cottonmouth moccasin venom (CMV) as a test antigen, since it is made up of components that are very similar to those of HBT, but would differ antigenically because it comes from a vertebrate animal. In these experiments, roaches were immunized with either inactivated HBT or CMV, rested 2 weeks to allow a maximum response to develop, and then challenged with either homologous or heterologous active venoms. We observed that animals immunized with inactivated HBT and challenged with active HBT had a survival rate of 84.2%, which was in agreement with our previous findings. Similarly, animals immunized with inactivated CMV and challenged with active CMV had a survival rate of 65%. However, when immunized animals received challenges with heterologous venoms, they did very poorly, since only 10% of the roaches immunized with inactivated HBT survived a challenge with active CMV, and conversely, only 5% of the animals immunized with inactivated CMV survived a challenge with active HBT. Thus, the initial results we observed using HBT as an antigen were not due to nonspecific detoxification mechanisms being activated, but rather to the generation of a factor that could discriminate between two similar toxins that differed antigenically. This finding was yet another characteristic that differentiated the roach immune factor from all of the mediators of antibacterial responses which do not show this type of antigenic discrimination.

Two key characteristics that typify an adaptive immune response, as defined in mammals, are the presence of specificity and immunologic memory. These features of immunity work hand in hand in allowing animals to selectively respond to a repeated threat by, or exposure to, a foreign agent. Our early experiments had already indicated that the roach immune factor displayed specificity for the inducing antigen, but was there also a memory component to the response? To test for the presence of immunologic memory, groups of adult male roaches were injected with primary and secondary doses of the inactivated HBT, spaced 2 weeks apart. The animals were then challenged with active HBT following various periods of rest post secondary immunization. The kinetics of responsiveness changed quite drastically after a second exposure to the bee venom (Karp and Rheins 1980). The response now was greatly enhanced, such that even at day 3 there was a significant difference (P=0.005) between the number of animals able to survive a lethal challenge following a primary immunization and the number able to survive following a secondary immunization. In addition, the peak response (80% survivors) came a week earlier at 7 days, as opposed to 14 days, and the response was more prolonged since it did not decline until 6 to 7 weeks post secondary immunization. These are all the hallmarks of what is considered to be immunologic memory. Additional experiments reinforced our conclusions since it was observed that: (1) Animals receiving a primary immunization, rested 7 to 8 weeks, then injected with a second dose of the inactivated HBT showed peak activity after only three days post secondary immunization, thus indicating that the memory was long-term in nature; (2) secondary responsiveness could be stimulated in primed animals with a subimmunizing dose of antigen; and (3) secondary responsiveness could not be stimulated in animals primed with the HBT toxoid and then given the heterologous CMV toxoid as the second immunization. Thus, it was quite obvious that roaches were capable of generating immunologic memory to soluble protein antigens.

At this point, we had discovered an adaptive response to the injection of soluble protein toxins, but did not know whether the protective response was mediated by cellular or humoral factors, or both. We turned to passive transfer experiments as a way of making such a determination. Groups of untreated animals were injected with either cell-free hemolymph from animals injected with BTS or inactivated HBT, or various concentrations of hemocytes from control and immune animals. Following a brief rest, all, groups were challenged with a lethal dose of HBT. The results indicated that only the transfer of immune cell-free hemolymph to naive animals brought about differential protection (as compared to the injection of control materials) against a lethal challenge of toxin (Rheins et al. 1980). Thus, we concluded that we had induced a humoral response in the roach to these antigens.

We became interested in studying some of the biological parameters of the humoral response in roaches in order to determine how comparable this response was to known responses in higher animals. It is well known that the females of most vertebrate species enjoy superior immune responses as compared to males (Lahita 1982). The rationale for this circumstance is that since females must bear and nurture the young, they developed enhanced immunological capacity not only to maintain themselves, but also to have the capability to passively protect their offspring. Since it has been estimated that a single female roach may produce 600 offspring during her life span, we reasoned that similar selective pressures may have been at work during the evolutionary development of cockroaches. To test this hypothesis, adult female roaches were immunized with inactivated HBT, rested for various periods of time, and then challenged with a lethal dose of HBT. The results (Rheins and Karp 1985a) indicated that female roaches did indeed have a superior immune response to that of males, since females attained a protection rate of 92% after only 7 days post immunization as compared to 62.9% for males; reached peak activity a week sooner than males; and generated a response that lasted twice as long as that of males. Thus, the general phenomenon amongst vertebrates in which females of the species are more immunocompetent than males was also found to be true among roaches.

Another characteristic that is usually associated with vertebrate immune responses is the association between age and degree of immunocompetence (Belisle and Strausser 1981). It is generally true that an animal is less immunocompetent at the two extremes of its life span. Very young animals are continuing to develop their immune capacity, while older members of the population begin to lose their ability to cope with environmental challenges. To determine if this trend was also true of the roach humoral response, animals representing distinct developmental phases were tested for their ability to respond to HBT. Immature nymphs (body weight 200–500 mg) were injected with inactivated HBT, rested for various periods of time, and then challenged with a lethal dose of HBT. We found that the nymphs displayed a significant lag of 2 weeks following immunization before they were able to respond to the challenge in the same manner as adults (Rheins and Karp 1985b). At the other end of the spectrum, older adults (at least 5 months into adulthood) immunized in the same fashion against HBT displayed a significant decline in reactivity particularly during the early phases of the response, and in general, demonstrated a less vigorous response overall (Rheins and Karp 1985b). It was interesting to note that both the nymphs and older adults demonstrated good secondary responsiveness following a second immunization, indicating that once cells were primed they could react just as well in these animals as those in mature adults. Taken with the results of the gender study, the effect of age on roach immune capacity led us to conclude that there were at least two very strong biological similarities between the roach humoral response and vertebrate immune responses indicating that environmental pressures had led very disparate animals along analogous evolutionary paths.

We subsequently determined that immune hemolymph would form strong precipitin bands in Ouchterlony gels with HBT, and that trypsin treatment prior to the incubation not only destroyed this activity, but also abrogated the ability of immune hemolymph to passively transfer protection to naive animals (Rheins and Karp 1982). Thus, the inducible humoral factor (IHF) in the roach had to be protein in nature.

We then began to analyze immune and control hemolymph samples using SDS-polyacrylamide gel electrophoresis (SDS-PAGE) in order to determine if there were any differences in the profiles of hemolymph proteins due to immunization. The results of those studies clearly indicated that a band with an estimated M_r of 102 kDa was always increased in concentration in immune animals (George et al. 1987b). Although other bands could also be found enhanced from time to time, later studies showed that it was only the 102 kDa band that was consistently increased no matter what antigen was used for immunization of the animals (Duwel-Eby et al. 1991). Phospholipase A_2 (a component of bee venom) now replaced HBT as the soluble protein toxin antigen, but other antigens such as creatine phosphokinase, and nontoxic antigens such as ovalbumin, bovine gamma globulin and cytochrome c, all induced the enhanced production of the 102 kDa band. Thus, we not only found that a single protein band was consistently associated with the immune state, but we also established that the IHF could be induced with nontoxic proteins. In fact, extensive absorption studies indicated that after incubating immune hemolymph with the inducing antigen, the only band missing in subsequent SDS-PAGE analysis of the supernatants was the 102 kDa band (George et al. 1987b).

Since we determined that the reduced 102 kDa protein was basically inactive, we subjected immune hemolymph to non reducing-PAGE (NR-PAGE) to determine if we could identify an enhanced fraction that would still be biologically active. We found that immune hemolymph was fractionated into 5 bands following treatment on 6% NR-PAGE gels, two of which (NR bands 2 and 4) showed enhanced levels of protein production as compared to hemolymph from BTS-injected control animals. When proteins were eluted from the various NR-PAGE bands and tested for their ability to passively protect naive animals against the lethal effects of PA2, we found that only NR bands 2 and 4 had the capability to pass on the immunity. Further experiments also determined that the ability to bind antigen in vitro was found to reside in NR bands 2 and 4 (Duwel-Eby et al. 1991). The first in vitro approach was to use affinity chromatography in which rabbit polyclonal antibodies against PA2 were covalently linked to the column. Immune hemolymph against PA2 was first incubated with the original antigen, and then introduced to the column. The idea was to allow the rabbit antibody to bind to the PA2 in the complexes, thus dragging the immune factor along with it. Following elution using a low pH wash, we found that NR bands 2 and 4 had specifically bound to the column as compared to similar analysis with hemolymph from BTS-injected controls. A second in vitro approach combined the use of Western blotting and autoradiography. NR-PAGE gels were run with both immune hemolymph and control hemolymph samples, and then electroblotted onto nitrocellulose. The blots were then incubated with ^{125}I-PA2 for 2 h at room temperature. Any binding of the radiolabeled antigen was then visualized using autoradiography. Once again, only NR bands 2 and 4 in the immune sample bound the iodinated PA2. It is interesting to note that SDS-PAGE analysis of NR bands 2 and 4 indicated that both contained the 102 kDa protein, which is consistent with our earlier findings associating the 102 kDa band with immunity in the roach.

At this point, we are attempting to determine how many species of proteins may be present in NR bands 2 and 4, and then identify which one is responsible for the immune activity. This will allow us to finally characterize and sequence the IHF so that we can compare it to other known immune mediators.

4 Concluding Remarks

Extensive studies, particularly among the Lepidopteran insects, clearly indicate that these animals have evolved very effective humoral mediators for the purpose of defense against bacteria. The strategy for these animals was to develop some very generalized effector molecules that could deliver a broad base of protection.

Thus, we have the lysozymes, defensins and sapecins which are lethal to many Gram-positive bacteria; the attacins, sarcotoxin II and diptericins which are lethal to Gram-negative bacteria; and the cecropins and sarcotoxin I which have strong activity against Gram-negative bacteria and some activity against Gram-positives as well. Thus, any notion of specificity for these molecules is based on a very broad subdivision of bacteria into Gram-positive and Gram-negative organisms. Unfortunately, there are very few studies in which holometabolous insects have been injected with either soluble proteins or toxins to determine how they respond to non particulate foreign material. The few studies that have been done (Bettini 1965; Kamon and Shulov 1965) proved to be inconclusive.

On the other hand, the American cockroach, although considered to be more primitive than holometabolous insects, seems to have evolved much more sophisticated immune defense mechanisms. The roach possesses various responses to soluble proteins, bacteria and foreign tissue, that demonstrate specificity at a much more refined level. In addition, unlike the holometabolous defense mechanisms, the roach demonstrates the development of immunologic memory during responses to soluble proteins (Karp and Rheins 1980) and to foreign allografts (Hartman and Karp 1989).

How do we explain these vast differences between two types of insects? First, we must remember that insects can be extremely different from one another physiologically as well as in life style. In other words, if you have seen one insect, you have not seen them all! In fact, the differences between insect taxa can be so great that comparing a moth to a roach could be considered the same as comparing a frog to a mouse, since both the moth and the frog go through very dramatic metamorphic changes during their life cycles, whereas the roach and the mouse develop from one basic morphic form. In addition, life spans of holometabolous insects are far shorter than that of the roach. The typical 1-year (or less) life span of the holometabolous insect can be thought of as being divided into very finite stages, in which three different animals, larva, pupa and adult, must find a way to survive for a relatively short period of time in order to make it to the next stage. It would be logical to assume that these short-lived species would need an extremely quick response to a foreign invader, such as a bacterium, and to rely on molecules that have broad specificity so as to not get too complicated, since time is of essence. Also, due to the time frame in question, long-term responses would be unnecessary, since: (1) they might not get the time required to developed and be fully effective, since each stage lasts for only a short period of time; (2) given the short life span of the animal, it might actually be more efficient to start over with a quickly mobilized nonspecific type of response; and (3) if points 1 and 2 are true, then why clutter the genome with unusable or ineffective genes?

In the case of insects like the roach, which enjoys a relatively long life span, the strategy has to be different because the animal is guaranteed of repeated exposures to infectious, noxious or other foreign agents throughout its existence. Although the roach certainly must have acute responses like holometabolous insects to avoid being overwhelmed, it would actually be more efficient for a longer lived animal to not have to start over every time it encounters the same antigen, but to develop more specific molecules that discriminate between antigens, and develop the capacity in the cells that generate these factors to be modified such that they will be even quicker in their response to a second exposure to that antigen. The same logic would apply to an explanation of why the roach has evolved the ability to recognize and react to allografts (George et al. 1987a). With a longer life span would come the increased threat of cells in the roach being adversely altered by environmental agents such as viruses and toxins, or by spontaneous mutation. Thus, it would be of great survival advantage to this animal to posses a mechanism for recognizing altered-self and disposing of aberrant cells before they became a threat.

Our understanding of adaptive immunity would require that clonal expansion must occur in stimulated immunocytes before one can conjur up the notion of immunologic memory. Thus, the possible existence of immune memory in invertebrates has been dismissed by many workers since such animals lack organized lymphoid tissue that could facilitate such activity. However, we have found that when roaches were injected with glutaraldehyde-killed *Enterobacter cloacae,* and the subsequent hemocyte division rate analyzed using flow cytometry, there was a significant increase in cell division and in the number of new hemocytes following antigenic stimulation (Ryan and Karp 1993). When one considers the physical size of the roach, and that it has an open hemocoel, one could almost consider the animal a walking lymph node. If the circulating hemocytes are able to respond to antigenic stimulation by cell division, then there really is no need for a specialized organ to carry this out in the way that it is in more complex life forms.

This review has described a rather rich catalogue of humoral immune factors that serve as potent molecules in the defense of insects against environmental threats. Cellular mechanisms such as phagocytosis (see review by Wago 1991), nodule formation, and encapsulation (see review by Huws Davies and Siva-Jothy 1991), as well as the prophenoloxidase system (see review by Sugumaran and Kanost 1993), which have not been part of the focus of this review, play an enormous role in the overall effectiveness of defense reactivity in insects. Taken together, one cannot come away with any conclusion other than insects have developed a very impressive set of mechanisms to deal with environmental threats. The future for insect immunology is a bright one, with the expectation of exciting revelations generated by understanding the significance of what has already been ascertained, as well as the prospect of new surprises that are awaiting our discovery.

Acknowledgments. The work presented in this review concerning the studies on the American cockroach was supported by NSF Research Grants PCM 8316140, DCB 8702382, NIH Research Grant GM GM 39398, and the University of Cincinnati Foundation Fund for Comparative Immunology.

References

Anderson RS, Cook ML (1979) Induction of lysozymelike activity in the hemolymph and hemocytes of an insect, *Spodoptera eridania.* J Invertebr Pathol 33: 197–203

Ando K, Okada M, Natori S (1987) Purification of sarcotoxin II, antibacterial proteins of *Sarcophaga peregrina* (flesh fly) larvae. Biochemistry 26: 226–230

Baba K, Okada M, Kawano M, Kormano T, Natori S (1987) Purification of sarcotoxin III, a new antibacterial protein of *Sarcophaga peregrina*. J Biochem (Tokyo) 102: 69–74

Belisle EH, Strausser HR (1981) Sex-related immunocompetence of Balb/c mice. II. Study of immunologic responsiveness of young, adult and aged mice. Dev Comp Immunol 5: 661–670

Bettini S (1965) Acquired immune response of the housefly *Musca domestica* (Linnaeus), to injected venom of the spider *Lactrodectus mactans tredecimagluttatus* (Rossi). J Invertebr Pathol 7: 378–383

Boman HG, Nilsson-Faye I, Paul K, Rasmuson T (1974) Insect immunity. I. Characterstics of an inducible cell-free antibacterial reaction in hemolymph of *Samia cythia* pupae. Inf Immunol 10: 136–145

Bulet P, Cociancich S, Dimarcq J-L, Lambert J, Reichhart J-M, Hoffmann D, Hetru C, Hoffmann JA (1991) Isolation from a coleopteran insect of a novel inducible antibacterial peptide and of new members of the insect defensin family. J Biol Chem 266: 24520–24525

Casteels P, Ampe C, Jacobs F, Vaeck M, Tempst P (1989) Apidaecins – antibacterial peptides from honeybees. EMBO J 8: 2387–2391

Casteels P, Ampe C, Riviere L, Van Damme J, Elicone C, Fleming M, Jacobs F, Tempst P (1990) Isolation and characterization of abaecin, a major antibacterial response peptide in the honeybee (*Apis mellifera*). Eur J Biochem 187: 381–386

Carlsson A, Engström P, Palva ET, Bennich H (1991) Attacin, an antibacterial protein from *Hyalophora cecropia*, inhibits synthesis of outer membrane proteins in *Escherichia coli* by interfering with OMP gene transcription. Inf Immunol 59: 3040–3045

Castro VM, Boman HG, Hammarström S (1987) Isolation and characterization of a group of isolectins with galactose/N-acetygalactoseamine specificity from hemolymph of the giant silk moth *Hyalophora cecropia*. Insect Biochem 17: 513–523

Chadwick JS (1970) Relation of lysozyme concentration to acquired immunity against *Pseudomonas aeruginosa* in *Galleria mellonella*. J Invertebr Pathol 15: 455–456

Chen C, Ratcliffe NA, Rowley AF (1993) Detection, isolation and characterization of multiple lectins from the haemolymph of the cockroach *Blaberus discoidalis*. Biochem J 294: 181–190

Christensen B, Fink J, Merrifield RB, Mauzerall D (1988) Channel-forming properties of cecropins and related model compounds incorporated into planar lipid membranes. Proc Natl Acad Sci USA 85: 5072–5076

Dickinson L, Russell V, Dunn PE (1988) A family of bacteria-regulated, cecropin D-like peptides from *Manduca sexta*. J Biol Chem 263: 19424–19429

Dimarcq JL, Keppi E, Dunbar B, Lambert J, Reichhart JM, Hoffmann D, Rankine S, Fothergill JE, Hoffmann JA (1988) Insect immunity. Purification and characterization of a family of novel inducible antibacterial proteins from immunized larvae of the dipteran *Phormia terranovae* and complete amino-acid sequence of the predominant member, diptericin A. Eur J Biochem 171: 17–29

Dimarcq JL, Zachary D, Hoffmann JA, Hoffmann D, Reichhart JM (1990) Insect immunity – Expression of the two major inducible antibacterial peptides, defensin and diptericin, in *Phormia terranovae*. EMBO J 9: 2507–2515

Drif L, Brehélin M (1989) Agglutinin mediated immune recognition in *Locusta migratoria* (Insecta). J Insect Physiol 35: 729–736

Dunn PE (1986) Biochemical aspects of insect immunology. Annu Rev Entomol 31: 321–339

Dunn PE, Dai W, Kanost MR, Geng D (1985) Soluble peptidoglycan fragments stimulate antibacterial protein synthesis by fat body from larvae of *Manduca sexta*. Dev Comp Immunol 9: 559–568

Duwel-Eby LE, Faulhaber LM, Karp RD (1991) The inducible humoral response in the cockroach. In: Gupta AP (ed) Immunology of insects and other arthropods. Academic Press, New York, pp 385–402

Faulhaber LM, Karp RD (1992) A diphasic immune response against bacteria in the American cockroach. Immunology 75: 378–381

Faulhaber LM, Karp RD (1995) An inducible antibacterial protein in the American cockroach that demonstrates killing activity in vitro. J Insect Physiol (submitted)

Faye I, Hultmark D (1993) The insect immune proteins and the regulation of their genes. In: Beckage NE, Thompson SN, Federici BA (eds) Parasites and pathogens of insects. Volume 2: Pathogens. Academic Press, New York, pp 25–53

Faye I, Pye A, Rasmuson T, Boman HG, Boman IA (1975) Insect immunity. II. Simultaneous induction of antibacterial activity and selective synthesis of some hemolymph proteins in diapausing pupae of Hyalophora cecropia and Samia cynthia. Inf Immunol 12: 1426–1438

Fujiwara S, Imai J, Fujiwara M, Yaeshima T, Kawashima T, Kobayashi K (1990) A potent antibacterial protein in royal jelly – purification and determination of the primary structure of royalisin. J Biol Chem 265: 11333–11337

George JF, Howcroft TK, Karp RD (1987a) Primary integumentary allograft reactivity in the American cockroach Periplaneta americana. Transplantation 43: 514–519

George JF, Karp RD, Rellahan BL, Lessard JL (1987b) Alteration of the protein composition in the hemolymph of American cockroaches immunized with soluble proteins. Immunology 62: 505–509

Gilliam M, Jeter WS (1970) Synthesis of agglutinating substances in adult honeybees against Bacillus larva. J Invertebr Pathol 16: 69–70

Gudmundsson GG, Lidholm D-A, Åsling B, Gan R, Boman HG (1991) The cecropin locus: cloning and expression of a gene cluster encoding three antibacterial peptides in Hyalophora cecropia. J Biol Chem 266: 11510–11517

Hapner KD, Jermyn MA (1981) Haemagglutinin activity in the haemolymph of Teleogryllus commodus (Walker). Insect Biochem 11: 287–295

Harrelson AL, Goodman CS (1988) Growth cone guidance in insects: fascilin II is a member of the immunoglobulin superfamily. Science 242: 700–708

Hartman RS, Karp RD (1989) Short-term immunologic memory in the allograft response of the American cockroach. Transplantation 47: 920–922

Hoffmann D (1980) Induction of antibacterial activity in the blood of the migratory locust Locusta migratoria L. J Insect Physiol 26: 539–549

Hoffmann D, Hultmark D, Boman HG (1981) Insect immunity: Galleria mellonella and other Lepidoptera have cecropia-P9-like factors active against gram negative bacteria. Insect Bochem 11: 537–548

Hultmark D, Steiner H, Rasmusön T, Boman HG (1980) Insect immunity. Purification and properties of three inducible bactericidal proteins from hemolymph of immunized pupae of Hyalophora cecropia. Eur J Biochem 106: 7–16

Hultmark D, Engström A, Bennich H, Kapur R, Boman HG (1982) Insect immunity: Isolation and structure of cecropin D and four minor antibacterial components from cecropia pupae. Eur J Biochem 127: 207–217

Hultmark D, Engström A, Andersson K, Steiner H, Bennich H, Boman HG (1983) Insect immunity. Attacins, a family of antibacterial proteins from Hyalophora cecropia. EMBO J 2: 571–576

Hurlbert RE, Karlinsey JE, Spence KD (1985) Differential synthesis of bacteria-induced proteins of Manduca sexta pupae and larvae. J Insect Physiol 31: 205–215

Huws Davies D, Siva-Jothy T (1991) Encapsulation in insects: polydnaviruses and encapsulation-promoting factors. In: Gupta AP (ed) Immunology of Insects and other arthropods. CRC Press, Boca Raton, pp 119–132

Ibrahim EAR, Ingram GA, Molyneux DH (1984) Haemagglutinins and parasite agglutinins in haemolymph and gut of Glossina. Tropenmed Parasitol 35: 151–156

Ishikawa M, Kubo T, Natori S (1992) Purification and characterization of diptericin homologue from Sarcophaga peregrina (flesh fly) Biochem J 287: 573–578

Jollès J, Schoentgen F, Croizier G, Croizier L, Jollès P (1979) Insect lysozymes from three species of lepidoptera: Their structural relatedness to the C (chicken) type lysozyme: J Mol Evol 14: 267–271

Jomori T, Natori S (1991) Molecular cloning of cDNA of lipopolysaccharide-binding protein from the hemolymph of the American cockroach, Periplaneta americana – Similarity of the protein with animal lectins and its acute phase expression. J Biol Chem 266: 13318–13323

Jomori T, Kubo T, Natori S (1990) Purification and characterization of lipopolysaccharide-binding protein from the hemolymph of the American cockroach. Eur J Biochem 190: 201–206

Jurenka R, Manfredi K, Hapner KD (1982) Haemagglutinin activity in Acrididae (grasshopper) haemolymph. J Insect Physiol 28: 177–181

Kaaya GP, Flyg C, Boman HG (1987) Induction of cecropin and attacin-like antibacterial factors in the haemolymph of *Glossina morsitans morsitans*. Insect Biochem 17: 309–315

Kamon E, Shulov A (1965) Immune response of locusts to venom of the scorpion. J Invertebr Pathol 7: 192–198

Kanai A, Natori S (1989) Cloning of gene cluster for sarcotoxin I, antibacterial proteins of *Sarcophaga peregrina*. FEBS Lett 258: 199–202

Kanai A, Natori S (1990) Analysis of a gene cluster for sarcotoxin II, a group of antibacterial proteins of *Sarcophaga peregrina*. Mol Cell Biol 10: 6114–6122

Kanost MR, Dai W, Dunn PE (1988) Petidoglycan fragments elicit antibacterial protein synthesis in larvae of *Manduca sexta*. Arch Insect Biochem Physiol 8: 147–164

Karp RD, Rheins LA (1980) Induction of specific humoral immunity to soluble proteins in the American cockroach (*Periplaneta americana*). II. Nature of the secondary response. Dev Comp Immunol 4: 629–639

Komano H, Natori S (1985) Participation of *Sarcophaga peregrina* humoral lectin in the lysis of sheep red blood cells injected into the abdominal cavity of larvae. Dev Comp Immunol 9: 31–40

Komano H, Mizuno D, Natori S (1980) Purification of lectin induced in the hemolymph of *Sarcophaga peregrina* larvae on injury. J Biol Chem 255: 2919–2924

Kubo T, Natori S (1987) Purification and some properties of a lectin from the hemolymph of *Periplaneta americana* (American cockroach). Eur J Biochem 168: 75–82

Kubo T, Komano H, Okada M, Natori S (1984) Identification of hemagglutinating protein and bactericidal activity in the hemolymph of adult *Sarcophaga peregrina* on injury of the body wall. Dev Comp Immunol 8: 283–291

Kubo T, Kawasaki K, Natori S (1990) Sucrose-binding lectin in regenerating cockroach (*Periplaneta americana*) legs: Purification from adult hemolymph. Insect Biochem 20: 585–591

Kylsten P, Kimbrell DA, Daffre S, Samakovlis C, Hultmark D (1992) The lysozyme locus in *Drosophila melanogaster*. Different genes are expressed in midgut and salivary glands. Mol Gen Genet 232: 335–343

Ladendorff NE, Kanost MR (1990) Isolation and characterization of bacteria-induced protein-P4 from hemolymph of *Manduca sexta*, Arch Insect Biochem Physiol 15: 33–41

Ladendorff NE, Kanost MR (1991) Becteria-induced protein P4 (hemolin) from *Manduca sexta*: A member of the immunoglobulin superfamily which can inhibit hemocyte aggregation. Arch Insect Biochem Physiol 18: 285–300

Lahita RG (1982) In: Stites DP, Stobo JD, Fudenberg H, Wells JV (eds) Basic and clinical immunology, 4th edn. Lange, Los Altos, CA

Lambert J, Keppi E, Dimarcq JL, Wicker C, Reichhart JM, Dunbar B, Lepage P, Van Dorsselaer A, Hoffmann J, Fothergill J, Hoffmann D (1989) Insect immunity: isolation from immune blood of the dipteran *Phormia terranovae* of two insect antibacterial peptides with sequence homology to rabbit lung macrophage bactericidal peptides. Proc Natl Acad Sci USA 86: 262–266

Lee J-Y, Edlund T, Ny T, Faye I, Boman HG (1983) Insect immunity. Isolation of cDNA clones corresponding to attacins and immune protein P4 from *Hyalophora cecropia*. EMBO J 2: 577–581

Lee J-Y, Boman A, Chuanxin S, Andersson M, Jornvall H, Mutt V, Boman HG (1989) Antibacterial peptides from pig intestine: Isolation of a mammalian cecropin. Proc Natl Acad Sci USA 86: 9159–9162

Lehrer RI, Ganz T, Selsted ME (1991) Defensins – endogenous antibiotic peptides of animal cells. Cell 64: 229–230

Ludwig D, Tracey KM, Burns ML (1957) Ratio of ions required to maintain the heartbeat of the American cockroach, *Periplaneta americana* Linnaeus. Ann Entomol Soc Am 50: 244–246

Matsuyama K, Natori S (1988a) Purification of three antibacterial proteins from the culture medium of NIH-Sape-4, an embryonic cell line of *Sarcophaga peregrina*. J Biol Chem 263: 17112–17116

Matsuyama K, Natori S (1988b) Molecular cloning of cDNA for sapecin and unique expression of the sapecin gene during the development of *Sarcophaga peregrina*. J Biol Chem 263: 17117–17121

Matsuyama K, Natori S (1990) Mode of action of sapecin, a novel antibacterial protein of *Sarcophaga peregrina* (flesh fly) J Biochem (Tokyo) 108: 128–132

Minnick MF, Rupp RA, Spence KD (1986) A bacterial-induced lectin which triggers hemocyte coagulation in *Manduca sexta*. Biochem Biophys Res Commun 137: 729–735

Morishima I, Suginaka S, Ueno T, Hirano H (1990) Isolation and structure of cecropins, inducible antibacterial peptides from the silkworm, *Bombyx mori*. Comp Biochem Physiol B 95: 551–554

Mulnix AB, Dunn PE (1994) Structure and induction of a lysozyme gene from the Tobacco Hornworm, *Manduca sexta*. Insect Biochem Mol Biol 24: 271–281

Okada M, Natori S (1983) Puriification and characterization of an antibacterial protein from haemolymph of *Sarcophaga peregrina* (flesh fly) larvae. Biochem J 211: 727–734

Okada M, Natori S (1984) Mode of action of a bactericidal protein induced in the haemolymph of *Sarcophaga peregrina* (flesh-fly) larvae. Biochem J 222: 119–124

Okada M, Natori S (1985) Ionophore activity of sarcotoxin I, a bactericidal protein of *Sarcophaga peregrina*. Biochem J 229: 453–458

Pendland JC, Boucias DG (1985) Hemagglutinin activity in the hemolymph of *Anticarsia gemmatalis* larvae infected with the fungus *Nomuraea rileyi*. Dev Comp Immunol 9: 21–30

Pendland JC, Heath MA, Boucias DG (1988) Function of a galactose-binding lectin from *Spodopetera exigua* larval haemolymph: opsonization of blastospores from entomopathogenous hypomycetes. J Insect Physiol 34: 533–540

Postlethwait JH, Saul SH, Postlethwait JA (1988) The antibacterial immune response of the medfly *Ceratitis capitata*. J Insect Physiol 34: 91–96

Powning RF, Irzykiewicz H (1967) Lysozyme-like action of enzymes from the cockroach *Periplaneta americana* and from some other sources. J Insect Physiol 13: 1293–1299

Powning RF, Davidson WJ (1973) Studies on insect bacteriolytic enzymes. I. lysozyme in the haemolymph of *Galleria mellonella* and *Bombyx mori*. Comp Biochem Physiol B 45B: 669–681

Rheins LA, Karp RD (1982) Cockroach inducible humoral factor: Precipitin activity that is sensitive to a proteolytic enzyme. J Invertebr Pathol 40: 190–196

Rheins LA, Karp RD (1985a) Effect of gender on the inducible humoral immune response to Honeybee venom in the American cockroach (*Periplaneta americana*). Dev Comp Immunol 9: 41–49

Rheins LA, Karp RD (1985b) Ontogeny of the invertebrate humoral immune response: Studies on various developmental stages of the American cockroach. Dev Comp Immunol 9: 395–406

Rheins LA, Karp RD, Butz A (1980) Induction of specific humoral immunity to soluble proteins in the American cockroach (*Periplaneta americana*). I. Nature of the primary response. Dev Comp Immunol 4: 447–458

Robertson M, Postlethwait JH (1986) The humoral antibacterial response of *Drosophila* adults. Dev Comp Immunol 10: 167–179

Ryan NA, Karp RD (1993) Stimulation of hemocyte proliferation in the American cockroach (*Periplaneta americana*) by injection of *Enterobacter cloacae*. J Insect Physiol 39: 601–608

Samakovlis C, Kimbrell DA, Kylsten P, Engström A, Hultmark D (1990) The immune response in *Drosophila*: pattern of cecropin expression and biological activity. EMBO J 9: 2969–2976

Samakovlis C, Kylsten P, Kimbrell DA, Engström A, Hultmark D (1991) The *Andropin* gene and its product, a male-specific antibacterial peptide in *Drosophila melanogaster*. EMBO J 10: 163–169

Seeger MA, Haffley L, Kaufman TC (1988) Characterization of *amalgam:* a member of the immunoglobulin superfamily from *Drosophila*. Cell 55: 589–600

Steiner H, Hultmark D, Engström A, Bennich H, Boman HG (1981) Sequence and specificity of two antibacterial proteins involved in insect immunity. Nature 292: 246–248

Steiner H, Andreu D, Merrifield RB (1988) Binding and action of cecropin and cecropin analogues: antibacterial peptides from insects. Biochim Biophys Acta 939: 260–266

Sugumaran M, Kanost MR (1993) Regulation of insect hemolymph phenoloxidases. In: Beckage NE, Thompson SN, Federici BA (eds) Parasites and pathogens of insects, vol 2. Parasites. Academic Press, New York, pp 317–342

Sun SC, Lindström I, Boman HG, Faye I, Schmidt O (1990) Hemolin – an insect immune protein belonging to the immunoglobulin superfamily. Science 250: 1729–1732

Sun S-C, Åsling B, Faye I (1991a) Organization and expression of the immunoresponsive lysozyme gene in the giant silk moth, *Hyalophora cecropia*. J Biol Chem 266: 6644–6649

Sun S-C, Lindström I, Faye I (1991b) Structure and expression of the attacin genes in *Hyalophora cecropia*. Eur J Biochem 196: 247–254

Suzuki T, Natori S (1983) Identificiation of a protein having hemagglutinating activity in the hemolymph of the silkworm, *Bombyx mori*. J Biochem 93: 583–590

Takahashi H, Komano H, Kawaguchi N, Kitamura N, Nakanishi S, Natori S (1985) Cloning and sequencing of cDNA of *Sarcophaga peregrina* humoral lectin induced on injury of the body wall. J Biol Chem 260: 12228–12233

Takahashi H, Komano H, Natori S (1986) Expression of the lectin gene in *Sarcophaga peregrina* during normal development and under conditions where the defence mechanism is activated. J Insect Physiol 32: 771–779

Wago H (1991) Phagocytic recognition in *Bombyx mori*. In: Gupta AP (ed) Immunology of insects and other arthropods. CRC Press, Boca Raton, pp 215–235

Wang Y, Willott E, Kanost MR (1995) Organization and expression of the hemolin gene, a member of the immunoglobulin superfamily in an insect, *Manduca sexta*. Insect Mol Biol 4: 113–123

Zachary D, Hoffmann D (1984) Lysozyme is stored in the granules of certain haemocyte types in *Locusta*. J Insect Physiol 30: 405–411

Blood Clotting in Invertebrates

S. SRIMAL

1 Introduction

Well before clotting evolved, innate mechanisms served to prevent loss of body fluid from an injured site. Injury to an animal can result in a loss of body fluid and favour attack and entry of opportunistic pathogens. Irrespective of the level of organization, each organism is equipped with defense mechanisms to counteract this threat.

The chief wound-sealing processes encountered in the animal kingdom are: (1) contractile retraction of the tissues resulting in the closure of the wound, (2) constriction of blood vessels and plugging of the wound by aggregates of specialized cells, (3) clotting or coagulation of the escaping body fluid itself (Spurling 1981). Coagulation refers/relates to a transformation of extracellular fluid from a sol to gel state.

A review of the evolutionary trends in hemostatic mechanism indicates the victory of complexity over simplicity. The superficial simple mechanisms of the invertebrates (Needham 1970; Spurling 1981) were soon replaced by the complex mammalian clotting cascade (Davie et al. 1991). In invertebrates, alliance exists between the immune system and hemostasis. Tissue specialization and division of labor in higher vertebrates, however, severed this bond and complex mechanisms evolved to counteract injury and invasion by foreign organisms.

The fact that clotting has not been observed in many invertebrate species implies that it is either not indispensable for species' survival, or alternate hemostatic mechanisms, yet undiscovered, exist. In coelenterates and echinoderms, agglutination of cells followed by the formation of a syncytial network and the shrinking or contraction of a plug of aggregated cells has been observed. Cell aggregation, contraction of musculature with some increase in viscosity of the hemolymph occurs in mollusks. In annelids, conversion of fluid into thick gum mucilages upon contact with air and plasma clotting have been reported (Florkin 1962; Valembios et al. 1988). Dependence on both cellular aggregation and hemolymph coagulation involving enzymatic cleavage was first described in arthropods. This review describes in detail the clotting system of arthropods,

Molecular Biophysics Unit, Indian Institute of Science, Bangalore 560012, India especially that of the horseshoe crab (*Limulus, Tachypleus, Carcinoscorpious* spp.) which to date is the best-studied system in invertebrates.

1.1 Hemostasis and Immunity – a Marriage of Convenience

Invertebrates possess an open circulatory system and injury or infection in an animal may initiate similar biochemical events. This is best exemplified in *Limulus* (Bang 1956) which has to its credit 390 million years of successful living. Its triumph against nature's odds is partly attributed to its excellent defense strategy. The clotting cascade in *Limulus* is activated by Gram-negative bacteria or LPS (component of the outer wall of Gram-negative bacteria) and assists the animal in surmounting the bacterial challenge and maintaining a sterile, pathogen-free circulation system (Loeb 1902; Levin and Bang 1964a,b). Initiation of clotting by bacteria is other arthropods is relatively less well-understood than that of *Limulus* and has been reviewed (Barker and Bang 1966; Bang 1967; Levin 1967). In higher arthropods, different mechanisms have evolved to circumvent infection. These include the prophenol-oxidase activity (Söderhäll 1982) and the induction of antibacterial peptides (Boman and Hultmark 1987; Nakamura et al. 1988a).

2 The *Limulus* Clotting System

2.1 Amebocytes and Clotting

The role of amebocytes/hemocytes in coagulation was described in the last century (Halliburton 1885; Hardy 1892). Hemocytes are cells of mesenchymal origin circulating in the hemolymph and classified according to their function and morphology (Ravindranath 1980). In *Limulus*, only one type of circulating blood cell (amebocyte) has been identified in the adult animal (Loeb 1902; Levin and Bang 1964a; Dumont et al. 1966; Armstrong 1985). The *Limulus* amebocyte is a nucleated oval cell (15–20 μm) packed with heterogenous refractile exocytic granules (Armstrong 1977). Two types of granules, L and S, housing several proteins, exocytosed at different rates have been described in *Tachypleus* (Toh et al. 1991). Upon extravasation, the amebocyte exhibits ameboid movement, adhesiveness, pleiomorphism, and undergoes degranulation (Loeb 1920; Armstrong 1977, 1979, 1991; Armstrong and Rickles 1982; Ornberg and Reese 1981). All the clotting factors and clotting proteins so far identified in the horseshoe crab are sequestered within secretory granules of the amebocytes in an inactive state.

Activation of amebocytes is a prerequisite for clotting. Relatively little is known about the initial steps involved in recognition of self/nonself, binding of LPS/bacteria/foreign particles and their presentation to the cell. Circumstantial evidence suggests a role of lectins (carbohydrate-binding proteins) in this step.

Several lectins with specificity for different sugars (Sialic acid, 2-keto-deoxy-octonate acid, β-glycero-phosphate, N-acetyl-glucosamine, N-acetyl-galactos-amine, glucuronic acid) have been purified from the hemolymph of the different species of the horseshoe crab (Marchalonis and Edelman 1968; Vaith et al. 1979; Bishayee and Dorai 1980; Dorai et al. 1982a,b; Brandin and Pistole 1983; Shishikura and Sekiguchi 1983; White et al. 1995). Some have been shown to agglutinate Gram-negative and Gram-positive bacteria (Dorai et al. 1982a,b). Based on these observations, their role in recognition and presentation of bacteria to the cell has been the subject of speculation (Srimal and Bachhawat 1985).

Divalent cations are involved in the activation step, as shown by the obser-vation that agents such as EDTA or citrate retard the process (Kenney et al. 1972; Armstrong and Rickles 1982). Mg^{2+} is more effective than Ca^{2+} in degranu-lation of isolated cells (Cannon et al. 1986). A hemagglutinin, mediating aggregation of amebocytes in vitro, has been purified both from Limulus (Liu et al. 1991; Fuji et al. 1992) and *Carcinoscorpious* (Srimal et al. 1985). In *Carcinoscorpious*, the hemagglutinin is localized both on the cell surface and in the cytosol. It can bind coagulogen which has also been localized on the cell surface by immunofluorescence techniques (Srimal 1984). Thus, it is proposed that aggregation of the amebocytes is mediated by the interaction of the cell-surface hemagglutinin and the coagulogen bound to the cell surface (Srimal et al. 1985).

2.1.1 The Proteolytic Cascade

Aggregation of cells results in the formation of a cellular hemostatic plug, while exocytosis of the granules, release of clotting factors, and subsequent activation of the clotting cascade results in the formation of an extracellular clot (Armstrong 1991). The clotting time varies from 75–90 s (Srimal 1984; Iwanaga 1993). A schematic representation of the events occurring during clotting is shown in Fig. 1. The serine protease zymogen factor C is autocatalyticaly activated in the presence of LPS to active factor C^-, which converts zymogen factor B to active B^-. The proteolytic activity of B^- converts the proclotting enzyme into clotting enzyme which transforms soluble coagulogen (clotting protein) into coagulin. Polymerization of coagulin monomers results in the formation of an insoluble gel resembling the vertebrate fibrin clot. The proclot-ting enzyme is also activated by β-(1,3)-glucans (components of cell walls of yeasts and other fungi) by a less well-characterized glucan-sensitive cascade system.

Biochemical Properties of Clotting Factors in Limulus. Purification of several clotting factors and reconstitution experiments have indicated a cascade type of reaction (Iwanaga et al. 1986). Elegant work in the last decade, involving sequencing and cloning of several of these factors, has unveiled the sequence of events in clotting at the molecular level (Iwanaga et al. 1992; Iwanaga 1993).

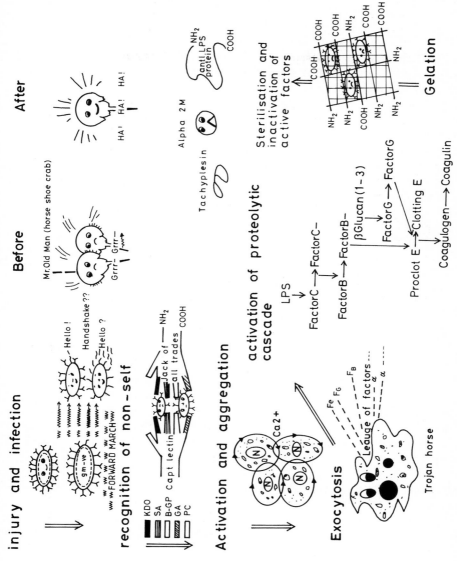

Fig. 1. Schematic representation of the sequence of events occurring upon injury or infection in the horseshoe crab

Factor C. Factor C is the initial activator of the clotting cascade triggered by LPS (Tokunaga et al. 1987). It is localized in the L-granules of the amebocytes. The zymogen factor C is a serine protease, and in the presence of LPS or Lipid A it is converted auto-catalytically to an activated form (C⁻) by cleavage of Phe-Ile bond (indicative of a chymotrypsin-like activity) in its L-chain (Tokunaga et al. 1987; Nakamura et al. 1988b). Purified factor C from *Limulus* and *Tachypleus* shows identical properties and is readily activated by chymotrypsin, but not by trypsin, and thus resembles mammalian α-thrombin in enzymatic properties. This initiation of a coagulation system is unique to the horseshoe crab, since all known mammalian coagulation, fibrinolytic, and complement systems are initiated by trypsin-like enzymes. The zymogen is a single-chain glycoprotein (9% carbohydrate) of Mr 123 kDa, which upon cleavage is converted to a two-chain intermediate form comprising of an 80 kDa heavy chain and a 43 kDa light chain. The latter is subsequently transformed to a 7.9 kDa A-chain and a 34 kDa B-chain. The active site of the protease is located in the B portion of the light chain and the LPS-binding domain is localized in the heavy chain (Tokunaga et al. 1991). This three-chain factor C⁻ has the potential to activate the next zymogen factor B of the cascade (Nakamura et al. 1986). Cloning and cDNA sequencing of factor C⁻ reveal several interesting structural features of this protein. The complete cDNA (3,474 bp long) encodes a putative signal sequence of 25 amino acids and a mature protein of 994 residues. There are five repeating units called "Sushi domains" of 60 amino acids each, which are found in many proteins, particularly in the mammalian complement system. A "C-type" lectin domain of 120 amino acid residues is present between the third and fourth sushi domains. An EGF-like domain which is found in several blood coagulation factors, complement proteins, and an LDL receptor, are also present in factor C. This type of mosaic structure of factor C is also contained in "Selectins" which are implied in cell recognition. Structurally, factor C is the first complement-like protein described in invertebrates, with a cysteine-rich amino terminus having homology with mammalian coagulation factor XII. Thus, it has been termed a "coagulation-complement factor" (Muta et al. 1991a).

Factor B. Factor B is a single chain glycoprotein with a molecular mass of 64 kDa. Upon activation by factor C⁻, the zymogen is converted into active factor B⁻ which is composed of an H chain (32 kDa) and an L chain (25 kDa) accompanied by the release of a short peptide (Nakamura et al. 1986). It is also a serine protease and activates the proclotting enzyme. Its cDNA sequence is known and the mature protein is composed of 377 amino acids (40.57 kDa). The protein shows sequence similarity, both at the amino terminus and carboxyl end, with *Limulus* proclotting enzyme. At the carboxyl terminal serine protease domain it exhibits 43.9% identity with the proclotting enzyme, implying that these proteins have arisen by gene duplication (Muta et al. 1991b; Iwanaga 1993).

Proclotting Enzyme. The proclotting enzyme, which is activated by factor B⁻ and also by factor G, is a single-chain glycosylated polypeptide of 54 kDa. Upon proteolysis at Arg-Ile (98–99 res) by factor B, it is converted into a two-chain active form (clotting enzyme) consisting of L (25 kDa) and H (31 kDa) chains (Nakamura et al. 1985). The clotting enzyme cleaves the clotting protein (coagulogen) at two specific sites (Arg-Gly and Arg-Thr) and is localized within the L-granules. Functionally, the proclotting/clotting enzyme resembles the vertebrate Prothrombin-thrombin complex. The carboxyl terminus of the heavy chain (containing the serine protease active site) has 34% homology with the mammalian factors Xa and XIa. Four disulfide linkages occur in the H chain at the same locations as these linkages occur in the chains of Prothrombin and clotting factor X. The light chain has a unique disulfide-knotted domain at its amino terminus, this domain is also found in a factor B and in *Drosophila* serine protease ester precursor. This unique "clip-like" domain may represent a common structural element in the serine protease zymogens of invertebrates (Iwanaga 1993).

Coagulogen. The clotting protein or coagulogen is one of the major cellular proteins and was the first *Limulus* clotting factor to be purified and sequenced (Nakamura et al. 1976a,b; Miyata et al. 1984). It is functionally analogous to vertebrate fibrinogen, but structurally bears no significant homology. Unlike fibrinogen, coagulogen is localised within the L-granules of amebocytes, and no plasma-soluble form has been reported yet. Exocytosis of the granules releases the precursor form of coagulogen [Mr 19.7 ± (50) kDa] into the plasma. The clotting enzyme cleaves coagulogen at specific sites (which are conserved in all three species) resulting in the release of a peptide C (28 amino acids) and coagulin (147 amino acids) which consists of two chains (A and B) linked by disulfide bridges. Peptide C bears some homology to primate fibrinopeptide A and B. Coagulin monomers polymerize noncovalently to form the extracellular clot (Nakamura et al. 1976a,b). In contrast to the situation in vertebrates, where fibrin monomers are cross-linked by transglutaminase to form a firm clot, *Limulus* transglutaminase does not cross-link or stabilize the clot (Tokunaga et al. 1993a). Coagulogen from all species of horseshoe crab have been sequenced and are well conserved especially in the A- and B- chain region (Srimal et al. 1985). Coagulogen cDNA from *Limulus* and *Tachypleus* have been cloned, and two mRNAs have been identified whose significance is not yet clear (Cheng et al. 1986; Miyata et al. 1986).

Factor G. The Proclotting enzyme can also be activated by factor G which is a serine protease activated by β-1,3-glucans (Mortia et al. 1981). Factor G is a glycoprotein of molecular mass 110 kDa. Upon proteolysis it is converted to active factor G, comprising two subunits, which can activate the proclotting enzyme and mediate clotting. Unlike the LPS activated cascade, this pathway is not well-characterized.

Transglutaminase. Calcium-dependent transglutaminase (Tgase) activity in the amebocytes was identified by Campbell Wilkes 1973 and has recently been purified, sequenced, and cloned. The purified protein (Mr 86 kDa) has properties of mammalian type 11 Tgase-like-enzyme and is localized in the cytosol. Tgase is expressed as a 3 kb mRNA and bears significant homology to the mammalian transglutaminase family (Tokunaga et al. 1993 a,b). Two major granule proteins of 8.6 and 80 kDa were identified as its potential substrate in the *Tachypleus* hemocytes. Unlike the transglutaminases of crustaceans and mammalians, its function with reference to stabilization of the clot is not very clear, although contradictory reports on N^e (γ-glutamyl) lysine cross-links in the blood clot exist (Wilson et al. 1992). It is possible that the abundant 8.6 kDa tgase substrate also becomes a constituent of the clot (by a mechanism yet to be identified) and is the reason for this discrepancy (Tokunaga et al. 1993a).

Regulators of Clotting. The function of the blood clot is to arrest bleeding and entrap/immobilize bacteria that would otherwise disseminate beyond the wound site and enter the hemocoel. Initiation of clotting results in the activation of several proteases and removal or inhibition of these protease and activating agents (LPS/bacteria), is equally important for regulation and homeostasis. Little is known about this aspect; however, several protease inhibitors and LPS-binding proteins have been identified in the hemolymph (Quigley and Armstrong 1983; Donavan and Laue 1991) and the hemocyte lysate (Aketagawa et al. 1986; Muta et al. 1987; Nakamura et al. 1987; Minetti et al. 1991). Of these, the best-characterized one are the Tachyplesins and LPS-binding protein from the amebocyte lysate whose amino acid sequence and crystal structure is known (Muta et al. 1987; Nakamura et al. 1988a; Wainwright et al. 1990; Hoess et al. 1993). The endoproteinase inhibitor α_2-macroglobulin is present in the plasma and is also secreted by the hemocytes by exocytosis (Armstrong et al. 1990). In mammals, α_2-macroglobulin entraps proteases and participates in a receptor mediated endocytic protease clearance in mammals. *Limulus* α_2-macroglobulin bears a striking similarity in the sequence at the thiol-ester domain with mammalian and other α_2-macroglobulin related proteins, and it is quite likely that it performs an analogous function (Armstrong and Quigley 1991).

Most of the clotting factors, except coagulogen, are glycoproteins; hence it is possible that upon release into the plasma they are bound to circulating lectins and are inactivated or removed.

3 Clotting in Crustaceans and Other Arthropods

In Crustaceans multiple coagulation mechanisms as described by Tait (1911) exist. Type A coagulation (where a dense hemocyte network is sufficient to seal the wound) is seen in the crab *Loxorhynchus grandis*; type B (hemocyte aggregation is followed by plasma coagulation) occurs in the marine lobster (*Homarus americunus*); and type C (involving the explosive cells) is present in the spiny lobster and shrimps (Hose et al. 1990). Clotting factors are localized both in the plasma and the hemocytes (Ravindranath 1980; Durliat 1985). Unlike clotting in

Limulus and vertebrates, a specific proteolytic cascade is not required for clot formation in crustaceans. Instead, the plasma-soluble clotting protein is cross linked by a cellular calcium dependent transglutaminase (Lorand 1972).

Hemocytes in crustaceans are heterogenous and have been classified based on their morphology and function, viz., hemostasis, hardening of exoskeleton, phagocytosis and encapsulation (Ravindranath 1980; Hose et al. 1990). In the penaid shrimp *Sicyonia* three types of hemocytes having three different functions have been described (Martin et al. 1987), whereas in *Homarus americunus, P. interrputus, and Loxorhynchus grandis* only two types exist (Hose et al. 1990). Hyaline cells, also known as explosive cells, are involved in the initiation of clotting, whereas granulocytes containing prophenol-oxidase and hydrolases are involved in phagocytosis and encapsulation.

Coagulogen in crustaceans is distributed both in plasma and within cells. Soluble plasma coagulogen appears to possess many diversified functions such as clotting, protein transport of tanning substances, lipid and sugar transport, etc. It is also localized in other tissues such as soft integument and cuticle, thus remotely resembling fibronectin (Fuller 1978; Durliat and Vranckx 1983b). The concentration varies from trace amounts to 20% of the total plasma protein in Lobsters (Durliat 1985). The only extensively studied crustacean clotting proteins are those of the spiny lobster (Fuller and Doolittle 1971a) and the fresh water crayfish (Kopacek et al. 1993). Lobster clotting protein is a lipo-glycoprotein of molecular mass 420 kDa, composed of two identical subunits (220 kDa) linked by disulfide bonds. Cross-linked polymers of higher molecular weight are known to occur in the plasma. The N-terminus of plasma coagulogen has some sequence similarity with vitellogenins from both invertebrates and vertebrates (Doolittle and Riley 1990). Crayfish clotting protein also has properties and sequence similarity to that of the lobster (Kopacek et al. 1993). Crustacean coagulogen bears no similarity to *Limulus* coagulogen or vertebrate fibrinogen.

Trace amounts of coagulogen have also been identified in the hemocytes of the sand crab and other crustraceans (Madaras et al. 1981; Durliat 1985; Durliat and Vranckx 1989). This protein, of 70 kDa, is distinct from the plasma coagulogen, and its exact function is not very clear. A clotting process triggered by β-1,3-glucan is present in the cryfish, sand crab and other crustacean hemocytes (Madaras et al. 1981; Söderhäll 1981; Durliat 1985; Söderhäll and Cerenius 1992). This process appears to be mediated to be mediated by serine proteases that hydrolyse Gly-Arg bonds and are activated by LPS, glucans, and other Gram-positive bacteria (Barker and Bang 1966; Bang 1967; Cornick and Stewart 1968; Söderhäll and Hall 1984). The β-1,3-glucan induced clotting involves the prophenol-oxidase which is absent in *Limulus* and *Carcinoscorpious* (Söderhäll et al. 1985; Srimal 1984).

Retardation of plasma clotting in crustaceans by amines (which inhibit cross-linking of vertebrate fibrinogen) indicates that polymerization requires cross-linking by transglutaminases (Lorand 1972; Lorand and Conrad 1984; Martin et al. 1991). No prior proteolytic cleavage of coagulogen occurs before polymerization, a feature which distinguishes it from vertebrate and *Limulus* clotting. The Ca^{2+}-dependent transglutaminase is chiefly localized in the hyaline hemocytes (Lorand and Conrad 1984; Fuller and Doolittle 1971b); however, trace amounts have been detected in the plasma of shrimp (Martin et al. 1991). Whether this plasma form is inherent or is itself from broken cells is not very clear.

The mechanism of clotting in insects resembles that of crustaceans. Amongst the heterogenous population of blood cells, the granular hemocytes and plasmatocytes appear to be involved in clotting. Both plasma and cellular forms of coagulogen have been described. The plasma coagulogen is a lipoglycoprotein (Mr 640 000) and carries multiple functions as in crustaceans. The enzymes required for clotting have not been very well characterized, but appear to be of hemocyte origin only, since cell-free plasma fails to clot. The fact that clotting is inhibited by amines and not by protease inhibitors, suggests that a mechanism is operative in insects similar to that of crustaceans involving transglutaminase, (Bohn 1986).

4 Conclusion

It is apparent that, with the exception of the *Limulus* clotting system, our understanding of invertebrate clotting systems is still lacking in detail and the biochemical characterization of such systems is in its infancy. The *Limulus* clotting system in evolutionary terms marks the beginning of the amplification of a signal by a cascade system (which ascertains specificity and sensitivity) that is conserved in vertebrate clotting and complement activation. The evolution of the response to bacteria, LPS, and a link between immunity and clotting is well-preserved in higher vertebrates, but several questions still remain unanswered about the "invertebrate clotting Pandora's box". LPS, by virtue of its physico-chemical properties, can interact nonspecifically with any protein, and confusion still exists regarding the true in vivo functional receptor for LPS. Little is known about the site of synthesis of clotting factors and their regulation. The elements of surprise in the elucidation of the clotting cascade, have been several and it seems there is still room for more. As Lewis puts "If you were looking about for an anecdote to illustrate the element of unexpectedness and surprise (in science) there is no better story than that of *Limulus polyphemus..*" (Lewis 1979). Further studies are bound to unravel the mystery and add to the above information to give a clearer picture of the sequence of events.

Acknowledgment. The author is supported by a grant from Department of Science and Technology (Grant number SP/SO/D14/91 to Prof A. Surolia), India.

References

Aketagawa J, Miyata T, Ohtsubo S, Nakamura T, Hayashida H, Miyata T, Iwanaga S (1986) Primary structure of *Limulus* anticoagulant anti-lipopolysaccharide factor. J Biol Chem 261: 7375–7365

Armstrong PB (1977) Interaction of the motile blood cells of the horseshoe crab, *Limulus polyphemus*. Studies on contact paralysis of pseudopodial activity and cellular overlapping in vitro. Exp Cell Res 107: 127–138

Armstrong PB (1979) Motility of the *Limulus* blood cell. J Cell Sci 37: 169–180

Armstrong PB (1985) Adhesion and motility of the blood cells of *Limulus*. In: Cohen WD (ed) Blood cells of marine invertebrates: experimental systems in cell biology and comparative physiology. Alan R Liss, New York

Armstrong PB (1991) Cellular and humoral immunity in the horseshoe crab. In: Gupta AP (ed) *Limulus polyphemus* immunology of insects and other arthropods. CRC Press, Boca Raton, pp 3–17

Armstrong PB, Quigley JP (1991) α_2-macroglobulin: a recently discovered defense system in arthropods. In: Gupta AP (ed) Immunology of insects and other arthropods. CRC Press, Boca Raton

Armstrong PB, Rickles FR (1982) Endotoxin-induced degranulation of the *Limulus* amebocyte. Exp Cell Res 140: 15–24

Armstrong PB, Quigley JP, Rickles FR (1990) The *Limulus* blood cell secretes α_2-macroglobulin when activated. Biol Bull 178: 137–143

Bang FB (1956) A bacterial disease of *Limulus polyphemus*. Bull Johns Hopkins 98: 325–350

Bang FB (1967) Serological responses among invertebrates other than insects. Fed Proc 26: 1680–1684

Barker WH Jr, Bang FB (1966) The effect of infection by Gram-negative bacteria and their endotoxins, on the blood clotting mechanisms of the crustacean *Sacculina carcini*, a parasite of the crab Carcinus Maenas. J Invertebr Pathol 8: 88–97

Bishayee S, Dorai DT (1980) Isolation and characterisation of a sialic binding lectin (carcinoscorpin) from Indian horseshoe crab *C. rotundicauda*. Biochem Biophys Acta 623: 89–97

Bohn H (1986) Hemolymph clotting in insects. In: Brehelin M (ed) Immunity in invertebrates. Springer, Berlin Heidelberg New York, pp 188–207

Boman HG, Hultmark D (1987) Cell free immunity in insects. Annu Rev Microbiol 41: 103–126

Brandin ER, Pistole TG (1983) Polyphemin: a techoic acid-binding lectin from the horseshoe crab *Limulus polyphemus*. Biochem Biophys Res Commun 113: 611–617

Cannon GW, Tsuchiya M, Rittschof D, Bonaventura J (1986) Magnesium dependence of endotoxin-induced degranulation of *Limulus* amebocytes. Biol Bull 143: 548–567

Campbell-Wilkes L (1973) PhD Dissertation, Northwestern Univ, Ann Arbor, MI, univ microfilms 73–30 763

Cheng SM, Suzuki A, Zon G, Liu TY (1986) Characterisation of a complementary deoxyribonucleic acid for the coagulogen of *Limulus polyphemus*. Biochem Biophys Acta 868: 1–8

Cornick JW, Stewart JE (1968) Interaction of the pathogen *Gaffkya homari* with natural defense mechanism of *Homarus americanus*. J Fish Res Board 25: 695–709

Davie EW, Fujikawa K, Kisiel W (1991) the coagulation cascade: initiation, maintenance and regulation. Biochemistry 30: 10363–10370

Doolittle RF, Riley M (1990) The amino-terminal sequence of lobster fibrinogen reveals common ancestry with vitellogenins. Biochem Biophys Res Commun 167: 16–19

Donavan MA, Laue TM (1991) A novel trypsin inhibitor from the hemolymph of the horseshoe crab *Limulus polyphemus*. J Biol Chem 266: 2121–2125

Dorai DT, Srimal S, Somasundaran M, Bachhawat BK, Balganesh TS (1982a) Recognition of 2-keto-deoxyoctonate in bacterial cell wall and lipopolysaccharides by the sialic acid binding lectin from the horseshoe crab. *C. rotundicauda*. Biochem Biophys Res Commun 104: 141–147

Dorai DT, Srimal S, Somasundaran M, Bachhawat BK, Balganesh TS (1982b) On the multispecifity of carcinoscorpin, the sialic acid binding lectin from the horseshoe crab. *C. rotundicauda*. FEBS Lett 148: 98–102

Dumont JN, Anderson E, Winner G (1966) Some cytologic characteristics of the hemocytes of *Limulus* during clotting. J Morphol 119: 181–208

Durliat M (1985) Clotting processes in crustacea decapoda. Biol Rev 60: 473–603

Durliat M, Vranckx R (1983) Analysis of clotting defects in diseased lobsters 2. Proteins of diseased lobster hemolymph. Comp Biochem Physiol 76A 103–108

Durliat M, Vranckx R (1989) Relationships between plasma and hemocyte proteins in decapoda. Comp Biochem Physiol 92B: 595–603

Fine structure, morphological changes during coagulation and localization of clotting factors and antimicrobial substances. Cell tissue Res 266: 137–147

Florkin M (1962) Comparative Biochemistry IV, a comprehensive treatise. Florkin M, Mason HS (eds) Academic Press, London, pp 435–482

Fuji N, Minetti CASA, Nakhasi H, Chen S, Barbehenn E, Nunes PH, Nguyen NY (1992) Isolation, cDNA cloning and characterisation of an 18 kDa hemagglutinin and amebocyte aggregation factor from *Limulus polyphemus*. J Biol Chem 267: 22452–22459

Fuller GM (1978) Are lobster fibrinogen and cold in soluble-immunoglobulin related molecules? Ann NY Acad Sci 312: 31–37

Fuller GM, Doolittle RF (1971a) Studies of invertebrate fibrinogen I. Purification and characterisation of fibrinogen from spiny lobster. Biochemistry 10: 1305–1310

Fuller GM, Doolittle RF (1971b) Studies of invertebrate fibrinogen. II. Transformation of lobster fibrinogen into fibrin. Biochemistry 10: 1311–1315

Halliburton WD (1885) On the blood of decapod crustacea. J Physiol 6: 300–335

Hardy WB (1892) The blood corpuscles of the crustacea together with a suggestion as to the origin of the crustacean fibrin ferment. J Physiol 13: 24–28

Hoess A, Watson S, Siber GR, Liddington R (1993) Crystal structure of an endotoxin-neutralizung protein from the horseshoe crab, *Limulus* anti LPS-factor, at 1.5 Å resolution. EMBO J 12: 3351–3356

Hose JE, Martin GG, Nguyen VA, Rosenstein T (1987) Cytochemical features of shrimp hemocytes. Biol Bull 173: 176–185

Hose JE, Martin GG, Gerard AS (1990) A decapod hemocyte classification scheme integrating morphology, cytochemistry, and function. Biol Bull 178: 33–45

Iwanaga S (1993) The *Limulus* clotting reaction. Curr Opin Immunol 5: 74–82

Iwanaga S, Morita T, Miyata T, Nakamura T, Aketagawa J (1986) The hemolymph coagulation system in invertebrate animals. J Protein Chem 5: 225–268

Iwanaga S, Miyata T, Tokunaga F, Muta T (1992) Molecular mechanism of hemolymph clotting system in *Limulus*. Thromb Res 68: 1–32

Kenney DM, Belamarich FA, Shepro D (1972) Aggregation of horseshoe crab (*Limulus polyphemus*) amebocytes and reversible inhibition of aggregation of EDTA. Biol Bull 143: 548–567

Kopacek P, Hall M, Soderhall K (1993) Characterisation of a clotting protein isolated from plasma of the freshwater crayfish *Pacifastacus leniusculus*. Eur J Biochem 213: 591–597

Lewis T (1979) In: Cohen E (ed) Biomedical applications of the horseshoe crab (Limulidae). Alan R Liss, New York, p 682

Levin J (1967) Blood coagulation and endotoxin in invertebrates. Fed Proc 26: 1707–1712

Levin J, Bang FB (1964a) The role of endotoxin in the extracellular coagulation of *Limulus* blood. Bull John Hopkins 115: 265–274

Levin J, Bang FB (1964b) A description of cellular coagulation in *Limulus*. Bull John Hopkins 115: 337–345

Liu T, Lin Y, Cislo T, Mineti CASA, Baba JMK, Liu TY (1991) Limunectin: A phosphochholine-binding protein from *Limulus* amebocytes with adhesion-promoting properties. J Biol Chem 266: 14813–14821

Loeb L (1902) On the blood lymph cells and inflammatory processes of *Limulus*. J Med Res 7: 145–158

Loeb L (1920) The movements of the amebocytes and the experimental production of amebocyte (cell fibrin) tissue. Wash Univ Stud 8:3

Lorand L (1972) Fibrinoligase: the fibrin-stabilising factor system of blood plasmas. Ann NY Acad Sci 202: 1–348

Lorand L, Conrad SM (1984) Transglutaminases. Mol Cell Biochem 58: 9–35

Madaras F, Chew MY, Parkin JD (1981) Purification and characterisation of the sand crab (*Ovalipes bipustulatus*) coagulogen (fibrinogen). Thromb Haemostasis 45: 77–81

Marchalonis JJ, Edelman GM (1968) Isolation and characterisation of a hemagglutinin from *L. ployphemus*. J Mol Biol 32: 265–275

Martin GG, Hose JE, Kim JJ (1987) Structure of hematopoietic nodules in the ridgeback prawn, *Sicyonia ingentis*: light and electron microscopic observations. J Morphol 192: 193–204

Martin GG, Hose JE, Omori S, Chong C, Hoodboy T, McKrell N (1991) Localization and roles of coagulogen and transglutaminase in hemolymph coagulation in decapod crustaceans. Comp Biochem Physiol 100B: 517–522

Minetti CASA, Lin Y, Cislo T, Liu TY (1991) Purification and characterization of an endotoxin-binding protein with protease inhibitory activity from amebocytes of *Limulus polyphemus*. J Biol Chem 266: 20773–20780

Miyata T, Hiranaga M, Umezu M, Iwanaga S (1984) Amino acid sequence of the coagulogen from *Limulus polyphemus* hemocytes. J Biol Chem 259: 8924–8933

Miyata T, Matsumoto H, Hattori M, Sakaki Y, Iwanaga S (1986) Two types of coagulogen mRNAs found in horseshoe crab (*Tachypleus tridentatus*) hemocytes: Molecular cloning and nuclotide sequences. J Biochem (Tokyo) 100: 213–220

Morita T, Tanaka S, Nakamura T, Iwanaga S (1981) D-glucan mediated coagulation pathway found in *Limulus* amebocytes. FEBS Lett 129: 318–321

Muta T, Miyata T, Tokunaga F, Nakamura T, Iwanaga S (1987) Primary structure of anti-lipopoly-saccharide factor from American horseshoe crab, *Limulus polyphemus*. J Biochem (Tokyo) 101: 1321–1330

Muta T, Miyata T, Misumi Y, Tokunaga F, Nakamura T, Toh Y, Ikehara Y, Iwanaga S (1991a) *Limulus* factor C: An endotoxin sensitive serine protease zymogen with a mosaic structure of complement-like, epidermal growth factor-like, and lectin-like domains. J Biol Chem 266: 6554–6561

Muta T, Hashimoto R, Oda T, Miyata T, Iwanaga S (1991b) *Limulus* clotting enzyme and factor B associated with endotoxin-sensitive coagulation cascade: Novel serine protease zymogens with a new type of "disulfide-knotted domain". Throm Haemostasis Abstr 65: 935

Nakamura S, Iwanaga S, Harada T, Niwa M (1976a) A clottable protein (coagulogen) from amebocyte lysate of Japanese horseshoe crab (*Tachypleus tridentatus*). J Biochem (Tokyo) 80: 1011–1021

Nakamura S, Iwanaga S, Harada T, Niwa M (1976b) A clottable protein (coagulogen) of horseshoe crab hemocytes: structural change of its polypeptide chain during gel formation. J Biochem (Tokyo) 80: 649–652

Nakamura T, Morita T, Harada-Suziki T, Iwanaga S (1985) Intracellular proclotting enzyme in (*Tachypleus tridentatus*) hemocytes: Its purification and properties. J Biochem (Tokyo) 97: 1561–1574

Nakamura S, Horuchi T, Morita T, Iwanaga S (1986) Purification and properties of intracellular clotting factor, factor B from horseshoe crab (*Tachypleus tridentatus*) hemocytes. J Biochem (Tokyo) 99: 847–857

Nakamura T, Hirai T, Tokunaga F, Kawabata S, Iwanaga S (1987) Purification and amino acid sequence of Kunitz-type protease inhibitor found in hemocytes of the horseshoe crab (*Tachypleus tridentatus*). J Biochem (Tokyo) 101: 1297–1306

Nakamura T, Tokunaga F, Morita T, Iwanaga S (1988a) Intracellular serine-protease zymogen, factor C from horseshoe crab hemocytes: its activation by synthetic lipid A analogues and acidic phospholipids. Eur J Biochem 176: 89–94

Nakamura T, Furunaka H, Miyata T, Tokunaga F, Muta T, Iwanaga S, Niwa M, Takao T, Shimonishi Y (1988b) Tachyplesin, a class of antimicrobial peptide from hemocytes of the horseshoe crab (*Tachypleus tridentatus*). J Biol Chem 263: 16709–16713

Needham AE (1970) Haemostatic mechanisms in the invertebrata. Symp Zool Soc Lond 27: 19–44

Ornberg RL, Reese TS (1981) Beginning of exocytosis captured by rapid-freezing of *Limulus* amebocyte. J Cell Biol 90: 40–54

Quigley JP, Armstrong PB (1983) An endopeptidase inhibitor similar to α_2-macroglobulin, detected in the hemolymph of an invertebrate *Limulus polyphemus*. J Biol Chem 258: 7903–7906

Ravindranath MH (1980) Hemocytes in hemolymph coagulation of arthropods. Biol Rev 55: 139–170

Shishikura F, Sekiguchi K (1983) Agglutinins in the horseshoe crab hemolymph: Purification of a potent agglutinin of horse erythrocytes from the hemolymph of *Tachypleus tridentatus*, the japanese horseshoe crab. J Biochem 93: 1539–1546

Söderhäll K (1981) Fungal cell wall *β*-1-3 glucans induce clotting and phenol-oxidase attachment to foreign surfaces of crayfish hemocyte lysate. Dev Comp Immunol 5: 565–573

Söderhäll K (1982) Prophenol-oxidase activating system and melanization – a recognition mechanism of arthropods? A review Dev Comp Physiol 6: 601–611

Söderhäll K, Cerenius L (1992) Crustacean immunity. Annu Rev Fish Dis 1:3–23

Söderhäll K, Hall L (1984) LPS induced activation of prophenol-oxidase activating system in crayfish hemocyte lysate. Biochem Biophys Acta 797: 99–104

Söderhäll K, Levin J, Armstrong PB (1985) The effect of *β*-1-3 glucans on the blood coagulation and amebocyte release in the horseshoe crab *Limulus polyphemus*. Biol Bull 169: 661–674

Spurling NW (1981) Comparative physiology of blood clotting. Comp Biochem Physiol 68A: 541–548

Srimal S (1984) The role of lipopolysaccharide binding proteins in the coagulation system of *Carcinoscorpious rotundicauda*. PhD Thesis, Univ Calcutta, India

Srimal S, Bachhawat B (1985) Blood Brother: a living fossil helps save human lives. 51–55 Science Age Nov-Dec: 51–55

Srimal S, Miyata T, Kawabata S, Iwanaga S (1985) the complete amino acid sequence of coagulogen isolated from the south-east Asian horseshoe crab *Carcinoscorpious rotundicauda*. J Biochem (Tokyo) 98: 305–318

Srimal S, Dorai DT, Somasundaran M, Bachhawat BK, Miyata T (1985) A new hemagglutinin from the amebocytes of the horseshoe crab *Carcinoscorpious rotundicauda*. Purification and role in cell aggregation. Biochem J 230: 321–327

Tait J (1911) Types of crustacean blood coagulation. J Mar Biol Assoc (UK) 9: 191–198

Toh Y, Mizutani A, Tokunaga F, Muta T, Iwanaga S (1991) Structure of hemocytes of the Japanese horseshoe crab *Tachypleus tridentatus*.

Tokunaga F, Miyata T, Nakamura T, Morita T, Kuma K, Miyata T, Iwanaga S (1987) Lipopolysaccharides-sensitive serine protease zymogen (Factor C) of horseshoe crab hemocytes: identification and alignment of proteolytic fragment produced during the activation show that it is a novel type of serine protease. Eur J Biochem 167: 405–416

Tokunaga F, Nakajima H, Iwanaga S (1991) Purification and characterisation of lipopolysaccharides-sesnsitive serine protease zymogen (factor C) isolated from *Limulus polyphemus* hemocytes: a newly identified intracellular zymogen activated by *α*-Chymotrypsin, not by trypsin, J Biochem (Tokyo) 109: 150–157

Tokunaga F, Yamada M, Miyata T, Ding YL, Hiranaga-Kawabata M, Muta T, Iwanaga S, Ichinose A, Davie EW (1993a) *Limulus* hemocyte transglutaminase: its purification, characterisation and identification of the intracellular substrates. J Biol Chem 268: 252–261

Tokunaga F, Muta T, Iwanaga S, Ichinose A, Davie EW, Kuma K, Miyata T (1993b) *Limulus* hemocyte transglutaminase: cDNA cloning, amino acid sequence and tissue localization. J Biol Chem 268: 262–265

Vaith P, Uhlenbruck G, Holz G (1979) Anti-glucoronyl activity *Limulus Polyphemus* agglutinin. Protides Biol Fluids Coll 27: 455–458

Valembios P, Roch PH, Lassegues M (1988) Evidence of plasma clotting in earthworm. J Invertebr Pathol 51: 221–228

Wainwright NR, Miller RJ, Paus E, Novitsky TJ, Fletcher MA, McKenna TM, Williams T (1990) In: Nowotny A, Spitzer JJ, Ziegler EJ (eds) Cellular and molecular aspects of endotoxin reactions. Elsevier New York, pp 315–325

White M, Wainwright N, Novitsky TJ (1995) Isolation of three lectin from the horseshoe rab *Limulus polyphemus*. (in prep)

Wilson J, Rickles FR, Armstrong PB, Lorand L (1992) N^e (*γ*-glutamyl) lysine crosslinks in the blood clot of the horseshoe crab *Limulus polyphemus*. Biophys Biochem Res Commun 188: 655–661

Immune Function of α_2-Macroglobulin in Invertebrates

P.B. Armstrong[1,2] and J.P. Quigley[2,3]

1 Introduction

Proteases play important roles in a variety of immune processes, including blood clotting and clot resolution (Furie and Furie 1992), complement activation (Reid and Porter 1981), inflammation (Cohn 1975; Haverman and Janoff 1978), and tissue remodeling (Werb 1993). Additionally, proteases contribute to a variety of pathological conditions such as tumor dissemination (Testa and Quigley 1990) and a variety of degenerative connective tissue diseases (Perlmutter and Pierce 1989). Proteases, whether of endogenous or of exogenous origin, have the potential for serious destructive effects on the surrounding tissues after their release into the tissue spaces. A variety of connective tissue disorders are directly traceable to the activities of proteases present in the wrong places and at the wrong times. Additionally, proteases are important agents facilitating the invasion of parasites (McKerrow et al. 1991; Breton et al. 1992). In response to this, higher animals have evolved a variety of protease inhibitors in the blood that limit the activities of the proteases of endogenous immune processes and the exogenous proteases of microbes and multicellular parasites. Circulating protease inhibitors are of two basic types; inhibitors that complex with and inhibit the active site of the target protease (Laskowski and Kato 1980; Travis and Salvesen 1983) and inhibitors of the α_2-macroglobulin family, which leave the active site intact and instead enfold the target protease to block its interaction with protein substrates (Starkey and Barrett 1977). Proteases bound to α_2-macroglobulin retain the ability to hydrolyze small amide and ester substrates (Barrett and Starkey 1973) but have lost the ability to hydrolyze proteins.

α_2-Macroglobulin from mammals has received extensive study, but we still know surprisingly little of its role in immunity even in mammals. Only more recently has it become clear that α_2-macroglobulin is also present in a variety of lower vertebrates (Starkey and Barrett 1982) and invertebrates (Armstrong and

[1]Department of Molecular and Cellular Biology, University of California, Davis, California 95616–8755, USA
[2]Marine Biological Laboratory, Woods Hole, Massachusetts 02543, USA
[3]Department of Pathology, Health Sciences Center, State University of New York, Stony Brook, New York 11794–8691, USA

Quigley 1991). This review will first present general features of our present understanding of the biochemistry of α_2-macroglobulin, and will then consider the application of this knowledge to the problems of the immune functions of α_2-macroglobulin in invertebrates.

2 Structure of α_2-Macroglobulin

Although it has recently become clear that α_2-macroglobulin can bind a diverse variety of proteins (Borth 1992), the binding of endopeptidases has received the most study (Sottrup-Jensen 1987). Protease binding by α_2-macroglobulin is unique in the annals of enzyme inhibition, since α_2-macroglobulin does not inhibit the active site of the enzyme but instead entraps the protease in a hydrophilic pocket within the α_2-macroglobulin molecule (Barrett and Starkey 1973). The entrapped protease is inhibited from reacting with macromolecules, including antibodies, protein substrates, and macromolecular active site protease inhibitors, but is still capable of hydrolyzing small amide and ester substrates and is still accessible to low molecular mass active site inhibitors (Beith et al. 1970; Barrett and Starkey 1973; Starkey and Barrett 1977). Entrapment of protease molecules involves a refolding and major compaction of α_2-macroglobulin (Barrett et al. 1979; Bjork and Fish 1982; Gonias et al. 1982) that follows upon the proteolytic cleavage of α_2-macroglobulin at a specialized domain of the molecule (Sottrup-Jensen et al. 1981; Harpel 1973; Hall and Roberts 1978).

The subunit polypeptide chain of α_2-macroglobulin from organisms as diverse as mammals, horseshoe crab, and octopus is 180–185 kDa. Proteolytic cleavage at the bait region yields peptides of 80 and 100 kDa. Forms of α_2-macroglobulin with one, two, and four polypeptide chains have been identified, with the two-chain form being the most widely distributed (Starkey and Barrett 1982; Armstrong et al. 1991). The polypeptide chains of dimeric forms are linked by diulfide bonds (Spycher et al. 1987; Armstrong et al. 1991), and tetrameric human α_2-macroglobulin consists of a noncovalently linked pair of disulfide-linked dimers (Harpel 1973).

Several functional domains have been identified in the monomer by the characterization of the products of limited proteolytic digestion (Thomsen and Sottrup-Jensen 1993) and the correlation of functional studies with the known sequences of α_2-macroglobulin and other members of the α_2-macroglobulin family of proteins (Sottrup-Jensen 1987, 1989). The best-characterized functional domains are, from amino- to carboxyl-terminus, the bait region, the internal thiol-ester domain, and the receptor-binding domain. The bait region of human α_2-macroglobulin is a flexible 25-residue stretch that includes a collection of peptide bonds that makes it a satisfactory substratum for proteolytic attack by proteases of diverse specificity and catalytic mechanism (Enghild et al. 1989c; Sottrup-Jensen et al. 1989). The presence of this diversity of protease-sensitive residues establishes the ability of α_2-macroglobulin to bind endopeptidases of all catalytic mechanisms.

The internal thiol-ester bond is activated following proteolysis of one or another of the susceptible peptide bonds of the bait region to generate a reactive glutamyl residue and a free sulfhydryl (Tack et al. 1980; Sottrup-Jensen et al. 1984):

-Gly-Cys-Gly-Glu-Glx-Asn-   -Gly-Cys-Gly-Glu-Glx-Asn-

    S————C=O   SH   C*=O

The reactive γ-carbonyl of the activated thiol ester subsequently reacts with nucleophiles in its immediate environment. Reaction of this glutamyl residue with ε-amino and hydroxyl residues on the target protease molecule allows α_2-macroglobulin to establish covalent γ-glutamyl isopeptide bonds linking the protease to the α_2-macroglobulin peptide bearing the thiol-ester domain (Sottrup-Jensen et al. 1990b; Chen et al. 1992). Inactivation of the thiol-ester domain by exposure of α_2-macroglobulin to small primary amines such as methylamine destroys the ability of α_2-macroglobulin to react subsequently with proteases. Methylamine-treated α_2-macroglobulin experiences bait region cleavage when subsequently treated with the protease, but is incapable of entrapping the protease molecule (Steinbuch et al. 1968; Swensen and Howard 1979). Under normal conditions, proteolysis at the bait region results in the activation of the thiol ester and the compaction of α_2-macroglobulin that results in the trapping of the reacting protease (Barrett and Starkey 1973).

The reactive internal thiol-ester bond of α_2-macroglobulin is unique to the members of the α_2-macroglobulin family of proteins (Tack 1983). Its presence can be documented by the exposure of a new thiol residue following reaction with protease, by the sensitivity of biological reactivity to incubation with small primary amines, and by the susceptibility of the polypeptide chain to heat-induced fragmentation at the peptide bond linking the glutamyl residue of the thiol ester to the glutamyl residue immediately before this glutamic acid. Additionally, there is a strong conservation of peptide sequence of the thiol-ester domain (Sottrup-Jensen et al. 1990a). The thermal fragmentation reaction is thought to be a consequence of the conversion of the glutamyl residue to a pyroglytamate residue by an intra-residue nucleophilic attack of the primary amino group on the γ-carbonyl group. The amino group breaks the peptide bond with the adjacent glutamyl residue to establish an amide bond with the γ-carbonyl (Howard 1983).

The structural rearrangement of α_2-macroglobulin following proteolytic activation also results in the exposure of the receptor-recognition domain, which is situated at the carboxyl terminus of the molecule (Enghild et al. 1989b; Thomsen and Sottrup-Jensen 1993). Protease-reacted α_2-macroglobulin is rapidly cleared from the circulation by receptor-mediated endocytosis initiated by the binding of this domain to receptors found on the surfaces of a variety of different cell types (Van Leuven 1984; Strickland et al. 1990). Receptor-bound α_2-macroglobulin is internalized into secondary lysosomes and degraded. Apparently, the receptor-recognition domain of unreacted α_2-macroglobulin is buried, since this form of the molecule does not bind to the α_2-macroglobulin receptor (Van Leuven 1984). Methylamine-reacted α_2-macroglobulin undergoes a compaction similar to that which follows reaction with proteases (Barrett et al. 1979; Bjork and Fish 1982; Gonias et al. 1982) and, if introduced back into the organism, methylamine-reacted α_2-macroglobulin is subsequently cleared by the same receptor system as protease-reacted α_2-macroglobulin (Van Leuven 1984).

3 The α_2-Macroglobulin Protein Family

α_2-Macroglobulin is the senior member of a family of immune defense proteins that share significant peptide sequence identity. Included in the α_2-macroglobulin family are α_2-macroglobulin itself, several closely-related proteins such as human pregnancy-zone protein (Christensen et al. 1989; Devriendt et al. 1991) and rodent $\alpha_1$inhibitor-3 (Enghild et al. 1989a; Rubenstein et al. 1991, 1992), the ovomacroglobulins, which are present in the albumins of the eggs of various birds and reptiles (Ikai et al. 1983, 1990; Nagase et al. 1983), and the effector proteins of the vertebrate complement systems C3, C4, and C5 (Tack 1983). As will be described below, the α_2-macroglobulin family also includes a variety of proteins found in the plasma of invertebrates. Most, but not all, of the members of the α_2-macroglobulin family possess an internal reactive thiol ester (Tack 1983). Exceptions include some of the ovomacroglobulins (Nagase and Harris 1983; Nielsen and Sottrup-Jensen 1993) and C5. In those forms that contain the thiol ester, the molecular function is susceptible to inactivation of this bond by low-molecular-mass primary amines. The thiol ester of C3 and C4 is responsible for the covalent linkage of these proteins to the surfaces of foreign particles (Law and Levine 1977), thereby targeting the particles for subsequent cytolysis or phagocytic destruction (Law and Reid 1988).

4 α_2-Macroglobulin in Invertebrates

The first indication that invertebrates possess α_2-macroglobulin came when we stumbled upon a protease inhibitor in the plasma of the American horseshoe crab, *Limulus polyphemus,* that had several of the features of α_2-macroglobulin (Quigley et al. 1982). This inhibitor was capable of inhibiting the proteolytic activity of a variety of proteases against [^{14}C]-casein (Table 1, Fig. 1A), but did not inhibit the activity against low-molecular-mass amide substrates (Fig. 1B). The anti-proteolytic activity was abolished by prior treatment with methylamine (Table 2; Quigley and Armstrong 1983). The inhibitor was also capable of protecting the amidolytic activity of trypsin against the macromolecular active site inhibitor, soya bean trypsin inhibitor (SBTI) (Armstrong et al. 1985; Fig. 2). The SBTI-protection assay has been used to quantify the amount of α_2-macroglobulin in the plasma of invertebrates (Armstrong et al. 1985) and to screen the plasma of various species for α_2-macroglobulin-like protease-binding molecules with this activity (Armstrong and Quigley 1992). Subsequently,

Table 1. Inhibition of the proteolytic activity of different proteases against [^{14}C]-casein and [^{125}I]-fibrin by *Limulus* plasma

Enzyme	Plasma[a]	Proteolytic activity (cpm)	
		[^{14}C]-Casein hydrolysis	[^{125}I]-Fibrin hydrolysis
Trypsin	–	1957	12559
Trypsin	+	194	823
Chymotrypsin	–	1384	
Chymotrypsin	+	132	
Plasmin	–		15250
Plasmin	+		620
Elastase	–		22359
Elastase	+		603
Subtilisin	–	3187	
Subtilisin	+	270	
Thermolysin	–	2541	
Thermolysin	+	173	
None	None	146	

[a]+ means that the protease was reacted with *Limulus* plasma that had been cleared of hemocyanin by ultracentrifugation before testing for its caseinolytic or fibrinolytic activity; – means that the protease was reacted with buffer and then tested.

Table 2. Effects of methylamine treatment of plasma on the caseinolytic activity of trypsin (0.5 μg)

Plasma fraction[a]	Protease activity (cpm [^{14}C]-casein hydrolyzed)
None	2385
Untreated plasma (10 μl)	10
Methylamine-treated plasma (40 μl)	2220

[a]The samples of plasma were treated as indicated and were then tested for their ability to inhibit the caseinolytic activity of trypsin.

α_2-macroglobulin has been identified in the plasmas of crustaceans (Armstrong et al. 1985; Hergenhahn and Söderhäll 1985; Spycher et al. 1987; Hergenhahn et al. 1988; Stöcker et al. 1991), and gastropod, bivalve, and cephalopod mollusks (Armstrong and Quigley 1992; Thøgersen et al. 1992).

So far, all of the forms of α_2-macroglobulin from invertebrates that have been characterized are dimeric (Spycher et al. 1987; Hergenhahn et al. 1988; Enghild et al. 1990; Armstrong et al. 1991; Thøgersen et al. 1992). This is nicely shown by transmission electron microscopy (Fig. 3) and scanning transmission electron microscopy (Fig. 4) studies of *Limulus* α_2-macroglobulin. *Limulus* α_2-macroglobulin undergoes an especially substantial compaction following reaction with proteases (Figs. 3 and 5). The magnitude of the protease-induced conformational

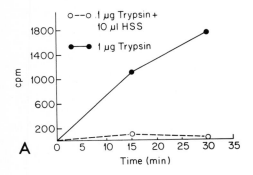

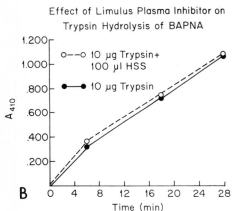

Fig. 1. Presence in *Limulus* plasma of a protease inhibitor that **A** eliminates the caseinolytic activity of trypsin but **B** fails to inhibit the activity of trypsin against the low molecular mass amide substrate BAPNA (N$^{\alpha}$-benzoyl-DL-arginine *p*-nitroanilide). Plasma was cleared of hemocyanin by ultracentrifugation. Hydrolysis of [^{14}C]-casein (**A**) was determined by the release of acid-soluble [^{14}C]. Hydrolysis of BAPNA (**B**) was determined by the increase in optical absorbance at 410 nm. Assay conditions were established for which both assays were approximately linear for the amount of enzyme and duration of incubation

change recommends *Limulus* α_2-macroglobulin for biophysical studies to characterize the nature of the molecular-trapping reaction of α_2-macroglobulin.

The sensitivity of α_2-macroglobulin from a variety of invertebrates to methylamine suggests that these variants, like human α_2-macroglobulin, contain a thiol ester that is required for function (Armstrong and Quigley 1987; Spycher et al. 1987; Thøgersen et al. 1992). This premise has been documented by the demonstration that, like other thiol-ester-containing proteins, *Limulus* α_2-macroglobulin experiences fragmentation of the polypeptide chain under mild denaturing conditions and by the exposure of a free sulfhydryl residue following proteolysis (Armstrong and Quigley 1987; Spycher et al. 1987; Hall et al. 1989). The thiol-ester domain of α_2-macroglobulin from invertebrates and vertebrates shows a striking conservation of peptide sequence (Spycher et al. 1987; Hall et al. 1989; Sottrup-Jensen et al. 1990a; Thøgersen et al. 1992). As mentioned before, the autolytic fragmentation during mild thermal denaturation, which yields fragments of 125 and 55 kDa, occurs by the conversion of the glutamyl residue of the thiol ester to a pyroglutamyl residue with the release of the peptide bond of this

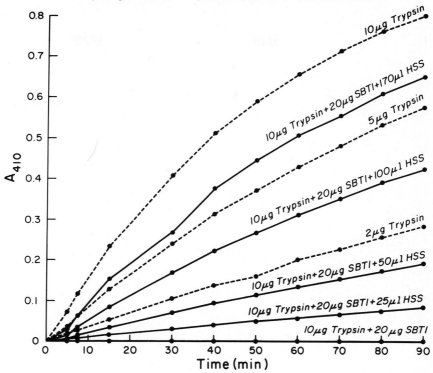

Fig. 2. Protection of trypsin from the macromolecular active site trypsin inhibitor, soybean trypsin inhibitor (SBTI) (Mr = 21 000) by *Limulus* plasma that has been cleared of hemocyanin by ultra-centrifugation (high-speed supernatant = HSS). This assay employed saturating amounts of trypsin and twice-saturating amounts of SBTI. The plasma sample was preincubated for 15 min with trypsin to allow binding by α_2-macroglobulin, and then was reacted with SBTI to inhibit any uncomplexed trypsin. The assay was initiated by adding the low-molecular-mass substrate, BAPNA. The only fraction of trypsin capable of hydrolyzing BAPNA is that fraction that is complexed with α_2-macroglobulin, and that, thereby, is protected from SBTI. As the amount of plasma in the assay is increased, the amount of trypsin that is in the protected state is likewise increased, as indicated by the increase in the rate of hydrolysis of BAPNA. In the absence of plasma, there is no hydrolysis of BAPNA, since all of the trypsin is inhibited by SBTI. The amount of α_2-macroglobulin-like activity in the sample, in μg trypsin bound/ml plasma, is estimated by comparing the hydrolytic rates in the presence of plasma with the rates of hydrolysis of BAPNA by different concentrations of trypsin in the absence of plasma and SBTI (*dashed line*), and from knowledge of the fractional activity of the preparation of trypsin, as determined by titration with *p*-nitrophenyl *p'*-guanidinobenzoate hydrochloride. (Chase and Shaw 1970)

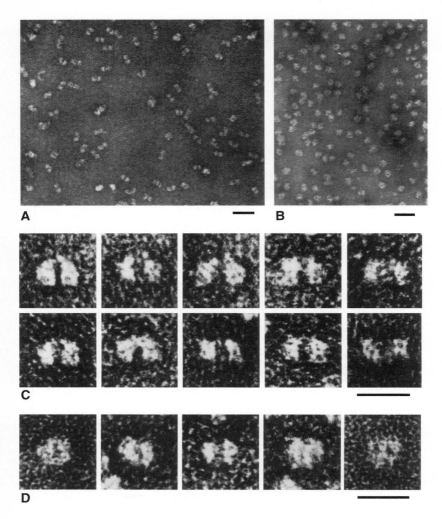

Fig. 3. Transmission electron microscopy of individual molecules of negatively stained *Limulus* α_2-macroglobulin. Low magnification views of **A** fields of native and **B** fields of chymotrypsin-reacted *Limulus* α_2-macroglobulin molecules. **C** High magnification views of individual molecules of native and **D** individual molecules of chymotrypsin-reacted *Limulus* α_2-macroglobulin. The native molecule of *Limulus* α_2-macroglobulin is relatively extended and resembles a butterfly, whereas the chymotrypsin-reacted molecule is significantly more compact

residue to the amino acid that was amino terminal to it in the polypeptide chain (Howard 1983). The new thiol residue that is exposed by the activation of the thiol-ester bond can be identified by reaction with an appropriate sulfhydryl titrant (Armstrong and Quigley 1987; Spycher et al. 1987).

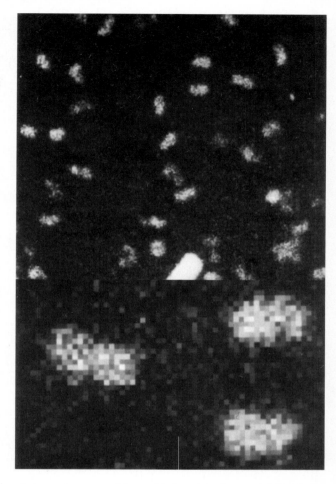

Fig. 4. Scanning transmission electron microscopy of freeze-dried, unstained, uncoated *Limulus* α_2-macroglobulin. The top panel shows a low magnification view of a field of molecules, the bottom panel shows a high magnification view of 3 α_2-macroglobulin molecules. The elongated structure of the individual α_2-macroglobulin molecules is consistent with a dimeric organization. The cylindrical structure at the *bottom edge of the top panel* is a tobacco mosaic virus particle, which was added to the sample as an internal calibration standard

The presence of a reactive internal thiol ester in α_2-macroglobulin of vertebrates and invertebrates indicates that the thiol ester appeared early in the evolution of higher animals and is characteristic of the ancestral form of α_2-macroglobulin. Although the thiol ester is a unique feature of proteins of the α_2-macroglobulin family, the functional significance of this bond in α_2-macro-globulin is not thoroughly understood. Certain forms of α_2-macroglobulin, notably certain of the ovomacroglobulins from the albumin of the eggs of egg-laying amniotes, lack the thiol-ester bond (Nagase et al. 1983; Nielsen and

GEL PERMEATION CHROMATOGRAPHY OF CHYMOTRYPSIN-
TREATED LIMULUS α_2 - MACROGLOBULIN

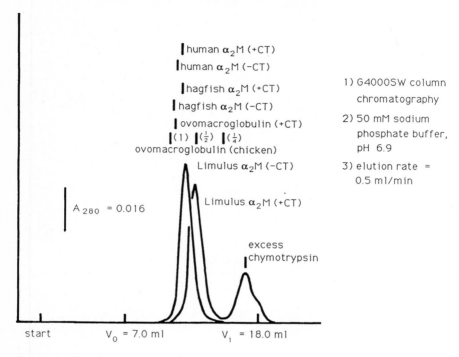

Fig. 5. Gel filtration chromatography of unreacted and chymotrypsin-reacted α_2-macroglobulin. Samples of unreacted and chymotrypsin-reacted *Limulus* α_2-macroglobulin were applied to a TSK G4000SW gel permeation column and were eluted with 0.05 M phosphate buffer. The protease-reacted sample was significantly retarded, indicative of a compaction of the molecule. The *calibration bars* above the figure indicate the elution positions of native and chymotrypsin-reacted tetrameric (human α_2M and ovomacroglobulin) and dimeric (hagfish α_2M) forms of α_2-macroglobulin. The column was further calibrated with tetrameric, dimeric, and monomeric ovomacroglobulin. The dimeric forms of α_2-macroglobulin (e.g., *Limulus* and hagfish α_2-macroglobulin) show a more pronounced compaction as estimated by the separation of the elution profiles of native and reacted forms, than do the tetrameric varieties

Sottrup-Jensen 1993), yet these forms of α_2-macroglobulin bind proteases perfectly well (Kitano et al. 1982; Nagase and Harris 1983; Feldman and Pizzo 1984). Human α_2-macroglobulin establishes covalent isopeptide bonds with the reacting protease via the reactive glutamyl of the thiol-ester bond (Sottrup-Jensen et al. 1990b; Chen et al. 1992), but covalent bonding is not required for the efficient trapping of the protease in the α_2-macroglobulin cage. Proteases, such as elastase, that lack reactive ε-amino groups, or trypsin in which the ε-amino groups have been acetylated, and are therefore rendered incapable of forming isopeptide N^ε (γ-glutamyl)-lysine bonds with the thiol ester glutamyl residue, nevertheless are efficiently trapped by human α_2-macroglobulin (Salvesen et al. 1981; Van Leuven et al. 1981; Wu et al. 1981). However, dimeric and monomeric forms of mammalian α_2-macroglobulin are incapable of binding proteases by noncovalent means (Christensen et al. 1989; Enghild et al. 1989a). In these forms, it has been suggested that the covalent γ-carbonyl-ε-amino isopeptide bonding of the target protease, which is mediated by the thiol-ester glutamyl residue, may be essential for function. In this regard, it is interesting that *Limulus* α_2-macro-globulin, which is dimeric, does not establish covalent bonds with the reacting protease (discussed below), but is highly efficient in protease trapping (Quigley et al. 1991). The functional role of the thiol ester is better characterized for C3 and C4 of complement, where it is essential for the covalent cross-linking of these proteins to the surfaces of target particles (Law and Reid 1988).

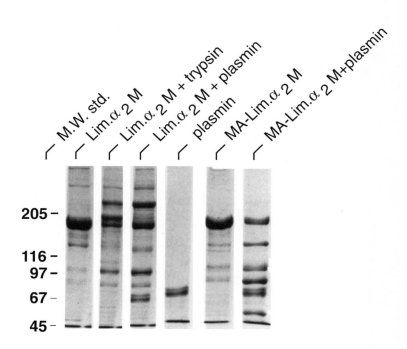

Fig. 6. SDS-polyacrylamide gel electrophoresis (reducing conditions) of unreacted and protease-reacted *Limulus* α_2-macroglobulin. The subunit molecular mass of unreacted *Limulus* α_2-macro-globulin (*lane 1*) is estimated to be 185 kDa. The formation of proteins with a greater molecular mass following exposure to trypsin (*lane 2*) and plasmin (*lane 3*) results from the establishment of isopeptide bonds linking formerly separate polypeptide chains into products with a molecular mass that is greater than 185 kDa. In addition, two lower molecular mass bands with approximate molecular masses of 100 and 85 kDa result from bait region cleavage without subsequent isopeptide bonding. High molecular mass products fail to form if the thiol ester of *Limulus* α_2-macroglobulin is eliminated by prior reaction with methylamine (*lane 6*)

Interestingly, although α_2-macroglobulin from *Limulus* has the thiol ester, it appears not to be involved in the cross linking of protease-reacted α_2-macroglobulin to other proteins. Instead, the γ-glutamyl isopeptide bonds that do form appear to cross-link entirely within the α_2-macroglobulin molecule itself. The formation of these isopeptide bonds is documented by the appearance of poly-peptides with a larger molecular mass than that of the unreacted α_2-macroglobulin monomer (Fig. 6, lanes 2, 3). Unlike human α_2-macroglobulin, which efficiently links trypsin into these high-molecular-mass polypeptides, dimeric *Limulus* α_2-macroglobulin fails to engage in isopeptide bonding to labeled protease (Quigley et al. 1991). The formation of high-molecular-mass products by protease-activated *Limulus* α_2-macroglobulin is dependent on the thiol ester, because inactivation of the thiol ester by prior treatment with methylamine prevents cross-linking upon subsequent protease treatment (Fig. 6, lane 6). Thus it is concluded that the binding of protease to *Limulus* α_2-macroglobulin is entirely noncovalent in character, and the isopeptide bonding that does occur establishes cross-links exclusively between the polypeptide chains of α_2-macroglobulin. Possibly, these isopeptide bonds function to strengthen the α_2-macroglobulin cage that encloses the bound protease molecule to ensure that escape is impossible. Interestingly, exogenous nucleophiles, such as glycerol, methyl-amine, and polylysine, do not affect isopeptide bonding if added at the same time as protease. This suggests that the thiol ester is protected from interaction with exogenous reactants upon proteolytic activation, and thus mediates exclusively intramolecular cross linking.

5 Receptor-Mediated Clearance of α_2-Macroglobulin-Protease Complex

Although α_2-macroglobulin is usually thought of as a protease inhibitor, it might better be considered as a protease-binding molecule that functions in a receptor-mediated endocytotic clearance pathway for proteases (Van Leuven 1984). In mammals, where the endocytotic clearance pathway has been extensively stud-ied, native, unreacted α_2-macroglobulin is stable in the plasma and fails to react with cell-surface receptors. After reaction with protease, the α_2-macroglobulin-protease complex is susceptible to rapid binding to surface-membrane receptors present on a variety of cell types and to subsequent internalization into secondary lysosomes. The clearance time of active proteases introduced into the circulation of mammals is short, typically 5–10 min (Ohlsson 1971). It appears that the receptor-mediated binding of the α_2-macroglobulin-protease complexes to hepa-tocytes and macrophages contributes to this rapid clearance pathway (Davidsen et al. 1985; Van Dijk et al. 1992; Ogata et al. 1993). Based on these observations, it has been suggested that one of the primary functions of α_2-macroglobulin in mammals is to serve as a protease-carrier molecule with the capabilities of binding endopeptidases of all possible enzymatic mechanisms and from all possible endogenous and exogenous sources to thereby render the bound pro-teases sensitive to removal from the body fluids by the receptor-mediated endocytotic pathway. In this sense, α_2-macroglobulin is thought to serve as a broad-spectrum recognition system for the identification and marking of proteases for subsequent endocytosis.

As mentioned above, the conformational changes that are induced by the proteolysis of mammalian α_2-macroglobulin, expose a domain at the carboxyl terminus of the polypeptide chains that is then recognized by the α_2-macroglobulin-protease receptor on the cell surface. The mammalian receptor for the α_2-macroglobulin-protease complex has been identified as low-density lipoprotein receptor-related protein (LRP) (Kristensen et al. 1990; Strickland et al. 1990). LRP is synthesized as a 600 kDa protein with a single transmembrane domain and is a member of the low-density lipoprotein receptor family. In addition to α_2-macroglobulin-protease complex, LRP appears also to be involved in the cell-surface binding of a variety of other ligands, including plasminogen activator-plasminogen activator inhibitor-1 complexes, lipoprotein lipase, lactoferrin, *Pseudomonas* exotoxin A, and apolipoprotein E-rich chylomicron remnant (Hussain et al. 1991; Nykjaer et al. 1993; Moestrup et al. 1993). LRP is, like α_2-macroglobulin itself, an ancient protein that was present prior to the great evolutionary radiation that established the divergent deuterostome and protostome invertebrate superphyla, since it has been found in modern representatives of lineages that diverged at the time of the Precambrian radiation, notably in verte-brates and the nematode, *Caenorhabditis elegans* (Yochem and Greenwald 1993).

It has not been established whether α_2-macroglobulin of invertebrates functions in a receptor-mediated clearance pathway. In our laboratory, we are currently exploring the possibility that *Limulus* and other arthropods (1) are capable of clearing active proteases from the circulation, and (2) if such a clearance capability does exist, that it is accomplished by the receptor-mediated uptake of α_2-macroglobulin-protease complex. To show specificity, it will be necessary to show that inactivated proteases are not cleared from the circulation[4] and that the clearance times of unreacted *Limulus* α_2-macroglobulin are very much longer than for reacted α_2-macroglobulin. The presence of cell-surface receptors for the α_2-macroglobulin-protease complex can be demonstrated by a specific binding of labeled α_2-macroglobulin-protease complex to a suitable population of cells. Methods of labeling include radioiodination, biotinylation, and fluoresceination. Specific binding is that which can be displaced by an excess of unlabeled α_2-macroglobulin-protease complex. Since only protease- or methylamine-reacted α_2-macroglobulin should bind to the α_2-macroglobulin-protease receptor (Debanne et al. 1975; Van Leuven et al. 1979; Gliemann et al. 1983), it is important to demonstrate that labeled, unreacted α_2-macroglobulin does not decorate cells.

These experiments are currently underway. *Limulus* has an important

---

[4]Since the binding of proteases requires the proteolytic processing of the bait region, inactive proteases fail to bind to α_2-macroglobulin (Barrett and Starkey 1973) and, thus, would not be expected to clear via the uptake of the α_2-macroglobulin-protease complex.

advantage over vertebrates for the analysis of the physiological functions of plasma α_2-macroglobulin because the composition of its plasma is much more simple. While mammals possess a plethora of different plasma protease inhibitors (Laskowski and Kato 1980; Travis and Salvesen 1983) that significantly complicate the determination of the importance of α_2-macroglobulin-mediated clear-ance (Sottrup-Jensen 1987), *Limulus* has only α_2-macroglobulin (Quigley and Armstrong 1983). There are just fewer potential players in *Limulus* plasma than are present in mammals.

6 α_2-Macroglobulin in the Blood Cells

Limulus has only a single cell type in the general circulation, the granular amebocyte (for review see Armstrong 1985a). All of the components of the blood-clotting system (the clottable protein, coagulogen, and the cascade of proteases that are responsible for its proteolytic modification) are contained in the amebocyte granules and are released by exocytosis at the initiation of the clotting response (Mürer et al. 1975; Armstrong and Rickles 1982). Interestingly, α_2-macroglobulin is also released during degranulation of the amebocyte (Armstrong et al. 1990). Functionally active α_2-macroglobulin can be detected by the SBTI-protection assay in the environment of washed degranulated cells, and α_2-macroglobulin antigen is detectable by immune staining of Western blots of the materials released by exocytosis from washed cells (Armstrong et al. 1990). Release requires exocytosis: α_2-macroglobulin is undetectable in the penultimate buffer wash of the blood cells and is not released if the cells are not stimulated to degranulate. Exocytosis does not involve cell lysis; although the cells contain an abundance of the cytoplasmic marker enzyme, lactate dehydrogenase, none is released when the cells are stimulated to degranulate (Armstrong and Quigley 1985). Immunoblotting analysis of the products released by activated amebocytes shows that the form of α_2-macroglobulin contained within the blood cells has a molecular mass identical to that of the plasma form. The blood cells also release active site protease inhibitors during degranulation (Armstrong and Quigley 1985; Nakamura et al. 1987; Donovan and Laue 1991). This situation is reminiscent of that of mammals, where the blood platelet, the thrombocytic cell of mammalian blood, releases α_2-macroglobulin and α_1-protease inhibitor during exocytotic degranulation (Nachman and Harpel 1976; Plow and Collen 1981). Our present hypothesis is that the α_2-macroglobulin and the active-site protease inhibitors released from the amebocyte may function in the cellular plug formed by aggregated amebocytes at sites of injury to inactivate the proteases that are also released from the blood cells and from bacteria entrapped within the cellular clot. The gradual degranulation of the blood cells within the clot would ensure a gradual release of α_2-macroglobulin in the depths of the cellular clot, which is presumably a site that is relatively inaccessible to plasma α_2-macroglobulin.

Although it has not been established that the blood cells of *Limulus* synthesize the α_2-macroglobulin contained within the secretory granules, blood cells from the crayfish *Pacifastacus* clearly do synthesize α_2-macroglobulin (Liang et al. 1992). It has not been shown that the blood cell is the principal source of plasma α_2-macroglobulin in this species, nor has the intracellular localization within the blood cell been defined.

7 Interaction of α_2-Macroglobulin with the Proteases of the Clotting System of *Limulus*

The blood clotting system of *Limulus* involves activation of a soluble apo-form of the principal structural protein of the clot by a specific proteolytic cleavage administered by a particular protease (Iwanaga et al. 1992). The apo-form of the structural protein of the clot is known as coagulogen; the proteolyzed form is known as coagulin (Tai et al. 1977; Mosesson et al. 1979; Miyata et al. 1984). Following proteolysis, coagulin polymerizes to form the fibrillar elements of the clot (Holme and Solum 1973). The protease involved in the proteolysis of coagulin is known as clotting enzyme (Nakamura et al. 1985; Muta et al. 1990b; Roth and Levin 1992) and is the terminal member of a proteolytic cascade that includes at least two enzymes more proximate in the reaction sequence (Nakamura et al. 1986; Muta et al. 1991, 1993). The clotting reaction presumably occurs in the presence of relatively high concentration of α_2-macroglobulin derived both from plasma and from the exocytotic blood cells themselves. As discussed above, *Limulus* α_2-macroglobulin reacts almost instantaneously with proteases of diverse catalytic mechanisms. However, *Limulus* α_2-macroglobulin fails to inhibit clotting, suggesting that it is uniquely incapable of suppressing the proteolytic activity of clotting enzyme against coagulogen, even when clotting enzyme and α_2-macroglobulin are preincubated prior to exposure to coagulogen (Armstrong et al. 1984). This failure is surprising for two reasons. Firstly, although the *Limulus* and human forms of α_2-macroglobulin show extensive functional similarity, they differ in their interaction with *Limulus* clotting enzyme: human α_2-macroglobulin inhibits the activity of the enzyme against coagulogen while, as indicated above, *Limulus* α_2-macroglobulin fails to inhibit (Armstrong et al. 1984). Secondly, *Limulus* α_2-macroglobulin is active against a broad spectrum of proteases (e.g., representatives of all categories of serine proteases, metalloproteases, and thiol proteases; proteases from vertebrates, plants, and bacteria (Tables 1 and 2; Quigley and Armstrong 1983, 1985)). In fact, the only protease that we have examined whose proteolytic activity is not abolished by *Limulus* α_2-macroglobulin is *Limulus* clotting enzyme. A similar situation may obtain in the blood of the crayfish, *Pacifastacus,* in which the proteases involved in activation of the prophenol oxidase system, a component of the systems involved in humoral immunity in crustaceans, are unaffected by the α_2-macroglobulin found in the plasma of that organism (Hergenhahn et al. 1987).

8 Cytokine-Binding Activities of α_2-Macroglobulin

Although α_2-macroglobulin has traditionally been identified as a protease-binding protein, it has recently become clear that it is also capable of binding a variety of other proteins, including peptide mitogens (James 1990; Borth 1992; Gonias 1992), basic proteins (Boyde and Pryme 1968; McPherson et al. 1970; Barrett and Starkey 1973; Stoller and Rezuke 1978; Peterson and Venge 1987), and ferritin (Santambrogio and Massover 1989). For some of these nonproteolytic proteins, the binding to α_2-macroglobulin requires activation of the thiol ester and involves the establishment of a disulfide linkage via the exposed thiol of the cysteinyl residue of the activated thiol ester (Borth and Luger 1989). The binding of growth factors has proven interesting since certain mitogens in the serum of mammals have been found to be complexed with α_2-macroglobulin (Huang et al. 1984; O'Connor-Court and Wakefield 1987). In some situations, the binding to α_2-macroglobulin inhibited the mitogenic activities of the growth factor, in other cases it did not (Borth 1992). Also, it must be remembered that a variety of serine proteases are potent mitogens that interact with a dedicated G-protein-coupled surface receptor (Vu et al. 1991). Although the best-described interaction of the different growth factors with the target cells is via specific mitogen receptors (Sporn and Roberts 1990), the frequent association of circulating mitogens and mitogenic proteases with α_2-macroglobulin opens the possibility that productive mitogenic interaction might also occur via the LRP/α_2-macroglobulin receptor. Thus, when exploring the functions of α_2-macroglobulin in invertebrates, it is important to extend our vision beyond the specifics of α_2-macroglobulin-protease interaction and entertain the possibility that proteases may not be the only important molecules whose functions are modulated by binding to α_2-macroglobulin.

9 The Plasma-Based Cytolytic System of *Limulus*

One of the important immune defense strategies of higher animals is to kill invading pathogenic organisms by inducing their cytolysis in the blood. In vertebrates, cytolysis is mediated by the complement system with its associated regulators and receptors (Law and Reid 1988). The key factor in the mammalian complement system is the protein C3, which binds to the surfaces of target cells, marking them for destruction by cytolysis and phagocytosis. Binding involves the covalent bonding of the γ-carbonyl of the glutamyl residue of the reactive internal thiol ester of the C3 molecule, with hydroxyl and amino residues at the surface of the target particle (Law and Levine 1977). As mentioned previously, C3 is a member of the α_2-macroglobulin family of proteins, based on peptide sequence homology and presence of the reactive thiol ester (Tack 1983; Sottrup-Jensen 1987). A recently-emergent topic in the complement field is the evolution of this complex defense system (Dodds and Day 1993; Farries and Atkinson 1991).

Lower vertebrates show many but not all of the elements of the mammalian system (Dodds and Day 1993). Invertebrates also have plasma- or hemocyte-based cytolytic systems (Noguchi 1903; Day et al. 1970; Bertheussen 1983; Cenini 1983; Komano and Natori 1985; Tucková et al. 1986; Canicatti and Cuilla 1987, 1988), but none had been convincingly demonstrated to be related to the vertebrate complement system until Enghild et al. (1990) observed that the form of α_2-macroglobulin found in the blood of the American horseshoe crab *Limulus polyphemus* is a component of the plasma-based cytolytic system of that animal. Based on the molecular similarity between α_2-macroglobulin and C3 and the involvement of the thiol ester in the activities of both molecules, these results are consistent with the possibility that in *Limulus*, α_2-macroglobulin serves a function like that of C3 in vertebrates.

We have confirmed that α_2-macroglobulin is involved in the hemolytic reactivity of *Limulus* plasma, by demonstrating a methylamine-sensitivity of the hemolytic reaction. Methylamine treatment of plasma resulted in a reduction in its hemolytic activity (Armstrong et al. 1993a) which was reversed by the addition of purified *Limulus* α_2-macroglobulin (Fig. 7). Methylamine-treated *Limulus* α_2-macroglobulin was unable to restore the hemolytic activity of methylamine-treated plasma.

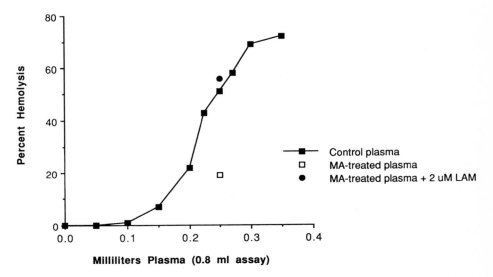

Fig. 7. Cytolytic activity of *Limulus* plasma that has been depleted of hemocyanin by ultracentrifugation ($120\,000 \times g$, 16 h). Washed sheep erythrocytes were incubated with *Limulus* plasma at room T for 4 h, and the extent of hemolysis was determined by measuring the quantity of hemoglobin released into the incubation buffer (optical absorbance at 412 nm). Untreated plasma shows a dose-responsive hemolytic activity that is diminished in plasma that has been pre-treated with methylamine to inactivate α_2-macroglobulin. The hemolytic activity of methylamine-treated plasma is restored by addition of $2\,\mu$M *Limulus* α_2-macroglobulin

The three most abundant proteins of *Limulus* plasma are α_2-macroglobulin (Quigley and Armstrong 1985), hemocyanin, and C-reactive protein (Robey and Liu 1981; Nguyen et al. 1986a, b; Amatayakul-Chantler et al. 1993; Tennent et al. 1993). α_2-Macroglobulin is present in *Limulus* plasma at 1–2 mg/ml (Enghild et al. 1990), C-reactive protein (CRP) at 1–5 mg/ml (Nguyen et al. 1986a), and hemocyanin at 30–50 mg/ml. The hemolytic activity of *Limulus* plasma is unaffected by the complete removal of hemocyanin by ultracentrifugation, indicating that this molecule plays no essential role in the process. The third abundant plasma protein, C-reactive protein, is present in two forms, a minor form that has lectin activity for sialyl-containing oligosaccharides, and an abundant form that lacks sialyl lectin activity. The CRP-related sialyl-binding lectin, which is probably the same protein as limulin (Kehoe and Seide 1986), is one of several sialyl-binding lectins in *Limulus* plasma (Srimal et al. 1993). CRP-related lectin is an essential component of the hemolytic reaction. Treatment of plasma with the C-reactive protein reactant, phosphorylethanolamine, resulted in a dose-dependent elevation of hemolytic activity and removal of C-reactive protein with phosphorylethanolamine-agarose or removal of sialyl-binding lectins with fetuin-agarose eliminated the hemolytic activity. Activity was restored to the phosphorylethanolamine-agarose-treated plasma by the addition of purified C-reactive protein. Purified CRP-related lectin was hemolytic in the absence of any other proteins. The hemolytic activity of purified C-reactive protein was unaffected by treatment with methylamine.

The possibility that invertebrates possess isolated elements of the vertebrate complement system has been suggested previously (Day et al. 1970, 1972; Bertheussen 1982, 1983; D'Cruz 1991), but the demonstration that α_2-macroglobulin, and a subfraction of C-reactive protein with sialyl lectin activity, participate in hemolysis in *Limulus*, is the first demonstration of a direct involvement of components related to elements of the vertebrate complement system in an invertebrate cytolytic system. However, important differences between complement-mediated hemolysis and hemolysis in *Limulus* are readily apparent. In the complement system, C3 is absolutely required, whereas in *Limulus* the C3 homologue, α_2-macroglobulin, is dispensable in purified systems containing only the CRP-related lectin. Also, although the action of C3 in complement-mediated cytolysis requires the formation of amide and ester bonds with macromolecules of the target cell (Law and Levine 1977), the binding of *Limulus* α_2-macroglobulin to erythrocytes does not appear to involve ester or amide bonds. We do not yet understand the role in hemolysis played by α_2-macroglobulin.

C-reactive protein is an activator of the complement system of mammals, but not a direct participant in the process of complement-mediated cytolysis (Jiang et al. 1991; Miyazawa and Inoue 1990; Volanakis 1982). In *Limulus*, the subfraction of C-reactivity protein with sialyl lectin activity appears to play the central role in cytolysis, since removal of this protein from plasma profoundly eliminated the hemolytic activity and purified CRP-related lectin could produce hemolysis in the absence of other plasma proteins.

10 α_2-Macroglobulin and the Inactivation of the Proteases of Invading Parasites

It has been thought that the plasma protease inhibitors of higher animals play an important role in inactivating the proteases involved in the invasive activities of potential pathogenic microbes and multicellular parasites (McKerrow et al. 1993). The most prominent ectoparasite of *Limulus* is the triclad turbellarian flatworm, *Bdelloura*, which is found in abundance on the cuticle of the gill leaflets and the legs (Rudloe 1971). The egg cases and immature stages of *Bdelloura* are invasive to the gill leaflets of *Limulus*, causing lesions that are portals for the entry of bacteria and other pathogens (Groff and Leibovitz 1982; Huggins and Waite 1993). *Bdelloura* releases a Ca^{+2}-dependent serine protease into its environment, that may contribute to the deleterious effects of the flatworm on its horseshoe crab host (Armstrong et al. 1993b). The reactivity of *Limulus* α_2-macroglobulin for the *Bdelloura* protease was shown by a shift of molecular weight of the protease from 40–60 kDa in the absence of α_2-macroglobulin to 370 kDa in the presence of α_2-macroglobulin. The protease was detected by its reactivity against the amide substrate S-2222, which is unaffected by binding to α_2-macroglobulin. The interaction of the α_2-macroglobulin of the host with the ectoprotease of the parasite, may serve to limit the extent of the damage wrought by the protease once an individual *Bdelloura* has penetrated the cuticle and has contacted the hemolymph, and may serve to restrict entry of the parasite and to facilitate repair of the damage to the integument resultant from penetration.

11 Interaction of the α_2-Macroglobulin with Other Systems of Immunity in Invertebrates

Individual metazoans are threatened by a nearly continual challenge from microbial and metazoan pathogens. In response, metazoans have evolved a varied array of defense strategies that limit the susceptibility to attack by potential pathogens. These include the activities of a variety of specialized cells and humoral factors that fight pathogens that have succeeded in gaining access into the body. Of particular interest are the interactions between α_2-macroglobulin and the other blood-borne systems of immunity.

In this regard, *Limulus* is an attractive system for analysis, because its various immune defense systems have been the subject of intense investigation, and are reasonably well understood (Armstrong 1985a, 1991; Iwanaga et al. 1992). The blood of *Limulus* is relatively simple in composition, with a single circulating blood cell, the granular amebocyte (Armstrong 1985a,b), and a plasma with only three abundant proteins, hemocyanin, which is the respiratory protein, C-reactive protein, and α_2-macroglobulin (Armstrong et al. 1993a). *Limulus* is one of the largest easily obtainable arthropods. The blood occupies approximately one-third of its body mass. It is relatively easy to obtain 100 ml or more of blood from an average-sized adult (Armstrong 1985b). *Limulus* is, thus, an ideal subject for hematologic studies requiring large volumes of material. *Limulus*, like many large arthropods, is relatively long-lived, requiring 9–12 years to reach maturity (Shuster 1950, 1954), and with a maximum life span estimated to be in excess of 14–19 years (Ropes 1961). Although *Limulus* lives in a septic environment and is subject to frequent wounding to the surface integument, the hemolymph of a majority of freshly collected animals is sterile (Brandin and Pistole 1985). Presumably, invading pathogens are effectively controlled by an array of internal defense systems that ensure an extended life span for an animal that is frequently beset by lethal pathogens.

The best characterized immune defense system of *Limulus* is the blood clotting system. Hemostasis is accomplished by two processes; the aggregation of the circulating blood cells (the granular amebocytes) (Armstrong 1985a), and the production of the fibrillar extracellular clot (Iwanaga et al. 1992). As described above, the extracellular clot forms when the clottable protein, coagulogen, is proteolytically modified by the serine protease, clotting enzyme. Of all of the proteases tested, only clotting enzyme was not inhibited in its proteolytic action by *Limulus* α_2-macroglobulin (Armstrong et al. 1984). Presumably, this is a functional adaptation to allow blood clotting to proceed in an environment containing high concentrations of α_2-macroglobulin.

The extracellular clot functions both as a mechanical barrier to prevent continued leakage of blood and intrusion of microorganisms that might attempt to gain access to the body through lesions in the cuticle, and as a solid substrate for the transglutaminase-mediated immobilization of pathogens (Wilson et al. 1992; Tokunaga et al. 1993a,b). Transglutaminase catalyzes the formation of N$^\varepsilon$ (γ-glutamyl) lysine isopeptide bonds between proteins (Greenberg et al. 1991). Human α_2-macroglobulin has a transglutaminase reactive site that is located approximately 20 residues from the bait region (Sottrup-Jensen et al. 1983) and α_2-macroglobulin is a major transglutaminase substrate in the blood (Van Leuven 1984). If *Limulus* α_2-macroglobulin does contain a similar transglutaminase reactive site as human α_2-macroglobulin, then it would probably be a major component to be immobilized onto the coagulin fibrils of the blood clot, and would thus be positioned to effect the immobilization of potentially destructive proteases. This proposed immobilization on the extracellular clot is a second possible mechanism for the sequestration of harmful proteases that might be expected to complement a cell-based receptor-mediated endocytotic pathway for protease clearance.

There is evidence for the presence of antimicrobial systems in *Limulus* that are associated with serum and the blood cells (Smith and Pistole 1985; Pistole and Graf 1986; Shirodkar et al. 1960; Johannsen et al. 1973; Furman and Pistole 1976; Pistole and Britko 1978; Nachum 1979; Nachum et al. 1980). The granular amebocytes release a variety of peptides and proteins with potent antimicrobial activities, including a low-molecular-mass protein known as anti-LPS factor, and a family of 17- and 18-residue peptides known as the tachyplesins. Lipopolysaccharide, which is the principal toxin of Gram-negative bacteria (Morrison and Ryan 1987), is bound and inactivated by anti-LPS factor (Tanaka et al. 1982;

Aketagawa et al. 1986). This protein also has antimicrobial activity for rough strains of Gram-negative bacteria (Morita et al. 1985) and is hemolytic for LPS-sensitized mammalian erythrocytes (Ohashi et al. 1984). The tachyplesins contain a preponderance of cationic and hydrophobic amino acids, giving them amphipathic characteristics (Kawano et al. 1990), and have potent antibiotic activities against Gram-positive and Gram-negative bacteria and fungi (Muta et al. 1987; Nakamura et al. 1988; Miyata et al. 1989; Muta et al. 1990a).

All of the factors described above are contained within the secretory granules of the granular amebocytes, and are released by exocytosis when the cells are exposed to a suitable secretagogue (Armstrong and Rickles 1981). In addition, there are humoral defense systems in the plasma, of which the α_2-macroglobulin-based system has been the principal subject of this review. The plasma contains a family of lectins, which include several lectins with sialyl and 2-keto-3-deoxy-octonate (KDO) sugar-binding activities (Marchalonis and Edelman 1968; Finstad et al. 1974; Kaplan et al. 1977; Vaith et al. 1979; Stebbins and Harper 1986) and a lectin, polyphemin, with galactosyl and teichoic acid specificities (Gilbride and Pistole 1979; Brandin and Pistole 1983). As was described above, one of the sialyl-binding lectins, CRP-related lectin, is the hemolytic protein in the plasma-based cytolytic system. It has been suggested that the plasma lectins may also function as opsonins to promote the phagocytic uptake and destruction of foreign cells (Pistole 1982).

It will in the future be of major interest to explore the interactions, if any, between these various systems for immune defense, and particularly to investigate the interactions of plasma and cell-released α_2-macroglobulin with the other defense proteins and peptides that are secreted by the granular amebocytes or that are free in the plasma.

12 Evolutionary Considerations

α_2-macroglobulin is clearly an ancient component of the immune systems of animals since it is present in all classes of vertebrates (Starkey and Barrett 1982; Osada et al. 1986), mandibulate and chelicerate arthropods (Quigley and Armstrong 1983; Armstrong et al. 1985; Spycher et al. 1987; Enghild et al. 1990), and cephalopod, gastropod, and bivalve mollusks (Armstrong and Quigley 1992; Thøgersen et al. 1992). The forms of α_2-macroglobulin found in invertebrates share many of the unique features of vertebrate α_2-macroglobulin, including the presence of the reactive internal thiol-ester bond (Armstrong and Quigley 1987; Spycher et al. 1987; Sottrup-Jensen et al. 1990a), the ability to interact with proteases of various catalytic mechanisms (Quigley and Armstrong 1985), and an interaction involving the physical envelopment of the protease molecule by α_2-macroglobulin (Quigley and Armstrong 1983, 1985; Armstrong et al. 1985; Armstrong et al. 1991). Since the vertebrate, arthropod, and molluskan lineages diverged in evolution approximately 550 million years ago (Sepkoski 1978; Runnegar 1982), the presence of α_2-macroglobulin in modern representatives of these 3 classes of animals indicates that the molecule must have evolved prior to this evolutionary divergence. The preservation of this unique molecule for so extended a period of evolution in organisms as distinctively different as arthropods, mollusks, and vertebrates, indicates that it plays an essential role in the survival of the organism. Additionally, the existence of members of the α_2-macroglobulin protein family with conserved functional characters that have been separated by 550 million years of divergent evolution presents the comparative biochemist with an excellent opportunity to explore relations of molecular structure and biochemical and physiological function of this unique protein.

Acknowledgments. The studies from our laboratory are supported by Grant MCB-9218460 from the National Science Foundation.

References

Aketagawa J, Miyata T, Ohtsubo S, Nakamura T, Morita T, Hayashi H, Miyata T, Iwanatga S, Takao T, Shimonishi Y (1986) Primary structure of *Limulus* anticoagulant anti-lipopolysaccharide factor. J Biol Chem 261: 7357–7365

Amatayakul-Chantler S, Dwek RA, Tennent GA, Pepys MB, Rademacher TW (1993) Molecular characterization of *Limulus polyphemus* C-reactive protein. 1. Subunit composition. Eur J Biochem 214: 91–97

Armstrong PB (1985a) Adhesion and motility of the blood cells of *Limulus*. In: Cohen WD (ed) Blood cells of marine invertebrates. AR Liss, New York, pp 77–124

Armstrong PB (1985b) Amebocytes of the American "horseshoe crab" *Limulus polyphemus*. In: Cohen WD (ed) Blood cells of marine invertebrates. AR Liss, New York, pp 253–258

Armstrong PB (1991) Cellular and humoral immunity in the horseshoe crab, *Limulus*. In: Gupta AP (ed) Immunology of insects and other arthropods. CRC Press, Boca Raton, pp 1–17

Armstrong PB, Quigley JP (1985) Proteinase inhibitory activity released from the horseshoe crab blood cell during exocytosis. Biochim Biophys Acta 827: 453–459

Armstrong PB, Quigley JP (1987) *Limulus* α_2-macroglobulin. First evidence in an invertebrate for a protein containing an internal thiol ester bond. Biochem J 248: 703–707

Armstrong PB, Quigley JP (1991) α_2-Macroglobulin: a recently discovered defense system in arthropods. In: Gupta AP (ed) Immunology of insects and other arthropods. CRC Press, Boca Raton, pp 291–310

Armstrong PB, Quigley JP (1992) Humoral immunity: α_2-macroglobulin activity in the plasma of mollusks. Veliger 35: 161–164

Armstrong PB, Rickles FR (1982) Endotoxin-induced degranulation of the *Limulus* amebocyte. Exp Cell Res 140: 15–24

Armstrong PB, Levin J, Quigley JP (1984) Role of endogenous proteinase inhibitors in the regulation of the blood clotting system of the horseshoe crab, *Limulus polyphemus*. Thromb Haemostasis (Stuttgart) 52: 117–120

Armstrong PB, Rossner MT, Quigley JP (1985) An α_2-macroglobulin-like activity in the blood of chelicerate and mandibulate arthropods. J Exp Zool 236: 1–9

Armstrong PB, Quigley JP, Rickles FR (1990) The *Limulus* blood cell contains α_2-macroglobulin and releases it upon exocytosis. Biol Bull 178: 137–143

Armstrong PB, Mangle WF, Wall JS, Hainfield JF, Van Holde KE, Ikai A, Quigley JP (1991) Structure of *Limulus* α_2-macroglobulin. J Biol Chem 266: 2526–2530

Armstrong PB, Armstrong MT, Quigley JP (1993a) Characterization of a complement-like hemolytic system in the arthropod, *Limulus polyphemus:* involvement of α_2-macroglobulin and C-reactive protein. Mol Immunol 30: 929–934

Armstrong PB, Selzer PM, Ahlborg N, Morehead K, Perregaux M, Komuniecki P, Komuniecki R, Srimal S, Hotez PJ (1993b) Identification and partial characterization of an extracorporeal protease activity secreted by the triclad turbellarid worm, *Bdelloura candida*. Biol Bull 185: 326

Barrett AJ, Starkey PM (1973) The interaction of α_2-macroglobulin with proteinases. Characteristics and specificity of the reaction and a hypothesis concerning its molecular mechanism. Biochem J 133: 709–724

Barrett AJ, Brown MA, Sayers CA (1979) The electrophoretically "slow" and "fast" forms of the α_2-macroglobulin molecule. Biochem J 181: 401–418

Beith J, Pichoir M, Metais P (1970) The influence of α_2-macroglobulin on the elastolytic and esterolytic activity of elastase. FEBS Lett 8: 319–321

Bertheussen K (1982) Receptors for complement on echinoid phagocytes II. Purified human complement mediates echinoid phagocytosis. Dev Comp Immunol 6: 635–642

Bertheussen K (1983) Complement-like activity in sea urchin coelomic fluid. Dev Comp Immunol 7: 637–640

Bjork I, Fish WW (1982) Evidence for similar conformational changes in α_2-macroglobulin on reaction with primary amines or proteolytic enzymes. Biochem J 207: 347–356

Borth W (1992) α_2-Macroglobulin, a multifunctional binding protein with targeting characteristics. FASEB J 6: 3345–3353

Borth W, Luger TA (1989) Identification of α_2-macroglobulin as a cytokine binding plasma protein. Binding of interleukin-1β to "F" α_2-macroglobulin. J Biol Chem 264: 5818–5825

Borth W, Dunky A, Viehberger G (1983) α_2-Macroglobulin in joint disease. Ann NY Acad Sci 421: 377–381

Boyde TRC, Pryme IF (1968) Alpha$_2$-macroglobulin binding of trypsin, chymotrypsin, papain, and cationic aspartate aminotransferase. Clin Chim Acta 21: 9–14

Brandin ER, Pistole TG (1983) Polyphemin: a teichoic acid-binding lectin from the horseshoe crab, *Limulus polyphemus*. Biochem Biophys Res Commun 113: 611–617

Brandin ER, Pistole TG (1985) Presence of microorganisms in the hemolymph of the horseshoe crab, *Limulus polyphemus*. Appl Environ Microbiol 49: 718–720

Breton CB, Blisnick T, Jouin H, Barale JC, Rabilloud T, Langsley, G, Pereira da Silva LH (1992) *Plasmodium chabaudi* p68 serine protease activity required for merozoite entry into mouse erythrocytes. Proc Natl Acad Sci USA 89: 9647–9651

Canicatti C, Cuilla D (1987) Studies on *Holothuria polii* (Echinodermata) coelomocyte lysate I. Hemolytic activity of coelomocyte hemolysins. Dev Comp Immunol 11: 705–712

Canicatti C, Cuilla D (1988) Studies on *Holothuria polii* (Echinodermata) coelomocyte lysate II. Isolation of coelomocyte hemolysins. Dev Comp Immunol 12: 55–64

Cenini P (1983) Comparative studies on hemagglutinins and hemolysins in an annelid and a primitive crustacean. Dev Comp Immunol 7: 637–640

Chase T, Shaw E (1970) Titration of trypsin, plasmin, and thrombin with p-nitrophenyl p'-guanidinobenzoate HCl. Methods Enzymol 19: 20–27

Chen BJ, Wang D, Yuan AI, Feinman RD (1992) Structure of α_2-macroglobulin-protease complexes. Methylamine competition shows that proteases bridge two disulfide-bonded half molecules. Biochemistry 31: 8960–8966

Christensen U, Simonsen M, Harritt N, Sottrup-Jensen L (1989) Pregnancy zone protein, a proteinase-binding macroglobulin. Interactions with proteinases and methylamine. Biochemistry 28: 9324–9331

Cohen E, Vasta GR, Korytnyk W, Petrie CR, Sharma M (1984) Lectins of the Limulidae and hemagglutination-inhibition by sialic acid analogs and derivatives. Prog Clin Biol Res 157: 55–69

Cohn Z (1975) The role of proteases in macrophage physiology. In: Reich E, Rifkin DB, Shaw E (eds) Proteases and biological control. Cold Spring Harbor Conf on Cell proliferation, vol 2. Cold Spring Harbor Press, Cold Spring Harbor, pp 483–893

Davidsen O, Christensen EI, Gliemann J (1985) The plasma clearance of human α_2-macroglobulin-trypsin complexes in the rat is mainly accounted for by uptake into hepatocytes. Biochim Biophys Acta 846: 85–92

Day NKB, Gewurz H, Johannsen R, Finstad J, Good RA (1970) Complement and complement-like activity in lower vertebrates and invertebrates. J Exp Med 132: 941–950

Day N, Geiger H, Finstad J, Good RA (1972) A starfish hemolymph factor which activates vertebrate complement in the presence of a cobra venom factor. J Immunol 109: 164–167

D'Cruz OJ (1991) Identification and characterization of insect hemolymph proteins interacting with the mammalian complement cascade. In: Gupta PA (ed) Immunology of insects and other arthropods. CRC Press, Boca Raton, pp 372–384

Debanne MT, Bell R, Dolovich J (1975) Uptake of protease-α-macroglobulin complexes by macrophages. Biochem Biophys Acta 411: 295–304

Devriendt K, Van den Berghe H, Cassiman J-J, Marynen P (1991) Primary structure of pregnancy zone protein. Molecular cloning of a full-length PZP cDNA clone by the polymerase chain reaction. Biochim Biophys Acta 1088: 95–103

Dodds AW, Day AJ (1993) The phylogeny and evolution of the complement system. In: Whaley K, Loos M, Weiler JM (eds) Complement in health and disease, 2nd edn. Kluwer Dordrecht, pp 39–88

Donovan MA, Laue TM (1991) A novel trypsin inhibitor from the hemolymph of the horseshoe crab *Limulus polyphemus*. J Biol Chem 266: 2121–2125

Enghild JJ, Salvesen G, Thøgersen IB, Pizzo SV (1989a) Proteinase-binding and inhibition by the monomeric α-macroglobulin rat α_1-inhibitor-3. J Biol Chem 264: 11428–11435

Enghild JI, Thøgersen IB, Roche PA, Pizzo SV (1989b) A conserved region in α-macroglobulins participates in binding to the mammalian α-macroglobulin receptor. Biochemistry 28: 1406–1412

Enghild JJ, Salvesen G, Brew K, Nagase H (1989c) Interaction of human rheumatoid synovial collagenase (matrix metalloproteinase 1) and stromolysin (matrix metalloproteinase 3) with human α_2-macroglobulin and chicken ovostatin. Binding kinetics and identification of matrix metalloproteinase cleavage sites. J Biol Chem 264: 8779–8785

Enghild JJ, Thøgersen IB, Salvesen G, Fey GH, Figler NL, Gonias SL, Pizzo SV (1990) α_2-Macroglobulin from *Limulus polyphemus* exhibits proteinase inhibitory activity and participates in a hemolytic system. Biochemistry 29: 10070–10080

Farries TC, Atkinson JP (1991) Evolution of the complement system. Immunol Today 12: 295–300

Feldman SR, Pizzo SV (1984) Comparison of the binding of chicken α_2-macroglobulin and ovomacroglobulin to the mammalian α_2-macroglobulin receptor. Arch Biochem Biophys 235: 267–275

Finstad CL, Good RA, Litman GW (1974) The erythrocyte agglutinin from *Limulus polyphemus* hemolymph: molecular structure and biological function. Ann NY Acad Sci 234: 170–182

Furie B, Furie BC (1992) Molecular and cellular biology of blood coagulation. N Engl J Med 326: 800–806

Furman RM, Pistole TG (1976) Bactericidal activity of hemolymph from the horseshoe crab, *Limulus polyphemus*. J Invertebr Pathol 28: 239–244

Gilbride KJ, Pistole TG (1979) Isolation and characterization of a bacterial agglutinin in the serum of *Limulus polyphemus*. Prog Clin Biol Sci 29: 525–535

Gliemann J, Larsen TR, Sottrup-Jensen L (1983) Cell association and degradation of α_2-macroglobulin-trypsin complexes in hepatocytes and adipocytes. Biochim Biophys Acta 756: 230–237

Gonias SL (1992) α_2-macroglobulin: a protein at the interface of fibrinolysis and cellular growth regulation. Ex Hematol 20: 302–311

Gonias SL, Reynolds JA, Pizzo SV (1982) Physical properties of human α_2-macroglobulin following reaction with methylamine and trypsin. Biochim Biophys Acta 705: 306–314

Greenberg CS, Birckbichler PJ, Rice RH (1991) Transglutaminases: multifunctional cross-linking enzymes that stabilize tissues. FASEB J 5: 3071–3077

Groff JM, Leibovitz L (1982) A gill disease of *Limulus polyphemus* associated with triclad turbellarid worm infection. Biol Bull 163: 392

Hall M, Söderhäll K, Sottrup-Jensen L (1989) Amino acid sequence around the thiolester of α_2-macroglobulin from plasma of the crayfish, *Pacifastacus leniusculus*. FEBS Lett 254: 111–114

Hall PK, Roberts RC (1978) Physical and chemical properties of human plasma α_2-macroglobulin. Biochem J 173: 27–38

Harpel PC (1973) Studies on human plasma α_2-macroglobulin-enzyme interactions. Evidence for proteolytic modification of the subunit chain structure. J Exp Med 138: 508–521

Harpel PC, Hayes MB, Hugli TE (1979) Heat-induced fragmentation of human α_2-macroglobulin. J Biol Chem 254: 8669–8678

Havermann K, Janoff A (1978) Neutral proteases of human polymorphonuclear leukocytes. Urbran and Schwarzenberg, Baltimore, MD

Hergenhahn H-G, Söderhäll K (1985) α_2-Macroglobulin-like activity in plasma of the crayfish Pacifastacus leniusculus. Comp Biochem Physiol 81B: 833–835

Hergenhahn HG, Aspan A, Söderhäll K (1987) Purification and characterization of a high-Mr proteinase inhibitor of prophenoloxidase activation from crayfish plasma. Biochem J 248: 223–228

Hergenhan HG, Hall M, Söderhäll K (1988) Purification and characterization of an α_2-macroglobulin-like proteinase inhibitor from plasma of the crayfish Pacifastacus leniusculus. Biochem J 255: 801–806

Holme R, Solum NO (1973) Electron microscopy of the gel protein formed by clotting of Limulus polyphemus hemocyte extracts. J Ultrastruct Res 44: 329–338

Howard FJB (1983) Reactive centers in α_2-macroglobulin. Ann NY Acad Sci 421: 69–80

Huang JS, Huang SS, Deuel TF (1984) Specific covalent binding of platelet derived growth factor to human plasma α_2-macroglobulin. Proc Natl Acad Sci USA 81: 342–347

Huggins LG, Waite JH (1993) Eggshell formation in Bdelloura candida, an ectoparasitic turbellarian of the horseshoe crab Limulus polyphemus. J Exp Zool 265: 549–557

Hussain MM, Maxfield FR, Más-Oliva J, Tabas I, Ji Z-S, Innerarity TL, Mahley RW (1991) Clearance of chylomicron remnants by the low density lipoprotein receptor-related protein/α_2-macroglobulin receptor. J Biol Chem 266: 13936–13940

Ikai A, Ditamoto T, Nishigai M (1983) Alpha-2-macroglobulin-like protease inhibitor from the egg white of Cuban crocodile (Crocodylas rhombifer). J Biochem (Tokyo) 93: 121–127

Ikai A, Kiruchi M, Nishigai M (1990) Interval structure of ovomacroglobulin studied by electron microscopy. J Biol Chem 265: 8280–8284

Iwanaga S, Miyata T, Tokunaga F, Muta T (1992) Molecular mechanism of hemolymph clotting system in Limulus. Thromb Res 68: 1–32

James K (1990) Interactions between cytokines and α_2-macroglobulin. Immunol Today 11: 163–167

Jiang H, Siegel JN, Gewurz H (1991) Binding and complement activation by C-reactive protein via the collagen-like region of Clq and inhibition of these reactions by monoclonal antibodies to C-reactive protein and Clq. J Immunol 146: 2324–2330

Johannsen R, Anderson RS, Good RA, Day NK (1973) A comparative study of the bactericidal activity of horseshoe crab (Limulus polyphemus) hemolymph and vertebrate serum. I Jnvertebr Pathol 22: 372–376

Kaplan R, Li SSL, Kehoe JM (1977) Molecular characterization of Limulin, a sialic acid binding lectin from the hemolymph of the horseshoe crab, Limulus polyphemus. Biochemistry 16: 4297–4303

Kawano K, Yoneya T, Miyata T, Yoshikawa K, Tokunaga F, Terada Y, Iwanaga S (1990) Antimicrobial peptide, Tachyplesin I, isolated from hemocytes of the horseshoe crab (Tachypleus tridentatus). J Biol Chem 265: 15365–15367

Kehoe JM, Seide RK (1986) Comparative structural studies of limulin. In: Gupta AP (ed) Hemocytic and humoral immunity in arthropods. Wiley, New York, pp 345–358

Kitano T, Nakashima M, Ikai A (1982) Hen egg white ovomacroglobulin has a protease inhibitory activity. J Biochem 92: 1679–1682

Komano H, Natori S (1985) Participation of Sarcophaga peregrina humoral lectin in the lysis of sheep red blood cells injected into the adbominal cavity of larvae. Dev Comp Immunol 9: 31–40

Kristensen T, Moestrup SK, Gliemann J, Bendtsen L, Sand O, Sottrup-Jensen L (1990) Evidence that the newly cloned low-density-lipoprotein receptor-related protein (LRP) is the α_2-macroglobulin receptor. FEBS Lett 276: 151–155

Laskowski M, Kato I (1980) Protein inhibitors of proteases. Annu Rev Biochem 49: 593–626

Law SK, Levine RP (1977) Interaction between the third complement protein and cell surface macromolecules. Proc Natl Acad Sci USA 74: 2701–2705

Law SK, Reid KBM (1988) Complement. IRL Press, Oxford, 72 pp

Liang Z, Lindblad P, Beauvais A, Johansson MW, Latage J-P, Hall M, Cerenius L, Söderhäll K (1992) Crayfish α_2-macroglobulin and 76 kDa protein; their biosynthesis and subcellular localization of the 76 kDa protein. J Insect Physiol 38: 987–995

Marchalonis JJ, Edelman GM (1968) Isolation and characterization of a hemagglutinin from *Limulus polyphemus*. J Mol Biol 32: 453–465

McKerrow JH, Newport G, Fishelson Z (1991) Recent insights into the structure and function of a larval proteinase involved in host infection by a multicellular parasite. Proc Soc Exp Biol Med 197: 119–124

McKerrow JH, Sun E, Rosenthal PJ, Bouvier J (1993) The proteases and pathogenicity of parasitic protozoa. Annu Rev Microbiol 47: 821–853

McPherson TA, Marchalonis JJ, Lennon V (1970) Binding of encephalitogenic basic protein by serum α-globulins. Immunology 19: 929–933

Miyata T, Hiranga M, Umezu M, Iwanaga S (1984) Amino acid sequence of the coagulogen from *Limulus polyphemus* hemocytes. J Biol Chem 259: 8924–8933

Miyata T, Tokunaga F, Yoneya T, Yoshikawa K, Iwanaga S, Niwa M, Takao T, Shimonishi Y (1989) Antimicrobial peptides isolated from horseshoe crab hemocytes, Tachyplesin II, and Polyphemusins I and II: chemical structures and biological activity. J Biochem 106: 663–668

Miyazawa K, Inoue K (1990) Complement activation induced by human C-reactive protein in mildly acidic conditions. J Immunol 145: 650–65

Moestrup SK, Holtet TL, Etzerodt M, Thøgersen HC, Nykjaer A, Andreasen PA, Rasmussen HH, Sottrup-Jensen L, Gliemann J (1993) α_2-Macroglobulin-proteinase complexes, plasminogen activator inhibitor type-1 plasminogen activator complexes, and receptor-associated protein bind to a region of the α_2-macroglobulin receptor containing a cluster of eight complement-type repeats. J Biol Chem 268: 13691–13696

Morita T, Ohtsubo S, Nakamura T, Tanaka S, Iwanaga S, Ohashi K, Niwa M (1985) Isolation and biological activities of *Limulus* anticoagulant (anti-LPS factor) which interacts with lipopolysaccharide (LPS). J Biochem 97: 1611–1620

Morrison DC, Ryan JL (1987) Endotoxins and disease mechanisms. Annu Rev Med 38: 417–432

Mosesson MW, Wolfenstein-Todd C, Levin J, Bertrand O (1979) Characterization of amebocyte coagulogen from the horseshoe crab *Limulus polyphemus*. Thromb Res 14: 765–779

Mürer EH, Levin J, Holme R (1975) Isolation and studies of the granules of the amebocytes of *Limulus polyphemus*, the horseshoe crab. J Cell Physiol 86: 533–542

Muta T, Miyata T, Tokunaga F, Nakamura T, Iwanaga S (1987) Primary structure of anti-lipopoly-saccharide factor from American horseshoe crab, *Limulus polyphemus*. J Biochem 101: 1321–1330

Muta T, Fujimoto T, Nakajima H, Iwanaga S (1990a) Tachyplesins isolated from hemocytes of Southeast Asian horseshoe crabs (*Carcinoscorpius rotundicauda*) and *Tachypleus gigas*): identification of a new Tachyplesin, Tachyplesin III, and a processing intermediate of its precursor J Biochem 108: 261–266

Muta T, Hashimoto R, Miyata T, Nishimura H, Toh Y, Iwanaga S (1990b) Proclotting enzyme from horseshoe crab hemocytes. cDNA cloning, disulfide locations, and subcellular localization. J Biol Chem 265: 22426–22433

Muta T, Miyata T, Misumi T, Tokunaga F, Nakamura T, Toh Y, Ikehara Y, Iwanaga S (1991) *Limulus* factor C. An endotoxin-sensitive serine protease zymogen with a mosaic structure of complement-like, epidermal growth factor-like, and lectin-like domains. J Biol Chem 266: 6554–6617

Muta T, Oda T, Iwanaga S (1993) Horseshoe crab coagulation factor B. A unique serine protease zymogen activated by cleavage of an dIle-Ile bond. J Biol Chem 268: 21384–21388

Nachman RL, Harpel PC (1976) Platelet α_2-macroglobulin and α_1 antitrypsin. J Biol Chem 251: 4514–4521

Nachum R (1979) Antimicrobial defense mechanisms in *Limulus polyphemus*. Prog Clin Biol Res 29: 513–524

Nachum R, Watson SR, Sullivan JD, Siegel SE (1979) Antimicrobial defense mechanisms in the horseshoe crab, *Limulus polyphemus:* Preliminary observations with heat-derived extracts of *Limulus* amebocyte lysate. J Invertebr Pathol 33: 290–299

Nachum R, Watson SW, Siegel SE (1980) Antimicrobial defense mechanisms in the horseshoe crab, *Limulus polyphemus:* effect of sodium chloride on bactericidal activity. J Invertebr Pathol 36: 382–388

Nagase H, Harris ED (1983) Ovostatin: a novel proteinase inhibitor from chicken egg white II. Mechanism of inhibition studied with collagenase and thermolysin. J Biol Chem 258: 7490–7498

Nagase J, Harris ED, Woessner JF, Brew K (1983) Ovostatin: a novel proteinase inhibitor from chicken egg white I. Purification, physicochemical properties and tissue distribution of ovostatin. J Biol Chem 258: 7481–7489

Nakamura T, Morita T, Iwanaga S (1985) Intracellular proclotting enzyme in Limulus (Tachypleus tridentatus) hemocytes: its purification and properties. J Biochem 97: 1561–1574

Nakamura T, Morita T, Iwanaga S (1986) Lipopolysaccharide-sensitive serine-protease zymogen (factor C) found in Limulus lysates. Eur J Biochem 154: 511–521

Nakamura T, Hirai T, Tokunaga F, Kawabata S, Iwanaga S (1987) Purification and amino acid sequence of Kunitz-type protease inhibitor found in the hemocytes of the horseshoe crab (Tachypleus tridentatus). J Biochem 101: 1297–1306

Nakamura T, Furunaka H, Miyata T, Tokunaga F, Muta T, Iwanaga S, Niwa M, Takao T, Shimonishi T (1988) Tachyplesin, a class of antimicrobial peptide from the hemocytes of the horseshoe crab, (Tachypleus tridentatus). J Biol Chem 263: 16709–16713

Nielsen K, Sottrup-Jensen L (1993) Evidence from sequence analysis that hen egg white ovomacroglobulin (ovostatin) is devoid of an internal β-Cys-γ-Glu thiol ester. Biochem Biophys Acta 1162: 230–232

Nguygen NY, Suzuki A, Cheng S-M, Zon G, Liu T-Y (1986a) Isolation and characterization of Limulus C-reactive protein genes. J Biol Chem 261: 10450–10455

Nguyen NY, Suzuki A, Boykins A, Liu TY (1986b) The amino acid sequence of Limulus C-reactive protein. Evidence of polymorphism. J Biol Chem 261: 10456–10459

Noguchi H (1903) A study of immunization – haemolysins, agglutinins, precipitins, and coagulins in cold-blooded animals. Zentralbel Bakteriol Parasitenkd Infektektionskr 33: 353–362

Nykjaer A, Bengtsson-Olivecrona G, Lookene A, Moestrup SK, Petersen CM, Weber W, Beisiegel U, Gliemann J (1993) The α_2-macroglobulin receptor/low density lipoprotein receptor-related protein binds lipoprotein lipase and β-migrating very low density lipoprotein associated with the lipase. J Biol Chem 268: 15048–15055

O'Connor-Court MD, Wakefield LM (1987) Latent transforming growth factor-β in serum. J Biol Chem 262: 14090–14099

Ogata H, Kouyoumdjian M, Borges DR (1993) Comparison between clearance rates of plasma kallikrein and of plasma kallikrein-α_2-macroglobulin complexes by the liver. Int J Biochem 25: 1047–1051

Ohashi D, Niwa M, Nakamura T, Morita T, Iwanaga S (1984) Anti-LPS factor in the horseshoe crab, Tachypleus tridentatus. Its hemolytic activity of the red blood cell sensitized with lipopolysaccharide. FEBS Lett 176: 207–210

Ohlsson K (1971) Elimination of [^{125}I]-trypsin α_2-macroglobulin complexes from blood by the reticuloendothelial cells in dog. Acta Physiol Scand 81: 269–272

Osada T, Nishigai M, Ikai A (1986) Open quaternary structure of the hagfish proteinase inhibitor with similar properties to human α_2-macroglobulin. J Ultrastruct Mol Struct Res 96: 136–145

Perlmutter DH, Pierce JA (1989) The α_1-antitrypsin gene and emphysema. Am J Physiol 257: L147–L162

Peterson CGB, Venge P (1987) Interaction and complex-formation between the eosinophil cationic protein and α_2-macroglobulin. Biochem J 245: 781–787

Phipps DJ, Chadwick JS, Leeder RG, Aston WP (1989) The hemolytic activity of Gasseria mellonella hemolymph. Dev Comp Immunol 13: 103–111

Pistole TG (1982) Limulus lectins: Analogues of vertebrate immunoglobins. Prog Clin Biol Res 81: 283–288

Pistole TG, Britko JL (1978) Bactericidal activity of amebocytes from the horseshoe crab, Limulus polyphemus. J Invertebr Pathol 31: 376–382

Pistole TG, Graf SA (1986) Antibacterial activity in Limulus. In: Gupta AP (ed) Hemocytic and humoral immunity in arthropods. Wiley, New York, pp 331–344

Plow EF, Collen D (1981) The presence and release of α_2-antiplasmin from human platelets. Blood 58: 1069–1074

Quigley JP, Armstrong PB (1983) An endopeptidase inhibitor, similar to mammalian α_2-macroglobulin, detected in the hemolymph of an invertebrate, *Limulus polyphemus*. J Biol Chem 258: 7903–7906

Quigley JP, Armstrong PB (1985) A homologue of α_2-macroglobulin purified from the hemolymph of the horseshoe crab *Limulus*. J Biol Chem 260: 12715–12719

Quigley JP, Armstrong PB, Gallant P, Rickles FR, Troll W (1982) An endopeptidase inhibitor, similar to vertebrate 2 macroglobulin, present in the plasma of *Limulus polyphemus*. Biol Bull 163: 402

Quigley JP, Ikai A, Arakawa H, Osada T, Armstrong PB (1991) Reaction of proteinases with α_2-macroglobulin from the American horseshoe crab, *Limulus*. J Biol Chem 266: 19426–19431

Reid KBM, Porter RR (1981) The proteolytic activation systems of complement. Annu Rev Biochem 50: 433–464

Robey FA, Liu T-Y (1981) Limulin: a C-reactive protein from *Limulus polyphemus*. J Biol Chem 256: 969–975

Roch P, Canicatti C, Valembois P (1989) Interactions between earthworm hemolysins and sheep red blood cell membranes. Biochim Biophys Acta 983: 193–198

Ropes JW (1961) Longevity of the horseshoe crab, *Limulus polyphemus*. Trans Am Fish Soc 90: 79–80

Roth RI, Levin J (1992) Purification of *Limulus polyphemus* proclotting enzyme. J Biol Chem 267: 24097–24102

Rubenstein DS, Enghild JJ, Pizzo SV (1991) Limited proteolysis of the α-macroglobulin rat α_1-inhibitor-3. Implications for a domain structure. J Biol Chem 266: 11252–11261

Rubenstein DS, Thøgersen IB, Pizzo SV, Enghild JJ (1992) Identification of monomeric α-macroglobulin proteinase inhibitors in birds, reptiles, amphibians and mammals, and purification and characterization of a monomeric α-macroglobulin proteinase inhibitor from the American bullfrog *Rana catesbeiana*. Biochem J 290: 85–95

Rudloe J (1971) The Erotic Ocean. Crowell, New York, pp 183–185

Runnegar B (1982) The cambrian explosion: animals or fossils? J Geol Soc Aust 29: 395–411

Salvesen GS, Sayers CA, Barrett AJ (1981) Further characterization of the covalent linking reaction of α_2-macroglobulin. Biochem J 195: 453–461

Santambrogio P, Massover WH (1989) Rabbit serum alpha-2-macroglobulin binds to liver ferritin: association causes a heterogeneity of ferritin molecules. Brit J Haematol 71: 281–290

Sepkoski JJ (1978) A kinetic model of phanerozoic taxonomic diversity I. Analysis of marine orders. Paleobiology 4: 223–251

Shirodkar MV, Warwick A, Bang FB (1960) The in vitro reaction of *Limulus* amebocytes of bacteria. Biol Bull 118: 324–337

Shuster CN (1950) Observations on the natural history of the American horseshoe crab, *Limulus polyphemus*. Woods Hole Oceanogr Inst Contr 564: 18–23

Shuster CN (1954) A horseshoe "crab" grows up. Ward's Nat Sci Bull (Rochester, NY) 28: 3–6

Smith RH, Pistole TG (1985) Bactericidal activity of granules isolated from amebocytes of the horseshoe crab, *Limulus polyphemus*. J Invertebr Pathol 45: 272–275

Scully MF (1992) The biochemistry of blood clotting: the digestion of a liquid to form a solid. Essays Biochem 27: 17–36

Sottrup-Jensen L (1987) α_2-Macroglobulin and related thiol ester plasma proteins. In: Putnam FW (ed) The plasma proteins. Structure, function, and genetic control, 2nd edn, vol 5. Academic Press, Orlando, pp 191–291

Sottrup-Jensen L (1989) α_2-Macroglobulin: structure, shape, and mechanism of proteinase complex formation. J Biol Chem 264: 11539–11542

Sottrup-Jensen L, Lonblad PB, Stephanik TM, Petersen TE, Magnusson S, Jornvall H (1981) Primary structure of the "bait" region for proteinases in α_2-macroglobulin. Nature of the complex. FEBS Lett 127: 167–173

Sottrup-Jensen L, Stepanik TM, Wierzbicki DM, Jones CM, Lonblad PB, Kristensen T, Mortensen SB, Petersen TE, Magnusson S (1983) The primary structure of α_2-macroglobulin and localization of a factor XIIIa cross-linking site. Ann NY Acad Sci 421: 41–60

Sottrup-Jensen L, Stepanik TM, Kristensen T, Wierzbicki DM, Jones CM, Lønblad PB, Magnusson S, Petersen TE (1984) Primary structure of human α_2-macroglobulin V. The complete structure. J Biol Chem 259: 8318–8327

Sottrup-Jensen L, Sand O, Dristensen L, Fey GH (1989) The α_2-macroglobulin bait region. Sequence diversity and localization of cleavage sites for proteinases in five mammalian α_2-macroglobulin. J Biol Chem 264: 15781–15789

Sottrup-Jensen L, Borth W, Hall M, Quigley JP, Armstrong PB (1990a) Sequence similarity between α_2-macroglobulin from the horseshoe crab, Limulus polyphemus, and proteins of the α_2-macroglobulin family from mammals. Comp Biochem Physiol 96B: 621–625

Sottrup-Jensen L, Hansen JF, Pedersen HS, Kristensen L (1990b) Localization of lysyl-γ-glutamyl cross-links in five human α_2-macroglobulin-proteinase complexes. Nature of the high molecular weight cross-linked products. J Biol Chem 265: 17727–17737

Sporn MB, Roberts AB (1990) Peptide growth factors, vols I, II. Springer, Berlin Heidelberg New York

Spycher SE, Arya S, Isenman DE, Painter H (1987) A functional thioester-containing α_2-macroglobulin homologue isolated from the hemolymph of the American lobster (Homarus americanus). J Biol Chem 262: 14606–14611

Srimal S, Quigley JP, Armstrong PB (1993) Limulin and C-reactive protein from the plasma of Limulus polyphemus are different proteins. Biol Bull 185: 325

Starkey PM, Barrett AJ (1977) α_2-Macroglobulin, a physiological regulator of proteinase activity. In: Barrett AJ (ed) Proteinases in mammalian cells and tissues. Elsevier, Amsterdam, pp 663–696

Starkey PM, Barrett AJ (1982) Evolution of α_2-macroglobulin. The demonstration in a variety of vertebrate species of a protein resembling human α_2-macroglobulin. Biochem J 205: 91–95

Stebbins MR, Harper KD (1986) Isolation, characterization and inhibition of arthropod agglutinins. In: Gupta AP (ed) Hemocytic and humoral immunity in arthropods. Wiley, New York, pp 463–491

Steinbuch M, Pejaudier L, Quentin M, Martin V (1968) Molecular alteration of α_2-macroglobulin by aliphatic amines. Biochim Biophys Acta 154: 228–231

Stöcker W, Breit S, Sottrup-Jensen L, Zwilling R (1991) α_2-Macroglobulin from the haemolymph of the freshwater crayfish Astacus astacus. Comp Biochem Physiol 98B: 501–509

Stoller BD, Rezuke W (1978) Separation of anti-histone antibodies from nonimmune histone-precipitating serum proteins, predominantly α_2-macroglobulin. Arch Biochem Biophys 190: 398–404

Strickland DK, Ashcom JD, Williams S, Burgess WH, Migliorini M, Argraves WA (1990) Sequence identity between the α_2-macroglobulin receptor and low density lipoprotein receptor-related protein suggests that this molecule is a multifunctional receptor. J Biol Chem 265: 17401–17404

Swenson RP, Howard JB (1979) Characterization of alkylamine-sensitive site in α_2-macroglobulin. Proc Natl Acad Sci USA 76: 4313–4316

Tack BF (1938) The β-Cys-γ-Glu thiolester bond in human C3, C4, and α_2-macroglobulin. Springer Semin Immunopathol 6: 259–282

Tack BF, Harrison RA, Janatova J, Thomas ML, Prahl JW (1980) Evidence for presence of an internal thiolester bond in third component of human complement. Proc Natl Acad Sci USA 77: 5764–5768

Tai JY, Liu T-Y (1977) Studies of Limulus amoebocyte lysate. Isolation of pro-clotting enzyme. J Biol Chem 252: 2178–2181

Tai JY, Seid RC, Huhn RD, Liu T-Y (1977) Studies on Limulus amoebocyte lysate II. Purification of the coagulogen and the mechanism of clotting. J Biol Chem 252: 4773–4776

Tanaka S, Nakamura T, Morita T, Iwanaga S (1982) Limulus anti-LPS factor, An anticoagulant which inhibits the endotoxin-mediated activation of Limulus coagulation system. Biochem Biophys Res Commun 105: 717–723

Tennent GA, Butler PJG, Hutton T, Woolfitt AR, Harvey DJ, Rademacher TW, Pepys MB (1993) Molecular characterization of Limulus polyphemus C-reactive protein. 1. Subunit composition. Eur J Biochem 214: 91–97

Testa JE, Quigley JP (1990) The role of urokinase-type plasminogen activator in aggressive tumor cell behavior. Cancer Metast Rev 9: 353–367

Thøgersen IB, Salvesen G, Brucato FH, Pizzo SV, Enghild JJ (1992) Purification and characterization of an α_2-macroglobulin proteinase inhibitor from the mollusc Octopus vulgaris. Biochem J 285: 521–527

Thomsen NK, Sottrup-Jensen L (1993) α_2-Macroglobulin domain structure studied by specific limited proteolysis. Arch Biochem Biophys 300: 327–334

Tokunaga F, Yamada M, Miyata T, Ding YL, Hiranaga-Kawabata M, Muta T, Iwanaga S, Ichinose A, Davie EW (1993a) *Limulus* hemocyte transglutaminase. Its purification and characterization, and identification of the intracellular substrates. J Biol Chem 268: 252–261

Tokunaga F, Muta T, Iwanaga S, Ichinose A, Davie EW, Kuma K, Miyata T (1993b) *Limulus* tissue transglutaminase. cDNA cloning, amino acid sequence, and tissue localization. J Biol Chem 268: 262–268

Travis J, Salvesen GS (1983) Human plasma proteinase inhibitors. Annu Rev Biochem 52: 655–709

Tucková L, Rejnek J, Síma P, Ondrejová R (1986) Lytic activities in coelomic fluid of *Eisenia foetida* and *Lumbricus terrestris*. Dev Comp Immunol 10: 181–189

Vaith P, Uhlenbruck G, Müller WEG, Cohen E (1979) Reactivity of *Limulus* polyphemus hemolymph with D-glucuronic acid containing glycosubstances. Prog Clin Biol Res 29: 579–587

Van Dijk MCM, Boers W, Linthorst C, van Berkel TJC (1992) Role of the scavenger receptor in the uptake of methylamine-activated α_2-macroglobulin by rat liver. Biochem J 287: 447–455

Van Leuven F (1984) Human α_2-macroglobulin. Primary amines and the mechanisms of endoprotease inhibition and receptor-mediated endocytosis. Mol Cell Biochem 58: 121–128

Van Leuven F, Cassiman J-J, Van Den Berghe J (1979) Demonstration of an α_2-macroglobulin receptor in human fibroblasts, absent in tumor-derived cell lines. J Biol Chem 254: 5155–5160

Van Leuven F, Cassiman J-J, Van Den Berghe J (1981) Functional modifications of α_2-macroglobulin by primary amines II. Inhibition of covalent binding of trypsin to α_2M by methylamine and other primary amines. J Biol Chem 256: 9023–9027

Volanakis JE (1982) Complement activation by C-reactive protein complexes. Ann NY Acad Sci 389: 235–250

Vu T-K, Hung DT, Wheaton VI, Coughlin SR (1991) Molecular cloning of a functional thrombin receptor reveals a novel proteolytic mechanism of receptor activation. Cell 64: 1057–1068

Werb Z (1993) Proteases and matrix degradation. In: Kelley WN, Harris ED, Ruddy S, Sledge CB (eds) Textbook of rheumatology, 4th edn. Saunders, Philadelphia, pp 300–321

Wilson J, Rickles FR, Armstrong PB, Lorand L (1992) N^{ε} (γ-glutamyl) lysine crosslinks in the blood clot of the horseshoe crab, *Limulus polyphemus*. Biochem Biophys Res Commun 188: 655–661

Wu K, Wang D, Feinman RD (1981) Inhibition of proteases by α_2-macroglobulin. The role of lysyl amino groups of trypsin in covalent complex formation. J Biol Chem 256: 10409–10414

Yochem J, Greenwald I (1993) A gene for a low density lipoprotein-related protein in the nematode *Caenorhabditis elegans*. Proc Natl Acad Sci USA 90: 4572–4576

Host–Parasite Interactions in Molluscs

S.E. Fryer and C.J. Bayne

1 Introduction

The phylum Mollusca is second only to the Arthropoda in both number and diversity of living species. Representatives include not only the readily recognized gastropods (snails, slugs and limpets) and bivalves (e.g., oysters, mussels and clams) but also the 'brainy' cephalopods including octopus and squid, and more primitive representatives such as chitons. These animals occur in a huge variety of terrestrial, freshwater and marine habitats. Interest in their pathogens and parasites arises primarily from the role of gastropods in the transmission of trematodes of medical and veterinary importance. Perhaps the most important of these are the human infecting schistosomes, estimated to parasitize 200 million people world wide. In addition, molluscs such as oysters, clams, mussels as well as gastropod "escargots" are increasingly being raised for human consumption, in both the developed and developing world. There are thus clear medical and economic needs to obtain a full understanding of molluscan pathogens, the capabilities of the molluscan immune systems, and how these interact to determine the outcome of a host–parasite encounter.

While knowledge of the general transmission and biology of many parasites that include molluscs in their life cycle has been relatively easy to obtain, our generally limited understanding of invertebrate internal defense systems (relative to those of vertebrates) has hindered elucidation of the interactions between molluscs and their parasites. Findings from laboratories throughout the world studying different host–parasite combinations are revealing a remarkable complexity of interactions at cellular, biochemical and molecular levels. This information, together with observed frequencies of natural resistance to specific parasites and pathogens, makes it clear that successful invasion, establishment, and survival to reproduce is no simple task for a molluscan parasite!

In this chapter, we summarise current knowledge on interaction between the internal defense systems of molluscan hosts and their parasites. The majority of work on this subject has involved studies of digenetic trematodes. Members of this large group of helminth parasites all include a mollusc as the first

Department of Zoology, Oregon State University, Cordley Hall 3029, Corvallis, Oregon 97331-2914, USA intermediate host in their life cycle. While the number of larval stages within this molluscan host varies between trematode species (Figs. 1 and 2), all undergo massive asexual reproduction within the snail tissues. Interactions between the tropical planorbid snail *Biomphalaria glabrata* and the causative agent of human intestinal schistosomiasis, *Schistosoma mansoni,* have been studied in a number of laboratories. In addition, investigations on echinostomes in the same host have yielded important data on mechanisms of parasite interference with host internal defenses. The 'terminal spined' schistosomes, including *S. haematobium* and several species of veterinary importance, have received much less attention, but recent studies are revealing information on their interactions with inter-mediate hosts in the genus *Bulinus* (Preston and Southgate 1994). *Trichobilharzia ocellata,* a schistosome parasite of birds, has been studied in its temperate molluscan host *Lymnaea stagnalis.* Progress with this model has been aided by the large volume of information available on the general biochemistry and physiology of this gastropod. Parasitic protozoans have a considerable economic impact on bivalve aquaculture throughout the world. Two extracellular proto-zoans, *Haplosporidium nelsoni* (MSX) and *Perkinsus marinus*, and the intra-cellular parasite *Bonamia ostreae* cause widespread pathology and mortality in

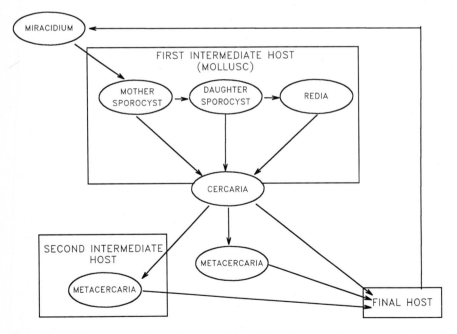

Fig. 1. Generalized representation of the larval stages of digenetic trematodes associated with mollus-can hosts. All digeneans parasitize molluscs as their first intermediate host. Life cycles vary in the number of intramolluscan stages, and the inclusion of second intermediate hosts which may also be molluscs. In all cases, the first contact between parasite and the molluscan internal defense system is as the miracidium transforms into a mother sporocyst

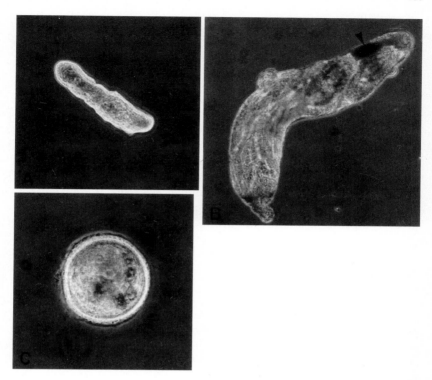

Fig. 2. Intramolluscan larval stages *Echinostoma paraensei*. **A** Mother sporocyst transformed axenically in vitro. **B** Mature daughter redia dissected from infected *B. glabrata*. Note the many developing cercariae within the body of the redia, and the dark gut (*arrowhead*) with ingested snail tissue. **C** Metacercaria retrieved from the tissues of *B. glabrata* acting as second intermediate host. (Photographs provided by E.S. Loker and coworkers)

oysters. Recent advances in the in vitro culture of such parasites is allowing more detailed studies of their interactions.

The examples that will be discussed here are restricted to parasites of gastropods and bivalves. Little if anything is known about internal defense systems and/or parasite interactions of other members of this extremely diverse phylum, and there is an obvious need for caution in extrapolating observations from the above models to other molluscan and parasitic groups.

2 Molluscan Internal Defenses

It has been argued (e.g., Klein 1989) that relatively small, short-lived invertebrates do not require a sophisticated "immune" response such as is acheived by the lymphoid arm of the vertebrate immune system. However, many invertebrates are larger than small vertebrates, surprisingly long-lived (many years) and constantly exposed to microbes and parasites, yet are infrequently infected, much less overcome, by these pathogens! As is clear from the subject matter of this book, invertebrates certainly possess effective and efficient internal defense systems.

The phenomenon of extreme host specificity seen in digenetic trematodes with respect to their molluscan hosts (e.g., Wright 1971; Basch 1976) indicates a sophisticated interplay between host and parasite. While some host restriction is due to the need for contact between hosts and infective stages of parasites in their microenvironments, and to the physiological suitability or unsuitability of potential hosts, the success or failure of recognition and killing of the invading pathogen by the host internal defense system plays a vital role (van der Knaap and Loker 1990). A potential molluscan host may therefore be physiologically unsuitable, susceptible or resistant to a given parasite (Fig. 3).

While lacking the classic components of vertebrate lymphoid systems, such as lymphocytes and immunoglobulins, and thus not showing specific immunological "memory", molluscs possess both cellular and humoral defense systems with remarkably effective capabilities. As more is discovered about the mediators and effectors of these systems, similarities at both cellular and molecular levels are being seen between them and the "innate" (nonlymphoid) components of vertebrate immunity. Molluscan circulatory systems, with the exception of the cephalopods, are of an 'open' type in which blood (hemolymph), passing out of open ends of arteries, bathes all the organs before returning to the heart by way of sinuses and respiratory structures (gills or lungs). The fluid portion, referred to as plasma since it lacks clotting capacity, contains a diversity of proteins usually including a dominant respiratory molecule, most frequently hemoglobin or

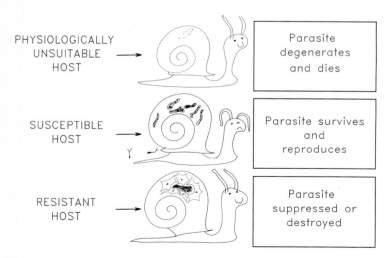

Fig. 3. The outcome of encounters between the parasite and potential molluscan hosts. Survival, reproduction, and continuation of the parasite life cycle only occurs in a physiologically suitable susceptible host

hemocyanin. Only rarely is the oxygen-carrying pigment located intracellularly as in vertebrates. In the majority of molluscs, circulating cells, usually termed hemocytes, are therefore all "leukocytes" (the few exceptions can be found described by Ratcliffe and Rowley 1981). Molluscan leukocytes show less structural diversity than is seen in either arthropods or vertebrates, with the main variable being in the degree of granularity. Granular hemocytes are usually the most numerous cell type, large and highly phagocytic. Experiments both in vivo and in vitro show that these cells phagocytose small particles, and encapsulate particles too large to be engulfed by an individual cell (frustrated phagocytosis) (Sminia et al. 1974). These molluscan granulocytes show other similarities to mammalian macrophages (McKerrow et al. 1985): they produce toxic metabolites such as reactive oxygen intermediates (reviewed by Adema et al. 1991; Bachère et al. 1991; Pipe 1992) and nitric oxide (Ottaviani et al. 1993), and possess varied lysosomal enzymes (Cheng et al. 1975; Sminia and Barendsen 1980; Cheng and Downs 1988). These highly mobile cells also have a high 'clumping' potential (helping in wound sealing; Chen and Bayne 1994), and in some cases have been demonstrated to have chemotaxic capacity towards pathogen products (Schmid 1975; Howland and Cheng 1982; Schneeweiß and Renwrantz 1983). Circulating granulocytes are clearly the cell type primarily responsible for defenses against invading parasites and pathogens in molluscs.

Humoral factors also play important roles in molluscan internal defense. While lacking immunoglobulins, as with other invertebrates, molluscs possess humoral agglutinins and opsonins (both often lectins; Renwrantz 1983; Olafsen 1986) and soluble lysosomal enzymes. Small pathogens may thus be agglutinated or opsonized for enhanced phagocytic clearance by the circulating hemocytes or lysed directly without the involvement of cells.

Detailed accounts of both cellular and humoral components of invertebrate and molluscan defense systems can be found in several reviews (e.g., Bayne 1983; Ratcliffe 1985; Sminia and van der Knaap 1987). It is our intention here to concentrate on how these are involved in interactions between molluscs and invading parasites.

3 The Fates of Invading Molluscan Parasites

One of the most striking characteristics of the majority of digenetic trematodes is the extreme specificity that is seen in infections of molluscan intermediate hosts. Wright, in his classic book, *Flukes and Snails* (1971), discussed this at length, and put forward hypotheses to account for observed host restrictions. Experimental evidence for many of these is emerging from recent studies that will be discussed later. Basch (1976) synthesized data from a large number of previously published studies of experimental infections of *Biomphalaria* species with *Schistosoma mansoni*. His analysis illustrates amazing host specificity; variations in susceptibility to infection occur between geographical strains of both host and parasite. He concluded in this and an earlier paper (Basch 1975) that the basis of mollusc-trematode compatibility resides in a concordance of genetically determined phenotypes in host and parasite, each of which is polymorphic with respect to the relevant traits. These genetically determined factors have been exploited in the establishment of laboratory strains of snail *B. glabrata*, and the trematode *S. mansoni* with known compatibility/incompatibility phenotypes (Richards and Shade 1987; Richards et al. 1992), and these strains have been used by several laboratories to examine the mechanisms of susceptibility and resistance in this system.

On entry into susceptible host snails, each trematode miracidium transforms within hours to a mother sporocyst (e.g., Meuleman et al. 1978; Pan 1980). Miracidia normally penetrate into the head-foot or tentacles of the host, where the ciliated plates of the parasite are rapidly shed by expansion of the tegument beneath them. In highly susceptible host snails there is no visible reaction to these developing larvae (e.g., Newton 1952). After about 2 weeks of development in tissues near the site of penetration, mother sporocysts release daughters that migrate to the host digestive gland. Here, once again the parasites survive with little or no host reaction to produce both more sporocysts and often thousands of cercariae.

In contrast to the situation with infections in compatible combinations, a marked cellular response is evident 15–18 h after miracidia penetrate a resistant snail (Newton 1952). Layers of concentrically arranged host cells accumulate around the parasites close to the point of penetration. Such host responses have been observed repeatedly (e.g., Lie et al. 1980) and ultrastructural aspects of this interaction have been described (Loker et al. 1982). By 24 h after infection, hemocytes of highly resistant hosts have destroyed the parasite tegument and only remnants of sporocyst material, much of it within hemocyte phagosomes, are seen up to 48 h after penetration. Acid phosphatase levels increase in encapsulating hemocytes (Cheng and Garrabbrant 1977) and increased levels of this and other lysosomal enzymes in the plasma (Granath and Yoshino 1983; Cheng and Dougherty 1989) may contribute to tegumental destruction. In addition, toxic reactive oxygen metabolites have been demonstrated at the host–parasite interface in some incompatible combinations (Dikkeboom et al. 1988; Adema et al. 1992). The ability of a host snail to attack and kill a larval schistosome can be transferred by means of allografted hematopoietic tissue (Sullivan and Spence 1994). *B. glabrata* of a susceptible phenotype for *S. mansoni* undergo a slight, but significant, reduction in infection rates when exposed to parasite miracidia between one and at least 7 weeks after receiving amoebocyte producing organs from resistant snails.

Some early parasite transplantation experiments by Chernin (1966) showed that the lack of host response to compatible *S. mansoni* was not restricted to snails that had been infected naturally. Daughter sporocysts within pieces of implanted digestive gland survived to produce cercariae in *B. glabrata* of the susceptible phenotype, yet the same material did not survive in resistant hosts. Further evidence that compatibility does not rely on a natural route of infection was found by Basch and DiConza (1974) who were able to infect susceptible *B.*

glabrata by implanting mother sporocysts produced by axenic in vitro transformation of *S. mansoni* miracidia, and daughter sporocysts cultured in vitro for
a period. Miracidia from *T. ocellata* develop normally after injection into
susceptible *L. stagnalis* (Meuleman et al. 1984), once again suggesting that the act
of penetration is not vital for parasite survival. In contrast to Chernin's (1966)
findings, *S. haematobium* daughter sporocysts survived after transplantation
from susceptible to resistant snails (Kechemir and Combes 1982).

While seemingly ignoring intact live parasites, susceptible snails are capable of
mounting cellular reactions against damaged parasite material. In vivo, contem-
poraneous infection by two trematode species may lead to predation of sporocysts of one by rediae (a carnivorous intramolluscan larval stage) of another.
Sporocysts damaged in this manner are rapidly surrounded by snail hemocytes
(Lim 1970), as are degenerating senescent sporocysts (Amen et al. 1991b). Similar
interactions are seen in vitro: healthy parasites are ignored by host hemocytes,
but even intact, dead larvae elicit a rapid encapsulation response. This has been
observed in various mollusc/trematode combinations including *Echinostoma
paraensei* larvae exposed to *B. glabrata* hemocytes (Loker et al. 1989) and various
cercariae exposed to hemolymph from the nonhost bivalve *Crassostrea virginica*
(Font 1980). A similar situation is observed when the protozoan parasite,
Haplosporidium nelsoni, is cultured with hemocytes from its susceptible oyster
host, *C. virginica* (Ford et al. 1993) with only dead parasites being phagocytosed.

The fates of parasites described here represent two ends of a continuum from
host resistance to susceptibility, or from parasite noninfectivity to infectivity.
Between these two extremes, intermediate levels of host–parasite compatibility
occur. For example certain "resistant" strains of *B. glabrata*, if kept long enough
after exposure to *S. mansoni* miracidia, will produce cercariae (Lewis et al. 1993).
In such snails, daughter sporocysts reside within the head-foot rather than in the
digestive gland, and cercarial production is never as high as in fully susceptible
individuals in which daughter sporocysts develop in the normal site. In schistosomes and molluscan hosts that are collected from endemic areas, there is a wide
variation in both the proportion of a host population that becomes infected, and
the number of cercariae shed from those that do develop patent infections. A
combination of these two features is considered to be an indicator of host–
parasite compatibility (Frandsen 1979), and data clearly demonstrate that a wide
range of "resistance" occurs in natural populations. In most of the protozoan
parasites of bivalves that have been studied, "resistant" stocks usually become
infected, but do not show the high mortality of "susceptible" populations. In
these cases, it appears that host survival depends on restriction of parasite
development, potentially involving components of the defense system, with
subsequent reduction in pathology (Ford 1988; Chintala and Fisher 1991; Ford
et al. 1993).

The compatibility of a mollusc-parasite combination can also alter with the
age of the host. Tests of laboratory strains of *S. mansoni* and *Biomphalaria spp.*
showed cases of juvenile susceptibility but adult resistance (Richards and Shade
1987). A similar situation is seen with infections of *B. glabrata* with *E. paraensei*

(Loker et al. 1987), where older snails recognize, encapsulate and kill invading miracidia which are able to survive in younger hosts (Fig. 4).

In some cases an aggressive host encapsulation response does not lead to the destruction of an intramolluscan parasite. A number of digenetic trematodes produce metacercariae, resistant postcercarial larvae, that reside in the tissues of a second intermediate host awaiting ingestion by the definitive host, often "encysted" within accumulations of host hemocytes that have presumably react-ed to the invader but failed to kill it. Dormant larvae of some nematodes are similarly encapsulated, such as is seen with the metastrongylid nematode *Angiostrongylus cantonensis*, in experimentally infected *B. glabrata* (Harris and Cheng 1975). A "paletot" seen around sporocysts of *Haplometra cylindracea* and other plagiorchid trematodes within molluscs (Cort and Olivier 1943) is com-posed of highly modified host hemocytes (Monteil and Martricon-Gondran 1991). This host cell envelope, with its high glycogen content, may be utilized as a nutrient source by the parasite, which, unlike the metacercariae and dormant nematode larvae mentioned above, is actively growing and replicating. Other pathogens, such as the protozoan *Bonamia* in oysters, actually live within host hemocytes. These parasites appear to gain entry into the host cell as a result of a normal phagocytic response (Chagot et al. 1992; Mourton et al. 1992).

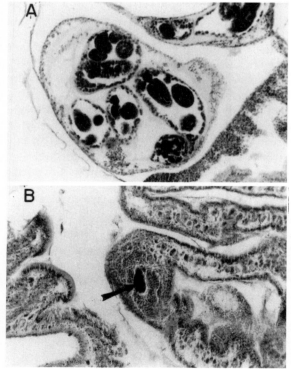

Fig. 4. Age-related resistance to *Echinostoma paraensei* in *Biompha-laria glabrata*. **A** Primary sporocysts (*dark objects*) developing within the ventricle of a young (6 mm) snail 8 days after infection. **B** Encapsulated primary sporocyst (*arrow*) in the periferal tissue of an older (12 mm) snail 2 days after infection. Note the large number of hemocytes sur-rounding the parasite. (Photographs provided by E.S. Loker, adapted from Loker et al. 1987)

Natural infection patterns of trematodes in molluscan hosts, and laboratory experiments on host–parasite interactions in vivo and in vitro, show that "resistance" to infection is the norm, and that it is the "compatible" combinations of snail and parasite that are the exception (Bayne and Yoshino 1989). While mechanisms of parasite recognition and destruction by molluscs constitute one intriguing aspect of this interaction, perhaps the more fascinating question remains; how, in a compatible combination, does an invading parasite escape destruction from the internal defense system of its molluscan host? It is vital to realize that in all host–parasite interactions, the two "players" can, and do, react to each other. This is emphasised by the evidence already given showing that escape from host recognition and subsequent destruction relies on the parasite being alive and intact. A parasitic life style may superficially seem a "lazy" alternative, with the host providing nutrition, protection and often transport for the parasite, yet most parasites survive in extremely hostile habitats!

4 How Do Molluscs Respond to Parasite Infections?

Relatively few studies have examined alterations in host defense functions within the first few hours after a snail is invaded by a parasite, yet it is at this early stage that the fate of most infections is determined. Our work has demonstrated that the phagocytic capacity of hemocytes from *B. glabrata* is affected by the trematode *S. mansoni* just three hours after contact with miracidia. In susceptible snails phagocytosis is depressed, while hemocytes from resistant snails show enhanced uptake of yeast (Fryer and Bayne 1990). In addition, the opsonic properties of plasma (its ability to facilitate phagocytosis) from resistant snails are significantly higher after exposure to the parasite compared to sham-treated individuals. *B. glabrata* exposed to *E. paraensei* show a slight increase in both hemocyte spreading (Noda and Loker 1989a) and phagocytosis (Noda and Loker 1989b) 24 h after miracidial penetration. In the case of *L. stagnalis* exposed to the bird schistosome, *Trichobilharzia ocellata*, hemocytes phagocytose more zymosan 90 min and 6 h after parasite penetration, indicating activation (Amen et al. 1992b).

A larger number of studies have examined the effects of compatible infections over longer time scales, after the parasites have become established. A variety of responses have been reported, both between different host species, and in the same host infected with different parasites. In *L. stagnalis* infected with the trematode *T. ocellata*, hemocytes show increased phagocytic capacity, DNA synthesis and endogenous peroxidase activity between 2 and 8 weeks after infection, apparently coinciding with mother sporocyst development and daughter sporocyst colonization of the digestive gland (Amen et al. 1991a,b, 1992a,b). However, phagocytic capacity has also been seen to decrease along with a reduction in agglutinin titers (van der Knaap et al. 1987). Similar decreases in agglutinin levels were seen when the same host species was infected with *Diplostomum spathaceum* (Riley and Chappell 1992), which also caused a decrease in phagocytic capacity. Infection of *Lymnaea* spp. with *Fasciola hepatica* appears to stimulate hemocytes, as seen by an increase in proliferative response of hemopoietic tissue (Rondelaud et al. 1988).

Different trematode species affect *B. glabrata* in different ways. Infection with *S. mansoni* causes relatively modest changes in defense-related parameters in susceptible host individuals, although miracidial penetration and migration does result in considerable pathology (Thompson et al. 1993), leading to alterations in host metabolism as seen by changes in the plasma protein composition, and levels of glucose and urea (e.g., Becker 1983; Rupprecht et al. 1989). Host and parasite seem to reach a metabolic equilibrium about 6 weeks after infections when the majority of sporocyst proliferation in the digestive gland has been completed, although disruption of the gonad resulting in parasitic castration is common (Crews and Yoshino 1989). A slight increase in hemagglutinin titers has been reported a few days after exposure of susceptible strains of *B. glabrata* to *S. mansoni* (Couch et al. 1990; Loker et al. 1994). This agglutinating activity may be due to a lectin that has been found to increase in plasma at the same time (Monroy et al. 1992). Changes in hemocyte function such as phagocytic capacity either do not occur, or are slight (e.g., Abdul Salam and Michelson 1980; Zelck and Becker 1992). Exposure of *B. glabrata* to echinostomes results in more dramatic changes in both humoral and cellular internal defense components. The spreading and locomotory ability of hemocytes is reduced dramatically shortly after infection (Noda and Loker 1989a; Adema et al. 1994). This suppression may be causally related to a reduction in phagocytic uptake of rabbit erythrocytes seen 8 and 30 days after infection (Noda and Loker 1989b). Jeong et al. (1981) reported increased titers of hemagglutinins after infection of *B. glabrata* with echinostomes, and this has since been described in more detail by Loker and colleagues. A fucose-inhibitable agglutinin increases gradually, peaking 8 days after infection of the snail (Couch et al. 1990; Loker et al. 1994). Two groups of molecules that bind to mixed carbohydrate affinity columns increase in concentration at a similar time after exposure of *B. glabrata* to *E. paraensei* (Fig. 5; Monroy and Loker 1993). These molecules represent several isolectins, with more than one being produced in individual animals.

The role of molluscan agglutinins in defense against parasites is uncertain. The observed changes in hemagglutinin titers in *B. glabrata* exposed to *E. paraensei* do not correlate with the fate of the parasite (Loker et al. 1994). These authors propose other roles, including precipitation of potentially toxic parasite products, and compensation for impaired hemocyte function. Recent work on an agglutinin purified from *Bulinus nasutus* has demonstrated binding to miracidia of an incompatible schistosome, *S. margrebowei* (Harris et al. 1993). While a similar 135-kDa molecule can be purified from all other bulinids tested, immunoreactivity differences suggest structural differences between taxonomic groups within this genus. This could indicate a role for this agglutinin in host-parasite interactions, since these taxonomic groups also differ in their susceptibility to various schistosome species.

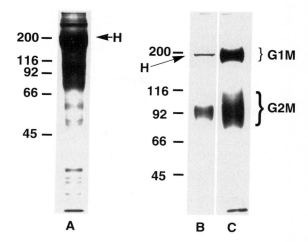

Fig. 5. Induction of lectins in *Biomphalaria glabrata* after exposure to *Echinostoma paraensei*. SDS-PAGE separation of **A** plasma from uninfected snails; **B** eluant from carbohydrate affinity column loaded with plasma from unexposed snails; **C** eluant from carbohydrate affinity column loaded with a similar volume of plasma from parasite-infected snails. Infected snails possess larger quantities of two groups of carbohydrate binding molecules (lectins) G1M and G2M. *H* Hemoglobin. See Monroy and Loker (1993) for further details

Infection of the bivalve *C. virginica* with the protozoan *H. nelsoni* leads to an increase in the number of circulating hemocytes and in their infiltration into tissues (Ford et al. 1993). It is postulated that these cells are involved in limiting parasite damage by plugging lesions, removing debris and repairing damaged tissue.

5 Are Susceptible Snails Simply Incompetent?

Natural and experimental infection patterns of snail hosts demonstrate that species, populations, and even individual snails, that are susceptible to one parasite are invariably resistant to others. This is particularly striking when looking at compatibility between members of the genus *Bulinus* and schistosomes belonging to the *S. haematobium* group (Preston and Southgate 1994). Susceptibility to a given parasite is not simply a result of some general lack of competence of the internal defense system. While some parasites clearly alter their host's defense capabilities, the recognition, killing, and clearance of other pathogens still occur. A clear example of this is evident in *B. glabrata* infected with the echinostome, *E. paraensei* (Lie et al. 1981); hemocytes from infected hosts were still able to phagocytose latex beads and ciliated plates from *S. mansoni*, encapsulate parasitic nematodes, and repair wounds.

6 How Does a Parasite Escape Destruction by a Molluscan Host?

Several hypotheses have been put forward concerning the mechanisms involved in larval trematode survival within compatible molluscan hosts. All of these were mentioned by Wright (1971), and have since been expanded and tested by others. The two main mechanisms proposed are: (1) evasion of host recognition by either (a) molecular mimicry or (b) active acquisition of host molecules; (2) interference with host defense mechanisms. After many years of descriptive studies that have given clues as to where to look, firm mechanistic evidence supporting each of the above hypotheses has now started to appear in the literature.

Studies in the 1980s using both in vitro and in vivo manipulations of *B. glabrata* hemolymph and *S. mansoni* larvae demonstrated three important characteristics of host components involved in determining the outcome of this snail–trematode interaction. (1) hemocytes from resistant strains can kill in the absence of plasma factors; (2) plasma on its own does not kill intact sporocysts (3) plasma from resistant snails elicits a cytotoxic response by normally benign susceptible strain hemocytes. These observations described first in vitro (Bayne et al. 1980a,b; Loker and Bayne 1982; LoVerde et al. 1984) and then in a series of inter strain plasma transfers in vivo (Granath and Yoshino 1984) led to a search for the "resistant factor" in plasma. The failure of any group to find this underlines the complexity of the in vivo interaction, which probably involves both evasion and interference by the parasite.

The idea that parasites may evade detection by their hosts by molecular mimicry was first proposed by Damian (1964). Salt (1970) suggested a role for such a mechanism in insect parasite interactions at almost the same time as Wright (1971) proposed its potential involvement in trematode survival within molluscs. Several studies, primarily using polyclonal antibodies, have found antigenically similar epitopes in molluscs and trematode larvae. The likelihood that such probes pick up common carbohydrate epitopes was pointed out by Bayne et al. (1987), and the presence of similar carbohydrates on host and parasite tissues has since been confirmed in lectin-binding studies (Zelck and Becker 1990). Common carbohydrate epitopes in *S. mansoni, B. glabrata* and even in nonhost snails have been demonstrated (Dissous et al. 1986). Evidence for a role for molecular mimicry has been considered tenuous (Yoshino and Boswell 1986), but Bayne and Yoshino (1989) revived interest in this as a potential mechanism of parasite survival after reanalysing data on compatibility between strains of *Oncomelania* and *S. japonicum* (Iwanaga and Tsuji 1985). In this, a positive correlation was seen between an increase in the number of shared antigens and higher rates of infectivity, indicating that the existence of common epitopes increases parasite survival.

Recent studies using molecular cloning techniques have provided the first evidence that antigenic similarities between *S. mansoni* and *B. glabrata* are genetically based. Dissous and Capron (1989) found that two *S. mansoni* miracidial components – Sm43 and Sm39 – could be immunoprecipitated by an antibody raised against *B. glabrata* proteins. Antisera raised against these *S.*

mansoni proteins recognised a 39-kDa translation product from *B. glabrata* mRNA which comigrated on 2D electrophoresis with in vitro translated Sm39. Evidence for a genetic basis for these shared antigens was obtained by molecular cloning of Bg39 (Dissous et al. 1990). The cDNA has 51–65% homology with tropomyosins of other species, with greatest similarity to schistosome tropomyosin previously cloned by Xu et al. (1989). The localization of Bg39 cross-reactive epitopes in tegumental vesicles and associated with epidermal ridges of *S. mansoni* miracidia (Dissous et al. 1990) is consistent with a role in schistosome–snail interactions.

Isoforms of tropomyosin from both *B. glabrata* and *S. mansoni* have also been cloned by Weston and Kemp (1993). Their analyses show closer similarity between an isoform from *B. glabrata* and one from *S. mansoni* (SMTM) than is seen between two schistosome isoforms. Monoclonal antibodies raised to recombinant SMTM have been used to probe antigens prepared from a wide variety of invertebrates (Weston et al. 1994). Some of these bound only to antigen prepared from other schistosome species and their molluscan hosts, indicating close similarity in epitopes on the tropomyosin. The lack of binding to material from invertebrates taxonomically closer to the schistosomes than molluscs suggests convergent evolution of this molecule in this host–parasite combination. Evidence is thus mounting for a selective pressure for the evolution of structural similarities in tropomyosins, and this may reflect a role for mimicry in protecting the parasite from host recognition.

Active acquisition of host components could also protect a parasite from recognition. The ability of *S. mansoni* larvae to acquire host plasma protein has been demonstrated in a number of studies. It should not be surprising that sporocysts incubated in host plasma rapidly become coated since, lacking a gut, the tegument is the main site of nutrient absorbtion (Bayne et al. 1986; Yoshino and Boswell 1986). However, in vitro incubations in plasma from a susceptible host does not protect parasites from subsequent recognition and killing by resistant hemocytes, and preexposure to resistant plasma does not lead to killing by susceptible cells (Loker and Bayne 1982; Granath and Yoshino 1984). While differences have been seen in the profile of components absorbed to *S. mansoni* larvae incubated in susceptible or resistant plasma (Bayne et al. 1986; Spray and Granath 1990) there is no evidence for their role in either protecting the parasite, or in marking it for destruction.

Evidence from both the host distribution of natural trematode infections, and changes in defense parameters after experimental exposure to miracidia, suggest direct interference with host defense as a mechanism of parasite survival. The classic example indicating the ability of a trematode to alter the capacity of its molluscan host to recognize and destroy other invaders is that of echinostome infections in *B. glabrata*, reviewed in detail by Lie et al. (1987). In this system, snails that are normally resistant to *S. mansoni* become susceptible to this schistosome after prior infection with *E. paraensei*. Similarly, paramphistome infections in *Bulinus tropicus* allow *S. bovis* to survive in this normally resistant host (Southgate et al. 1989). Many aspects of echinostome interference in

B. glabrata have now been described in more detail by Loker and coworkers. In vitro investigations using parasite excretory/secretory (ES) products showed alterations in hemocyte behavior, including reduced spreading and phagocytosis (Loker et al. 1992) confirming the previous observations on cells taken from snails infected in vivo (Noda and Loker 1989a,b). Most significantly, treatment of hemocytes from *S. mansoni* resistant strains of *B. glabrata* with echinostome ES products interfered with in vitro encapsulation of the schistosome sporocysts. Ultrastructural examination showed hemocytes binding weakly to the parasite with little spreading, and none of the tegumental damage that was observed with control hemocytes. Observations of in vitro preparations have demonstrated that host hemocytes round up and become immobile within minutes of the introduction of echinostome sporocysts to glass adherent hemocyte monolayers, without the need for physical contact (Fig. 6; Adema et al. 1994).

The ability of *S. mansoni* to interfere with hemocyte function is far less dramatic. Our observations of reduced phagocytosis of yeast in vitro by hemocytes taken from susceptible *B. glabrata* that had been exposed to miracidia 3 h previously (Fryer and Bayne 1990) indicate that interference may also play a role in parasite survival is this combination. Supporting in vitro evidence has now come from several studies. *S. mansoni* ES products (characterized by Lodes and Yoshino 1989) have been found to inhibit motility of hemocytes from susceptible snails, while stimulating those from resistant individuals. Addition of ES products to hemocytes in the absence of plasma stimulates cells from both strains to synthesize polypeptides that bind to the parasite (Lodes et al. 1991; Lodes and Yoshino 1993). However, if homologous plasma is present, secretion of two polypeptides by susceptible hemocytes is inhibited, while their production by resistant cells is unaltered (Lodes et al. 1991). This requirement for plasma, at the time of exposure to sporocyst products, for alteration of hemocyte function, reflects our own in vitro observations of inhibition of susceptible hemocyte phagocytic ability with whole parasites (Fryer and Bayne 1990), where parasites, hemocytes, and plasma all had to be present for modulation to occur. It may also explain apparently contradictory data (Connors and Yoshino 1990) showing decreased phagocytic ability of hemocytes from both susceptible and resistant snails treated with *S. mansoni* ES products, since all assays were carried out in the absence of plasma.

Superoxide, or other toxic oxygen metabolites, have been implicated in the killing of trematode larvae (Dikkeboom et al. 1988; Adema et al. 1992). There is now evidence that schistosomes may interfere with this defense mechanism. Production of superoxide following zymosan stimulation was reduced in *B. glabrata* hemocytes when cells were treated with *S. mansoni* sporocyst ES products (Conners and Yoshino 1990), although there was no difference between snail strains (again, no plasma was present). A superoxide scavenging molecule has been isolated from these ES products (Connors et al. 1991). While this clearly could protect the parasite from toxic oxygen metabolites, the ability of the parasite molecule to scavenge superoxide from both susceptible and resistant

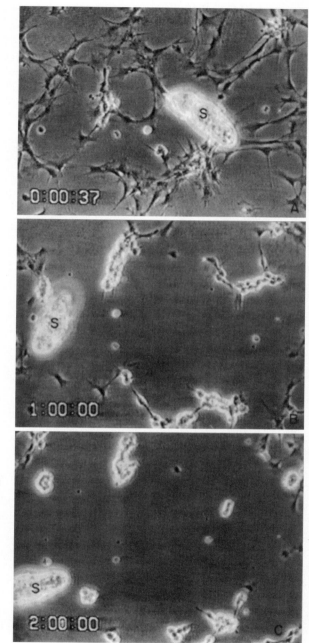

Fig. 6. Sporocyst induced rounding of *Biomphalaria glabrata* hemocytes. Time-lapse series of host glass adherent hemocytes in the presence of a single *Echinostoma paraensei* sporocyst (*S*). Hemocytes at **A** the initial point of contact with the parasite (37 s) show **B** gradual rounding after 1 h, with **C** some cells detaching from the substrate by 2 h. The sporocyst has moved across the field of view. (Photographs provided by C.M. Adema)

hemocytes indicates that this alone does not determine the outcome of the host–parasite encounter.

As evidence mounts for the role of interference in protecting trematode larvae from aggressive gastropod host responses, similar observations have been made for a protozoan parasite of bivalve molluscs (Ford et al. 1993). While hemocytes from bivalve species that are resistant to *H. nelsoni* move toward and engulf plasmodia, cells from susceptible oysters are actively repelled by the parasite

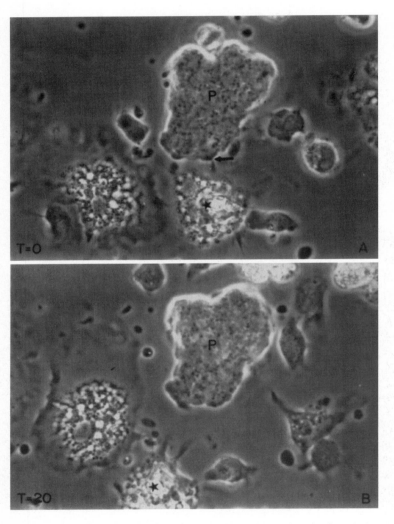

Fig. 7. Repulsion of oyster hemocytes by *Haplosporidium nelsoni* in vitro. **A** A granular hemocyte (*star*) has contacted a plasmodium (*P*) with an extended pseudopod (*arrow*) (T = O). **B** By 20 min, the same hemocyte has moved away from the parasite. In contrast, dead plasmodia are rapidly phagocytosed by similar hemocytes (not shown). (Photographs provided by S.E. Ford)

(Fig. 7). Dead plasmodia are rapidly engulfed, indicating that the living parasites secrete products that alter cell behavior. In contrast, the intracellular protozoan parasite *B. ostrae* gains entry into the host cell via phagocytosis (Chagot et al. 1992). In this case hemocytes from both susceptible *O. edulis* and a resistant species, *C. gigas,* rapidly internalize the parasites, forming morphologically normal phagolysosomes. Parasites within *O. edulis* hemocytes remained undamaged through the 4-h period of observations, while some of the engulfed protozoans were destroyed within *C. gigas* cells. There is some evidence of direct interference with host-cell function in that the normal respiratory burst is not elicited by phagocytosis of the parasite as is seen during phagocytosis of other particles (Hervio et al. 1989). However, since inhibition of the respiratory burst was seen in both bivalve species, this can not account for the differences in susceptibility. It is hoped that further development of longer term in vitro culture systems (Gauthier and Vasta 1993) will reveal more about the mechanisms used for survival by protozoan parasites of molluscs.

7 Conclusions

It is clear that molluscs possess verstatile internal defense systems that involve both humoral and cellular effectors. Similarly, parasites have many methods of dealing with such internal defense reactions of their hosts, whether they be vertebrates or invertebrates. The relatively high level of understanding of vertebrate, and particularly mammalian, immune systems, has led to fascinating discoveries of mechanisms of parasite survival (van der Ploeg et al. 1990; Toft et al. 1991). Studies of invertebrate host–parasite systems are aiding us in understanding both host internal defense systems and the methods of evasion, avoidance and interference used by their parasites (Bayne 1991). In addition to the work on molluscan systems reviewed here, studies on insect pathogens are revealing evidence of interference with host defense mechanisms (reviewed by Vinson 1993). Recent reports suggest the presence of molecules either known to be involved in vertebrate defense reactions, or thought to have characteristics of "primitive" defense molecules, in various invertebrate phyla. Cytokines such as IL-1 may be present in several phyla (Beck et al. 1989, 1993). Molluscs have now been shown to posess the acute phase reactants, C-reactive protein (Mandal et al. 1991) and α_2-macroglobulin (Bender et al. 1992; Thørgersen et al. 1992), a multifunctional binding protein with antiprotease activity which may have a role in recognition and clearance of foreign particles (Borth 1992). Continued studies of mollusc–parasite interactions should aid us in understanding how molluscs protect themselves from invading parasites and pathogens, and how parasites may use them to their own advantage to survive within their hosts.

Acknowledgements. We would like to thank Eric Loker, Coen Adema and Susan Ford for providing us with material for figures. Thanks are also due to Randy Bender for his comments on early drafts of this chapter.

References

Adbul-Salam JM, Michelson EH (1980) *Biomphalaria glabrata* amoebocytes: effect of *Schistosoma mansoni* infection on in vitro phagocytosis. J Invertebr Pathol 35: 241–248

Adema CM, van der Knaap WPW, Sminia T (1991) Molluscan hemocyte-mediated cytotoxicity: the role of reactive oxygen intermediates. Crit Rev Aquat Sci 4: 201–223

Adema CM, Harris RA, van Deutekom-Mulder EC (1992) A comparative study of hemocytes from six different snails: morphology and functional aspects. J Invertebr Pathol 59: 24–32

Adema CM, Arguello DF, Stricker SA, Loker ES (1994) A time-lapse study of interactions between *Echinostoma paraensei* intramollscan larval stages and adherent hemocytes from *Biomphalaria glabrata* and *Helix aspersa*. J Parasitol 80: 719–727

Amen RI, Baggen JMC, Meuleman EA, Wijsman-Grootendorst A, Boon ME, Bezemer PD, Sminia T (1991a) *Trichobilharzia ocellata*: quantification of effects on hemocytes of the pond snail *Lymnaea stagnalis* by morphometric means. Tissue Cell 23: 665–676

Amen RI, Tijnagel JMG, van der Knaap WPW, Meuleman EA, de Lange-de Klerk ESM, Sminia T (1991b) Effects of *Trichobilharzia ocellata* on hemocytes of *Lymnaea stagnalis*. Dev Comp Immunol 15: 105–115

Amen RI, Aten JA, Baggen JMC, Meuleman EA, de Lange-de Klerk ESM, Sminia T (1992a) *Trichobilharzia ocellata* in *Lymnaea stagnalis*: flow cytometric approach to study its effects on hemocytes. J Invertebr Pathol 59: 95–98

Amen RI, Baggen JMC, Bezemer PD, de Jong-Brink M (1992b) Modulation of the activity of the internal defence system of the pond snail *Lymnaea stagnalis* by the avian schistosome *Trichobilharzia ocellata*. Parasitology 104: 33–40

Bachère E, Hervio D, Mialhe E (1991) Luminol-dependent chemiluminescence by hemocytes of two marine bivalves, *Ostrea edulis* and *Crassostrea gigas*. Dis Aquat Org 11: 173–180

Basch PF (1975) An interpretation of snail-trematode infection rates: specificity based on concordance of compatible phenotypes. Int J Parasitol 5: 449–452

Basch PF (1976) Intermediate host specificity in *Schistosoma mansoni*. Exp Parasitol 39: 150–169

Basch PF, Di Conza JJ (1974) The miracidium-sporocyst transition in *Schistosoma mansoni*: surface changes in vitro with ultrastructural correlation. J Parasitol 60: 935–941

Bayne CJ (1983) Molluscan immunobiology. In: Saleuddin ASM, Wilbur KM (ed) The Mollusca vol 5 Physiology, part 2, Academic Press, New York, pp 407–486

Bayne CJ (1991) Invertebrate host immune mechanisms and host escapes. In: Toft CA, Aeschlimann A, Bolis L (eds) Parasite-host associations. Coexistence or conflict? Oxford Univ Press, New York, pp 299–315

Bayne CJ, Yoshino TP (1989) Determinants of compatibility in mollusc-trematode parasitism. Am Zool 29: 399–407

Bayne CJ, Buckley PM, Dewan PC (1980a) *Schistosoma mansoni*: cytotoxicity of hemocytes from susceptible snail hosts for sporocysts in plasma from resistant *Biomphalaria glabrata*. Exp Parasitol 50: 409–416

Bayne CJ, Buckley PM, Dewan PC (1980b) Macrophage-like hemocytes of resistant *Biomphalaria glabrata* are cytotoxic for sporocysts of *Schistosoma mansoni* in vitro. J Parasitol 66: 413–419

Bayne CJ, Loker ES, Yui MA (1986) Interactions between the plasma proteins of *Biomphalaria glabrata* (Gastropoda) and the sporocyst tegument of *Schistosma mansoni* (Trematoda). Parasitology 92: 653–664

Bayne CJ, Boswell CA, Yui MA (1987) Widespread antigenic cross-reactivity between plasma proteins of a gastropod and its trematode parasite. Dev Comp Immunol 11: 321–329

Beck G, O'Brien RF, Habicht GS (1989) Invertebrate cytokines: the phylogenetic emergence of interleukin-1. Bioessays 11: 62–67

Beck G, O'Brien RF, Habicht GS, Stillman DL, Cooper EL, Raftos DA (1993) Invertebrate cytokines III: Invertebrate interleukin-1-like molecules stimulate phagocytosis by tunicate and echinoderm cells. Cell Immunol 146: 284–299

Becker W (1983) Purine metabolism in *Biomphalaria glabrata* under starvation and infection with *Schistosoma mansoni*. Comp Biochem Physiol 76B: 75–79

Bender RC, Fryer SE, Bayne CJ (1992) Proteinase inhibitory activity in the plasma of a mollusc: evidence for the presence of α-macroglobulin in *Biomphalaria glabrata*. Comp Biochem Physiol 102B: 821–824

Borth W (1992) α_2-macroglobulin, a multifunctional binding protein with targeting characteristics. FASEB J 6: 3345–3353

Chagot D, Boulo V, Hervio D, Mialhe E, Bachère E, Mourton C, Grizel H (1992) Interactions between *Bonamia ostreae* (Protozoa: Ascetospora) and hemocytes of *Ostrea edulis* and *Crassostrea gigas* (Mollusca: Bivalvia): entry mechanisms. J Invertebr Pathol 59: 241–249

Chen J-H, Bayne CJ (1994) The roles of carbohydrates in aggregation and adhesion of hemocytes from the California mussel (*Mytilus californianus*). Comp Biochem Physiol 109A: 117–125

Cheng TC, Dougherty WJ (1989) Ultrastructural evidence for the destruction of *Schistosoma mansoni* sporocysts associated with elevated lysosomal emzyme levels in *Biomphalaria glabrata*. J Parasitol 75: 928–941

Cheng TC, Downs JCU (1988) Intracellular acid phosphatase and lysozyme levels in subpopulations of oyster, *Crassostrea virginica*, hemocytes. J Invertebr Pathol 52: 163–167

Cheng TC, Garrabbrant TA (1977) Acid phosphatase in granulocytic capsules formed in strains of *Biomphalaria glabrata* totally and partially resistant to *Schistosoma mansoni*. Int J Parasitol 7: 467–472

Cheng TC, Rodrick GE, Foley DA, Koehler SA (1975) Rlease of lysozyme form hemolymph cells of *Mercenaria mercenaria* during phagocytosis. J Invertebr Pathol 25: 261–265

Chernin E (1966) Transplantation of larval *Schistosoma mansoni* from infected to unifected snails. J Parasitol 52: 473–482

Chintala MM, Fisher WS (1991) Disease incidence and potential mechanisms of defense for MSX-resistant and -susceptible oysters held in Chesapeake Bay. J Shellfish Res 10: 439–443

Connors VA, Yoshino TP (1990) In vitro effect of larval *Schistosoma mansoni* excretory-secretory products on phagocytosis-stimulated superoxide production hemocytes from *Biomphalaria glabrata*. J Parasitol 76: 895–902

Connors VA, Lodes MJ, Yoshino TP (1991) Identification of a *Schistosoma mansoni* sporocyst excretory-secretory antioxidant molecule and its effect on superoxide production by *Biomphalaria glabrata* hemocytes. J Invertebr Pathol 58: 387–395

Courch L, Hertel LA, Loker ES (1990) Humoral response of the snail *Biomphalaria glabrata* to trematode infection: observations on a circulating hemagglutinin. J Exp Zool 255: 340–349

Cort WW, Olivier L (1943) The development of the larval stages of *Plagiorchis muris* in the first intermediate host. J Parasitol 29: 81–99

Crews AE, Yoshino TP (1989) *Schistosoma mansoni*: effect of infection on reproduction and gonadal growth in *Biomphalaria glabrata*. Exp Parasitol 68: 326–324

Damian RT (1964) Molecular mimicry: antigen sharing by parasite and host and its consequences. Am Nat 98: 129–149

Di Conza JJ, Basch PF (1974) Axenic cultivation of *Schistosoma mansoni* sporocysts. J Parasitol 60: 757–763

Dikkeboom R, Bayne CJ, van der Knaap WPW, Tijnagel JMGH (1988) Possible role of reactive forms of oxygen in in vitro killing of *Schistosoma mansoni* sporocysts by hemocytes of *Lymnaea stagnalis*. Parasitol Res 75: 148–154

Dissous C, Capron A (1989) *Schistosoma mansoni* and its intermediate host *Biomphalaria glabrata* express a common 39 kilodalton acidic protein. Mol Biochem Parasit 32: 49–56

Dissous C, Grzych JM, Capron A (1986) *Schistosoma mansoni* shares a protective oligosaccharide epitope with fresh water and marine snails. Nature 323: 443–445

Dissous C, Torpier G, Duvaux-Miret O, Capron A (1990) Structural homology of tropomyosins from the human trematode *Schistosoma mansoni* and its intermediate host *Biomphalaria glabrata*. Mol Biochem Parasitol 43: 245–256

Font WF (1980) Effects of hemolymph of the American oyster, *Crassostrea virginica*, on marine cercariae. J Invertebr Pathol 36: 41–47

Ford SE (1988) Host-parasite interactions in eastern oysters selected for resistance to *Haplosporidium nelsoni* (MSX) disease: survival mechanisms against a natural pathogen. Am Fish Soc Spec Publ 18: 206–224

Ford SE, Ashton-Alcox KA, Kanaley SA (1993) In vitro interactions between bivalve hemocytes and the oyster pathogen *Haplosporidium nelsoni* (MSX). J Parasitol 79: 255–265

Frandsen F (1979) Discussion of the relationship between *Schistosoma* and their intermediate hosts, assessment of the degree of host-parasite compatibility, and evaluation of schistosome taxonomy. Z Parasitenkd 58: 275–296

Fryer SE, Bayne CJ (1990) *Schistosoma mansoni* modulation of phagocytosis in *Biomphalaria glabrata*. J Parasitol 76: 45–52

Gauthier JD, Vasta GR (1993) Continuous in vitro culture of the Eastern oyster parasite *Perkinsus marinus*. J Invertebr Pathol 62: 321–323

Granath WO, Yoshino TP (1983) Lysosomal enzyme activities in susceptible and refractory strains *Biomphalaria glabrata* during the course of infection with *Schistosoma mansoni*. J Parasitol 69: 1018–1026

Granath WO, Yoshino TP (1984) *Schistosoma mansoni*: passive transfer of resistance by serum in the vector snail *Biomphalaria glabrata*. Exp Parasitol 58: 188–193

Harris KR, Cheng TC (1975) The encapsulation process in *Biomphalaria glabrata* experimentally infected with the metastrogylid *Angiostrongylus cantonensis*: light microscopy. Int J Parastiol 5: 521–528

Harris RA, Preston TM, Southgate VR (1993) Purification of an agglutinin from the haemolymph of the snail *Bulinus nasutus* and demonstration of related proteins in other *Bulinus* spp. Parasitology 106: 127–135

Hervio D, Chagot D, Miahle E, Grizel H (1989) Chemiluminescent responses of *Ostrea edulis* and *Crassostrea gigas* hemocytes to *Bonamie ostreae* (Ascetospora). Dev Comp Immunol 13: 449

Howland KH, Cheng TC (1982) Identification of bacterial chemoattractants for oyster (*Crassostrea virginica*) hemocytes. J Invertebr Pathol 39: 123– 132

Iwanaga Y, Tsuji M (1985) Studies on host-parasite relationship between *Schistosoma japonicum* and *Oncomelania* snails. 1. Antigenic communities between the Chinese strain of *Schistosoma japonicum* adult worm and *Oncomelania* snails. Jpn J Parasitol 34: 1–6

Jeong KH, Sussman S, Rosen SD, Lie KJ, Heyneman D (1981) Distribution and variation of hemagglutinating activity in the hemolymph of *Biomphalaria glabrata*. J Invertebr Pathol 38: 256–263

Kechemir N, Combes C (1982) Développment due trématode *Schistosoma haematobium* après transplanation microchirurgicale chez le gastéropode *Planorbis metidjensis*. CR Acad Sci Paris 295: 505–508

Klein (1989) Are invertebrates capable of anticipatory immune responses? Scan J Immunol 29: 499–505

Lewis FA, Richards CS, Knight M, Cooper LA, Clark B (1993) *Schistosoma mansoni*: analysis of an unusual infection phenotype in the intermediate host snail *Biomphalaria glabrata*, Exp Parasitol 77: 349–361

Lie KJ, Jeong KH, Heyneman D (1980) Tissue reactions induced by *Schistosoma mansoni* in *Biomphalaria glabrata*. Ann Trop Med Parasitol 74: 157–166

Lie KJ, Jeong KH, Heyneman D (1981) Selective interference with granulocyte function induced by *Echinostoma paraensei* (Trematoda) larvae in *Biomphalaria glabrata* (Mollusca). J Parasitol 67: 790–796

Lie KJ, Jeong KH, Heyneman D (1987) Molluscan host reactions to helminthic infections. In: Soulsby EJL (ed) Immune responses in parasitic infection. CRC Press, Boca Raton, pp 211–270

Lim SHK (1970) Parameters and mechanisms of antagonistic interactions between *Schistosoma mansoni* and *Paryphostomum segregatum* in the snail *Biomphalaria glabrata*. Doctoral Dissertation, Univ California, San Francisco

Lodes MJ, Yoshino TP (1989) Characterization of excretory-secretory proteins synthesized in vitro by *Schistosoma mansoni* primary sporocysts. J Parasitol 75: 853–862

Lodes MJ, Yoshino TP (1990) The effect of schistosome excretory-sectory products on *Biomphalaria glabrata* hemocyte motility. J Invertebr Pathol 56: 75–85

Lodes MJ, Yoshino TP (1993) Polypeptides synthesized in vitro by *Biomphalaria glabrata* hemocytes bind to *Schistosoma mansoni* primary sporocysts. J Invertebr Pathol 61: 117–122

Lodes MJ, Conners VA, Yoshino TP (1991) Isolation and functional characterization of snail hemocyte-modulating polypeptide from primary sporocyts of *Schistosoma mansoni*. Mol Biochem Parasitol 49: 1–10

Loker ES, Bayne CJ (1982) In vitro encounters between *Schistosoma mansoni* primary sporocysts and hemolymph components of susceptible and resistant strains of *Biomphalaria glabrata*. Am J Trop Med Hyg 31: 999–1005

Loker ES, Bayne CJ, Buckley PM, Kruse KT (1982) Ultrastructure of encapsulation of *Schistosoma mansoni* mother sporocysts by hemocytes of juveniles of the 10-R2 strain of *Biomphalaria glabrata*. J Parasitol 68: 84–94

Loker ES, Cimino DF, Stryker GA, Hertel LA (1987) The effect of size of M-line *Biomphalaria glabrata* on the course of development of *Echinostoma paraensei*. J Parasitol 73: 1090–1098

Loker ES, Boston ME, Bayne CJ (1989) Differential adherence of M line *Biomphalaria glabrata* hemocytes to *Schistosoma mansoni* and *Echinostoma paraensei* larvae, and experimental manipulation of hemocyte binding. J Invertebr Pathol 54: 260–268

Loker ES, Cimino DF, Hertel LA (1992) Excretory-secretory products of *Echinostoma paraensei* sporocysts mediate interference with *Biomphalaria glabrata* hemocyte function. J Parasitol 78: 104–115

Loker ES, Couch L, Hertel LA (1994) Elevated agglutinin titres in plasma of *Biomphalaria glabrata* exposed to *Echinostoma paraensei*: characterization and functional relevance of a trematode-induced response. Parasitoloty 108: 17–26

LoVerde PT, Shoulberg N, Gherson J (1984) Role of cellular and humoral components in the encapsulation response of *Biomphalaria glabrata* to *Schistosoma mansoni* sporocysts in vitro. In: Cohen E (ed) Recognition proteins, receptors, and probes: invertebrates. AR Liss, New York, pp 17–29

Mandal C, Biswas M, Nagpurkar A, Sailen M (1991) Isolation of a phosphoryl choline-bining protein from the hemolymph of the snail *Achatina fulica*. Dev Comp Immunol 15: 227–239

McKerrow JH, Jeong KH, Beckstead JH (1985) Enzyme histochemical comparison of *Biomphalaria glabrata* amoebocytes with human granuloma macrophages. J Leukocyte Biol 37: 341–347

Meuleman EA, Lyaruu DM, Khan MA, Holzmann PJ, Sminia T (1978) Ultrastructural changes in the body wall of *Schistosoma mansoni* during the transformation of the miracidium into the mother sporocyst in the snail host *Biomphalaria pfeifferi*. Z Parasitenkd 56: 227–242

Meuleman EA, Huyer AR, Luub TWJ (1984) Infection of *Lymnaea stagnalis* with miracidia of *Trichobilharzia ocellata*. Z Parasitenkd 70: 275–278

Monroy F, Loker ES (1993) Production of heterogeneous carbohydrate-binding proteins by the host snail *Biomphalaria glabrata* following exposure to *Echinostoma paraensei* and *Schistosoma mansoni*. J Parasitol 79: 416–423

Monroy F, Hertel LA, Loker ES (1992) Carbohydrate-binding plasma proteins from the gastropod *Biomphalaria glabrata*: strain specificity and the effects of trematode infection. Dev Comp Immunol 16: 355–366

Monteil J-F, Matricon-Gondran M (1991) Interactions between the snail *Lymnaea truncatula* and the plagiorchid trematode *Haplometra cylindracea*. J Invertebr Pathol 58: 127–135

Mourton C, Boulo V, Chagot D, Hervio D, Bachère E, Mialhe E, Grizel H (1992) Interactions between *Bonamia ostreae* (Protozoa: Ascetospora) and hemocytes of *Ostrea edulis* and *Crassostrea gigas* (Molluscs: Bivalvia) In vitro system establishment. J Invertebr Pathol 59: 235–240

Newton WL (1952) The comparative tissue reaction of two strains of *Australorbis glabratus* to infection with *Schistosoma mansoni*. J Parasitol 38: 362–366

Noda S, Loker ES (1989a) Effects of infection with *Echinostoma paraensei* on the circulating hemocyte population of the host snail *Biomphalaria glabrata*. Parasitology 98: 35–41

Noda S, Loker ES (1989b) Phagocytic activity of hemocytes of M-line *Biomphalaria glabrata*: effect of exposure to the trematode *Echinostoma paraensei*. J Parasitol 75: 261–269

Olafsen JA (1986) Invertebrate lectins: biochemical heterogeneity as a possible key to their biological function. In: Brehélin M (ed) Immunity in invertebrates. Springer, Berlin Heidelberg New York

Ottaviani E, Paemen LR, Cadet P, Stefano GB (1993) Evidence for nitric oxide production and utilization as a bacteriocidal agent by invertebrate immunocytes. Eur J Pharmacoe (Environ Toxicol Pharmacol) 248: 319–324

Pan SC-T (1980) The fine structure of the miracidium of *Schistosoma mansoni*. J Invertebr Pathol 36: 307–372

Pipe RK (1992) Generation of reactive oxygen metabolites by the hemocytes of the mussel *Mytilus edulis*. Dev Comp Immunol 16: 111–122

Preston TM, Southgate VR (1994) The species specificity of *Bulinus–Schistosoma* interactions. Parasitol Today 10: 69–73

Ratcliffe NA (1985) Invertebrate immunity – a primer for the non-specialist. Immunol Lett 10: 253–270

Ratcliffe NA, Rowley AF (eds) (1981) Invertebrate blood cells. Academic Press, London

Renwrantz L (1983) Involvement of agglutinin (lectins) in invertebrate defense reactions: the immuno-biological importance of carbohydrate-specific binding molecules. Dev Comp Immunol 7: 603–608

Richards CS, Shade PC (1987) The genetic variation of compatibility if *Biomphalaria glabrata* and *Schistosoma mansoni* J Parasitol 73: 1146–1151

Richards CS, Knight M, Lewis FA (1992) Genetics of *Biomphalaria glabrata* and its effect on the outcome of *Schistosoma mansoni* infection. Parasitol Today 8: 171–174

Riley EM, Chappell LH (1992) Effect of infection with *Diplostomum spathaceum* on the internal defense system of *Lymnaea stagnalis*. J Invertebr Pathol 59: 190–196

Rondelaud D, Bouix-Busson D, Barthe D (1988) Relationship between shell height and a proliferative response of the amoebocyte-producing organ in two species of *Lymnaea* (Gastropoda: Mollusca) infected by *Fasciola hepatica*. J Invertebr Pathol 51: 294–295

Rupprecht H, Becker W, Schwanbek A (1989) Alterations in hemolymph components in *Biomphalaria glabrata* during long-term infection with *Schistosoma mansoni*. Parasitol Res 75: 233–237

Salt G (1970) The cellular defence reactions of insects. Cambridge Monographs in Exp Biol 16 Cambridge Univ Press, Cambridge, 118 pp

Schmid LS (1975) Chemotaxis of hemocytes from the snail *Viviparus malleatus*. J Invertebr Pathol 25: 125–132

Schneeweiss H, Renwrantz L (1993) Analysis of the attraction of haemocytes from *Mytilus edulis* by molecules of bacterial origin. Dev Comp Immunol 17: 377–387

Sminia T, Barendsen L (1980) A comparative and enzyme histochemical study on blood cells of the freshwater snails *Lymnaea stagnalis*, *Biomphalaria glabrata* and *Bulinus truncatus*. J Morphol 165: 31–39

Sminia T, van der Knaap WPW (1986) Immunorecognition in invertebrates with special reference to molluscs. In: Brehelin M (ed) Immunity in invertebrates. Springer, Berlin Heidelberg New York, pp 113–124

Sminia T, van der Knaap WPW (1987) Cells and molecules in molluscan immunology. Dev Comp Immunol 11: 17–28

Sminia T, Borghart-Reinders E, van de Linde AW (1974) Encapsulation of foreign materials experimentally introduced into the freshwater snail *Lymnaea stagnalis*. An electron microscopic and autoradiographic study. Cell Tissue Res 153: 307–326

Southgate VR, Brown DS, Warlow A, Knowles RJ, Jones A (1989) The influence of *Calicophoron microbothium* on the susceptibility *Bulinus tropicus* to *Schistosoma bovis*. Parasitol Res 75: 381–391

Spray FJ, Granath WO (1990) Differential binding of hemolymph proteins from schistosome-resistant and -susceptible *Biomphalaria glabrata* to *Schistosoma mansoni* sporocysts. J Parasitol 76: 225–229

Sullivan JT, Spence JV (1994) Transfer of resistance to *Schistosoma mansoni* in *Biomphalaria glabrata* by allografts of amoebocyte-producing organ. J Parasitol 80: 449–453

Thompson SN, Lee RW-K, Mejia-Scales V, El-Din MS (1993) Biochemical and morphological pathology of the foot of the schistosome vector *Biomphalaria glabrata* infected with *Schistosoma manson*. Parasitology 107: 275–285

Thøgersen IB, Salvesen G, Brucato FH, Pizzo SV, Enghild JJ (1992) Purification and characterization of an α-macroglobulin proteinase inhibitor from the mollusc *Octopus vulgaris*. Biochem J 285: 521–527

Toft CA, Aeschlimann A, Bolis L (eds) (1991) Parasite-host associations. Coexistence of conflict? Oxford Univ Press, New York

Van der Knaap WPW, Loker ES (1990) Immune mechanisms in trematode-snail interactions. Parasitol Today 6: 175–182

Van der Knaap WPW, Meuleman EA, Sminia T (1987) Alterations in the internal defense system of the pond snail *Lymnaea stagnalis* induced by infection with the schistosome *Trichobilharzia ocellata*. Parasitol Res 73: 57–65

Van der Ploeg LHT, Cantor CR, Vogel HJ (eds) (1990) Immune recognition and evasion: molecular aspects of host-parasite interaction. Academic Press, San Diego

Vinson SB (1993) Suppression of the insect immune system by parasitic hymenoptera. In: Pathak JPN (ed) Insect immunity. Kluwer, Dordrecht, pp 171–187

Weston DS, Kemp WM (1993) *Schistosoma mansoni*: comparison of cloned tropomyosin antigens shared between adult parasites and *Biomphalaria glabrata*. Exp Parasitol 76: 358–370

Weston DS, Schmitz, Kemp MW, Kunz W (1993) Cloning and sequencing of a complete myosin heavy chain cDNA from *Schistosoma mansoni*. Mol Biochem Parasitol 58: 161–164

Weston DS, Allen B, Thakur A, LoVerde PT, Kemp WM (1994) Invertebrate host-parasite relationships: convergent evolution of a tropomyosin epitope between *Schistosoma* sp. *Fasciola hepatica*, and certain pulmonate snails. Exp Parasitol 78: 269–278

Wright CA (1971) *Flukes and Snails*. Allen and Unwin, London

Xu H, Miller S, Van Keulen H, Wawrzynski MR, Rekosh DM, LoVerde PT (1989) *Schistosoma mansoni* tropomyosin: cDNA characterization, sequence, expression and gene product localization. Exp Parasitol 69: 373–392

Yoshino TP, Boswell CA (1986) Antigen sharing between larval trematodes and their snail hosts: how real a phenomenon in immune evasion? Symp Zool Soc Lond 56: 221–238

Zelck U, Becker W (1990) Lectin binding to cells of *Schistosoma mansoni* sporocysts and surrounding *Biomphalaria glabrata* tissue. J Invertebr Pathol 55: 93–99

Zelck U, Becker W (1992) *Biomphalaria glabrata*: influence of calcium, lectins, and plasma factors on in vitro phagocytic behavior of hemocytes of noninfected or *Schistosoma mansoni* infected snails. Exp Parasitol 75: 126–136

Clotting and Immune Defense in Limulidae

T. Muta and S. Iwanaga

1 Introduction

The evolution of an effective system for microbial defense is central to the survival and perpetuity of higher organisms. Invertebrates, which lack typical immune systems, have developed unique systems to detect and respond to microbial surface antigens, such as lipopolysaccharide (LPS), peptideglycan, and β-(1,3)-glucan. Because both invertebrates and vertebrate animals respond to these substances, it is likely that a system recognizing these epitopes emerged at a very early stage in the evolution of these animals.

Invertebrate defense systems include hemolymph coagulation, melanization, cytolysis, cell agglutination, antimicrobial activities, and phagocytosis against pathogens. Hemolymph coagulation and phenoloxidase-mediated melanization is induced in some invertebrate species by foreign materials, that result in the engulfment of invading microorganisms (Bohn 1986; Söderhäll and Smith 1986). Their hemolymph plasma displays a cytolytic activity against exogenous cells, which is apparently similar to the mammalian complement-reactions (Day et al. 1970). Lectins in either plasma or in hemocytes aggregate foreign materials (Olafsen 1986). Those immobilized microorganisms are finally killed by antimicrobial substances (Boman 1994).

In the horseshoe crab, an arthropod, the hemolymph is extremely sensitive to bacterial endotoxin or LPS (Bang 1956; Levin and Bang 1964). A minute quantity of LPS activates the hemocytes to induce cell adhesion and aggregation, followed by degranulation of granules in the cell (Ornberg and Reese 1979). The released granular components include LPS-sensitive and β-(1,3)-glucan-sensitive coagulation factors, lectins, and antimicrobial substances (Iwanaga et al. 1994a). Although melanization activity of the horseshoe crab hemolymph is not evident, LPS-sensitive hemolymph coagulation is well known (Iwanaga et al. 1992; Iwanaga 1993). This reaction is now being widely utilized to detect or quantify trace amounts of LPS in clinical applications (Tanaka and Iwanaga 1993). On the other hand, the hemolymph plasma shows hemagglutinating activities and cytolytic activities on foreign cells (Enghild et al. 1990). The hemocytes and

Department of Biology, Faculty of Science and Department of Molecular Biology, Graduate School of Medical Science, Kyushu University 33, Fukuoka 812-81, Japan plasma also contain proteinase inhibitors with broad specificities that would block the activities of proteases secreted from invaded microorganisms (Sottrup et al. 1990). Table 1 summarizes proteins so far found in the horseshoe crab hemocytes and the hemolymph plasma.

The first and most important step of the defense mechanism is the recognition of foreign materials. Studies of the defense systems of invertebrates, which are different from the typical immune systems known in mammals, are of interest because they may lead to the elucidation of the basic mechanisms used by animals to distinguish self and nonself materials.

Table 1. Proteins in horseshoe crab hemolymph

Protein	Molecular mass (kDa)	Function	Localization
Coagulation factors			
Factor C	123	Serine protease	L-granule
Factor B	64	Serine protease	N.D.
Factor G	110	Serine protease	L-granule
Proclotting Enzyme	54	Serine protease	L-granule
Coagulogen	20	Gelation	L-granule
Protease inhibitors			
LICI-1	48	Serpin	L-granule
LICI-2	42	Serpin	L-granule
Trypsin inhibitor	6.8	BPTI-like	N.D.
Cystatin	13	Cystatin-like	N.D.
α_2-Macroglobulin	180	Complement-like	Plasma
Antimicrobial substances			
Anti-LPS factor	12	Antimicrobial	L-granule
Tachyplesin	2.3	Antimicrobial	S-granule
Big defensin	8.6	Antimicrobial	L- & S-granule
Factor D	42	Serprocidin	L-granule
Lectins			
L6	27	Lectin	L-granule
L10	27	Lectin	L-granule
Limunectin	54	Lectin	L-granule
18K-LAF	18	Lectin	L-granule
Limulin/CRP	460	Lectin	Plasma
Polyphemin	N.D.	Lectin	Plasma
Others			
Transglutaminase (TGase)	86	Cross-link	Cytosol
8.6 kDa protein	8.6	TGase substrate	L-granule
Pro-rich protein	80	TGase substrate	L-granule
L-SPC	70	Processing	N.D.
S-2	8	LPS-binding	S-granule

BPTI, bovine pancreatic trypsin inhibitor; L, large; S, small; N.D.: not determined.

2 Hemolymph

Horseshoe crab hemolymph, when freshly collected, is white. It turns gradually to a beautiful blue fluid owing to oxidation of divalent copper ions contained in hemocyanin. In contrast to mammalian blood, oxygen-carrying protein (hemocyanin) is present in the plasma, and coagulation factors are in the cells. Coagulation factors are released when the cells are stimulated. The hemolymph plasma contains four major proteins; hemocyanin, limulin (lectin), C-reactive protein, and α_2-macroglobulin (Quigley and Armstrong 1994). The plasma has hemagglutinin activities, caused by lectins, and shows cytolytic activities (Enghild et al. 1990), the mechanism of which remains to be clarified.

3 Hemocytes

The horseshoe crab hemolymph contains two types of hemocytes (Toh et al. 1991). The minor type of hemocytes contains no granules; however, the major type, which comprises 99% of total hemocytes, is filled with two types of granules and a single nucleus (Fig. 1a). Thus, this cell is often referred as granulocyte or amebocyte, and it plays an important role in the defense system of this animal. A scanning electron micrograph shows that it has a rugby ball-like oval shape with a longer axis of up to 20 μm (Fig. 1b). The surface of the cell is rugged because of the many granules inside. The degranulation of these granules is initiated within 30 s after the LPS stimulation. A cross section of the cells treated with LPS clearly shows membrane fusions between granule and plasma membranes, indicating that the degranulation is caused not by cell lysis but by exocytosis (Fig. 2; Ornberg and Reese 1979; Toh et al. 1991).

4 Hemocyte Granules

Two types of granules are evident in the hemocytes (Fig. 1a; Toh et al. 1991). One type of granules is larger (up to 1.5 μm in equatorial diameter) and less electron-dense, whereas the other is smaller (less than 0.6 μm in diameter) but dense. Population density of the larger granule is approximately 40/100 μm^2 cytoplasmic area, whereas that of the smaller is about 20/100 μm^2. This results in occupation ratios of 33% of cytoplasmic area versus 3%, respectively. These granules contain coagulation factors, protease inhibitors, antimicrobial peptides, lectins and other proteins, which are released into the hemolymph in response to LPS (Table 1). Although both of the granules are exocytosed, the larger granules show fast response and the exocytosis of the smaller ones is delayed.

In 1975, Mürer et al. reported the isolation of large granules and showed that an extract from the granules is gelled by LPS. More recently, we have succeeded in separately isolating the larger and smaller granules (Shigenaga et al. 1993). The

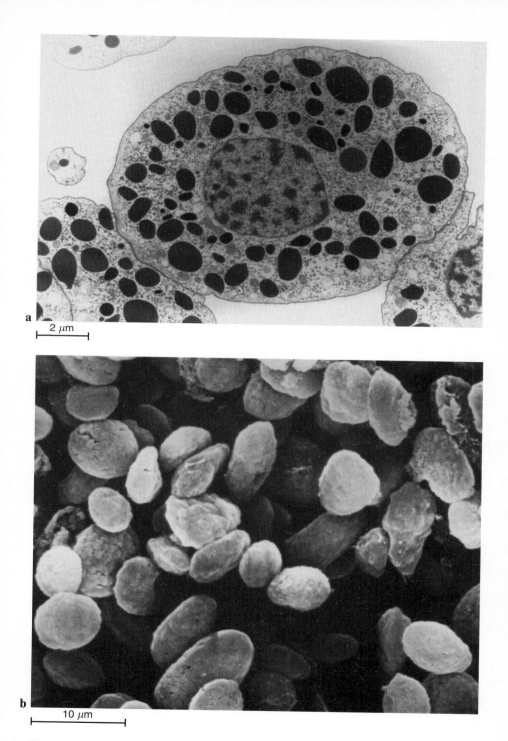

Fig. 1. Horseshoe crab hemocytes. **a** Electron micrograph; *bar* = 2 μm. **b** scanning electron micrographs; *bar* = 10 μm

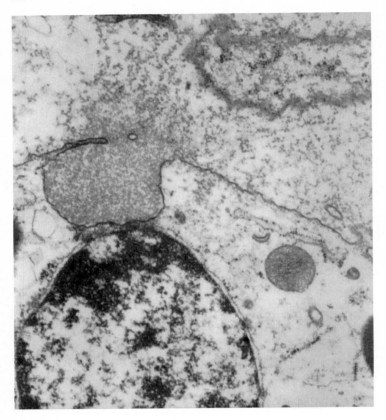

Fig. 2. Exocytosis of a granule of the hemocytes upon stimulation of LPS. The hemocyte was fixed
*l*min after the addition of LPS. (Toh et al. 1991)

separation of these granules made it possible to identify various granular components (Fig. 3a). SDS-PAGE and HPLC analyses show that they contain
distinct components (Fig. 3b; Shigenaga et al. 1993). The major proteins/peptides
in the larger and smaller granules are coagulogen and tachyplesin, respectively
(described later). In addition to these abundant proteins and peptides, HPLC
analyses of each granule component resulted in the identification and purification of at least twenty proteins from the larger and six proteins from the
smaller granules: these proteins, so far identified, are tentatively named L1–L10
and S1–S6.

5 Clotting Cascade

The horseshoe crab (or limulus) hemolymph is known to be very sensitive to LPS
on Gram-negative bacteria (Iwanaga et al. 1992; Iwanaga 1993). A trace amount
of LPS activates the hemocytes, releasing LPS-sensitive coagulation factors and

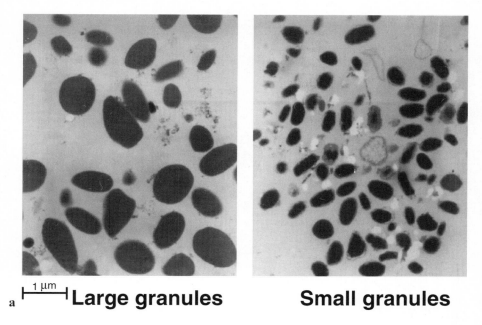

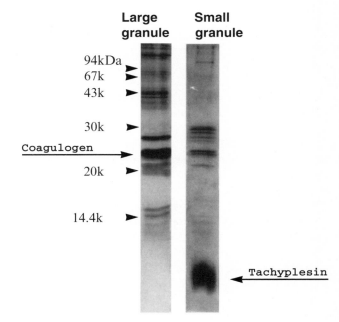

Fig. 3. Larger and smaller granules isolated from hemocytes of the horseshoe crab. Two granule fractions were separated by centrifugation on sucrose density gradient (Shigenaga et al. 1993). **a** Electron micrograph of larger and smaller granules. **b** SDS-PAGE analysis of the separated granules. (Shigenaga et al. 1993)

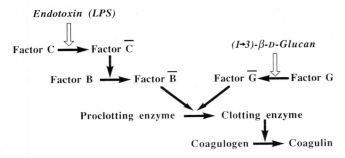

Fig. 4. Horseshoe crab coagulation cascade. Protease zymogens are activated by limited proteolyses

antimicrobial substances via degranulation. Among the proteins released from the cells, an LPS-sensitive serine protease zymogen, factor C, is autocatalytically converted to its active form, factor $\overline{C}$ (Nakamura et al. 1986a; Muta et al. 1991). The active factor $\overline{C}$ activates zymogen factor $\overline{B}$ to factor $\overline{B}$ (Nakamura et al. 1986b; Muta et al. 1993), which then activates proclotting enzyme to clotting enzyme (Nakamura et al. 1985; Muta et al. 1990b). The resulting clotting enzyme catalyzes the activation of coagulogen, resulting in the formation of an insoluble coagulin gel (Nakamura et al. 1976a; Miyata et al. 1984a; Fig. 4). In addition to the LPS-mediated coagulation pathway, the hemocyte lysate also responds to β-(1,3)-glucans, which are one of the major cell-wall components of fungi. A serine protease zymogen, factor G, is directly activated in the presence of β-(1,3)-glucan, and its active form activates proclotting enzyme to clotting enzyme, resulting also in the coagulin gel formation (Seki et al. 1994; Muta et al. 1995). Gram-negative bacteria and fungi which invade the horseshoe crab hemolymph are thus engulfed in the coagulin gel and are subsequently killed by antimicrobial substances released from the cells.

5.1 Coagulogen and Coagulin

Coagulogen is one of the major proteins in the large granules of the hemocytes (Mürer et al. 1975; Shigenaga et al. 1993). The protein band of coagulogen can be easily identified as a major band in the gel of SDS-PAGE of the hemocyte lysate (Fig. 3b). The amino acid sequences of coagulogens from all four species of horseshoe crab have been determined at the protein level (Fig. 5) (Nakamura et al. 1976b; Miyata et al. 1984a,b; Takagi et al. 1984; Srimal et al. 1985). cDNAs for coagulogens have also been cloned from Japanese (*Tachypleus tridentatus*) and American (*Limulus polyphemus*) horseshoe crabs (Cheng et al. 1986; Miyata et al. 1986). The coagulogens from the different species are all basic proteins composed of 175 amino acids with molecular masses of 20 kDa. The amino acid sequences of these proteins are highly conserved among the four species; the coagulogen from the American horseshoe crab has 70% identity with that from

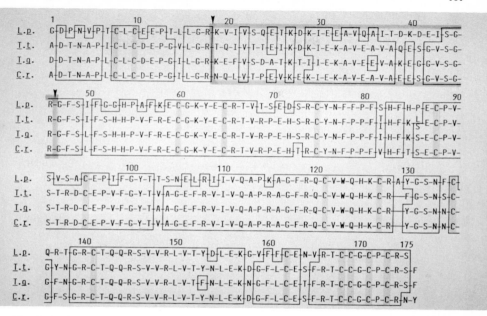

Fig. 5. Amino acid sequences of coagulogens isolated from the four species of horseshoe crab. *Arrowheads* indicate the cleavage sites in the conversion of coagulogen to coagulin gel. *L. p.* (*Limulus polyphemus,* American horseshoe crab), *T. t.* (*Tachypleus tridentatus,* Japanese), *T. g.* (*T. gigas,* Southeast Asian), *C. r.* (*Carcinoscorpius rotundicauda,* Southeast Asian)

the Japanese horseshoe crab, and the two Southeast Asian species (*T. gigas* and *Carcinoscorpius rotundicauda*) show 86% and 90% identity, respectively. In particular, 16 cysteine residues in the molecule are completely conserved among the species, suggesting the similarity of their tertiary structures.

Upon activation by clotting enzyme, two peptidyl bonds Arg^{18}–X^{19} and Arg^{46}–Gly^{47}, are cleaved in the coagulogen molecule. This results in conversion into insoluble coagulin gel, releasing a peptide of 28 residues, named peptide C (Fig. 6). The insoluble coagulin gel contains an NH_2-terminal A chain (18 residues) and a COOH-terminal B chain (129 residues), which are held by two intramolecular disulfide bonds. The resulting two-chain coagulin has a cell-agglutination activity towards rabbit erythrocytes or formalin-fixed horseshoe crab hemocytes (Fortes-Dias et al. 1993). The sequence identity of the peptide C region (50% between Japanese and American coagulogens) is lower than those of the A and B chains (73%). Similar results have been observed in fibrinogen; fibrinopeptides A and B, which are released from the NH_2-termini of Aα and Bβ chains when activated, show lower similarities among different species than the rest of the molecule. The coagulin gel is solubilized in lower (pH < 3.0) and higher pH (pH > 10.5) conditions or in the presence of 6 M urea, suggesting that there are no covalent bonds between the monomers. The exact mechanism for this gelation is still unknown. Recently the crystallization of coagulogen was accomplished (Fig. 7), and the X-ray crystallography is the subject of current research

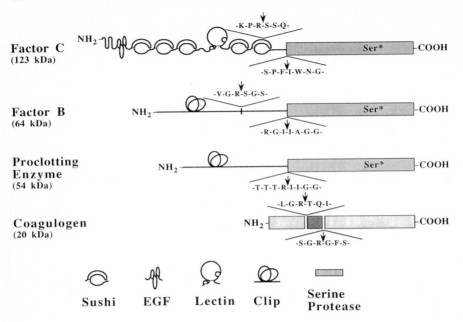

Factor C
(123 kDa)

-K-P-R-S-S-Q-

Ser*   —COOH

-S-P-F-I-W-N-G-

Factor B
(64 kDa)

-V-G-R-S-G-S-

Ser*   —COOH

-R-G-I-I-A-G-G-

**Proclotting
Enzyme**
(54 kDa)

Ser*   —COOH

-T-T-T-R-I-I-G-G-

-L-G-R-T-Q-I-

Coagulogen
(20 kDa)

-S-G-R-G-F-S-

Sushi   EGF   Lectin   Clip   Serine
Protease

Fig. 6. Gross structures of coagulation factors. Cleavage sites associated with activation are indicated by *arrows*

Fig. 7. Crystal of *T. tridentatus* coagulogen

which may provide insights into the elucidation of the molecular mechanism of the gelation.

5.2 Proclotting Enzyme and Clotting Enzyme

Synthetic chromogenic substrates having a –Gly–Arg-sequence at the P_2 and P_1 sites, such as Tos–Ile–Glu–Gly–Arg–pNA and Boc–Leu–Gly–Arg–pNA, have been developed based on the conserved sequence of the cleavage site during the transformation of coagulogen to coagulin (Iwanaga et al. 1978). By using these substrates, an active serine protease, clotting enzyme, was purified in 1982 from the hemocyte lysate activated by LPS (Nakamura et al. 1982). The zymogen form of the enzyme, proclotting enzyme, was purified three years later (Nakamura et al. 1985). Proclotting enzyme is a single-chain glycoprotein with a molecular mass of 54 kDa (Nakamura et al. 1985). Upon activation, the single-chain zymogen is cleaved to yield a light (L, 25 kDa) and a heavy (H, 31 kDa) chains held by a disulfide bond. This activation is catalyzed by factor $\bar{\text{B}}$. A higher concentration of trypsin also activates the zymogen in vitro. The resulting two-chain form clotting enzyme is an active serine protease with an active site in the H chain, which is sensitive to antithrombin-III, α_2-plasmin inhibitor, soybean trypsin inhibitor, or diisopropyl fluorophosphate (DFP). This enzyme also catalyzes the conversion of bovine prothrombin to α-thrombin, showing its similar substrate specificity to the mammalian coagulation factor Xa.

In 1990, a cDNA-encoding proclotting enzyme was cloned and sequenced (Muta et al. 1990b). The deduced amino acid sequence reveals that the COOH-terminal H chain is a typical serine protease domain with sequence similarity to the mammalian coagulation factors IXa (34.5%) or Xa (34.1%). The NH_2-terminal L-chain with pyroglutamic acid at its NH_2-terminus, contains a small compact domain with three disulfide bonds, called "clip domain", followed by a region with six O-linked carbohydrate chains. The "clip domain" has been found in horseshoe crab factor B and in *Drosophila snake* and *easter* protease precursors. *Easter* and *snake* are both indispensable proteins for the normal embryonic developments in flies (Chasan and Anderson 1989; DeLotto and Spierer 1986). Although the significance of this domain is not yet clear, the presence of this type of domain in *Drosophila* strongly suggests the existence of protease cascade systems similar to those found in the horseshoe crab in other animals. The cleavage site associated with the activation by factor $\bar{\text{B}}$, is an Arg–Ile bond in the sequence of –Thr–Thr–Thr–Thr–Arg–Ile. Threonine is located at the P_2 site of the cleavage site, which is consistent with the substrate specificity of factor $\bar{\text{B}}$ (Nakamura et al. 1986b). Furthermore, this region might be near the surface of the molecule since the Asn at the P_6 position is glycosylated. Following a hydrophobic signal sequence, the sequence preceding the mature NH_2-terminus has –Arg–X–Arg–Arg, which is unlikely to be processed by a signal peptidase. This sequence is similar to the conserved COOH-terminal sequence of propeptides found in vitamin K-dependent clotting factors or complement factors in mammals (Arg–X–Arg/Lys–Arg) (Barr 1991; Hosaka et al. 1991). The similar sequences are found in factor B or G. This fact indicates that a processing protease with Kex-2 or furin-like substrate specificity is involved in the biosynthesis of these proteins. Indeed, an mRNA encoding furin-like protease has been identified in the horseshoe crab hemocytes (unpubl. results).

5.3 Factor B and Factor $\bar{B}$

Factor B is a 64 kDa glycoprotein identified as a serine protease zymogen, whose active form (factor $\bar{B}$) activates proclotting enzyme to clotting enzyme (Nakamura et al. 1986b). The activation of factor B is catalyzed by factor $\bar{C}$. The zymogen factor B has been purified as a mixture of a single-chain form and a two-chain form composed of an L chain (26 kDa) and an H chain (40 kDa), which are held by a disulfide linkage. The activation by factor $\bar{C}$ causes a limited proteolysis in the H chain to yield the active form factor $\bar{B}$ which has a smaller H-chain (32 kDa; Fig. 6). The active form factor $\bar{B}$ is inhibited by α_2-plasmin inhibitor and is highly sensitive to DFP and benzamidine. The H chain in the factor $\bar{B}$ incorporates DFP, indicating that this chain contains the active serine residue. With synthetic substrates, this protease has a preference for an arginine residue and a hydroxyamino acid at the P_1 and the P_2 sites, respectively; it cleaves favorably Boc–Met–Thr–Arg–MCA, Bz–Thr–Thr–Arg–MCA, and Bz–Ser–Thr–Arg–MCA.

The cDNAs for factor B indicate that the amino acid sequence of this zymogen is homologous to that of proclotting enzyme, not only in the serine protease domain but also in the NH$_2$-terminal "clip domain" (Muta et al. 1993). The sequence similarity allows us to speculate that these two proteins evolved from a common ancestor by gene duplication. Among the mammalian serine proteases, human plasma prekallikrein, which is homologous to factor XI, shows the highest sequence similarity to factor B. Factor B differs from proclotting enzyme in that factor B has an insertion in the NH$_2$-terminus of the H-chain, which is released upon activation as an activation peptide with 21 amino acid residues. Like proclotting enzyme, the mature NH$_2$-terminus is preceded by a sequence of Lys–Val–Ser–Arg. This sequence also has basic residues at P_4 and P_1 positions, strongly suggesting that the protein is synthesized as a preproprotein and processed by a signal peptidase and a processing protease in the same manner that proclotting enzyme is. Most of the activations of serine protease zymogens in the mammalian coagulation and complement systems are known to occur by the cleavage of Arg–Ile/Val linkage. In contrast, the activation of factor B by factor $\bar{C}$ results in a cleavage between Ile–Ile. Due to this sequence, the zymogen factor B is activated neither by trypsin nor by chymotrypsin. This unusual activation site should prevent factor B from being activated by clotting enzyme, a typical trypsin-type enzyme, in the positive feedback manner. So far, factor $\bar{C}$ is the only known activator of factor B.

5.4 Factor C and Factor $\overline{C}$

Factor C is an LPS-sensitive serine protease zymogen. Purified factor C is a mixture of a single-chain form (123 kDa) and a two-chain form composed of an H-chain (80 kDa) and an L-chain (43 kDa) (Nakamura et al. 1986a). In the presence of LPS, both forms of factor C are autocatalytically activated to an active form, factor $\overline{C}$ composed of three chains, the H chain, an A (7.9 kDa) chain and a B (34 kDa) chain (Fig. 6). The three chains of the active form, the H, A, and B chains, are held by disulfide bonds. Being distinct from most of serine protease zymogens, the activation site in the L chain (between the A and B chains) is a Phe–Ile bond (Tokunaga et al. 1987; Muta et al. 1991). In fact, factor C is activated by chymotrypsin or chymase, a serine protease in mast cells with a chymotrypsin-like substrate specificity, but not by trypsin (Tokunaga et al. 1991). The B-chain of factor $\overline{C}$ constitutes a serine protease domain having a DFP-sensitive active serine residue. On the other hand, LPS binds to the H chain (Nakamura et al. 1988a). The active factor $\overline{C}$ has a similar substrate specificity to that of α-thrombin, efficiently hydrolyzing Boc–Val–Pro–Arg–pNA.

Factor C is also activated by lipid A, which has a core structure of LPS and is important for the expression of the endotoxin activity of LPS (Nakamura et al. 1988b). Acidic phospholipids such as phosphatidylinositol, phosphatidylserine, and cardiolipin, which have structures partly similar to lipid A, are also capable of activating the zymogen factor C. The negatively charged surfaces that activate the intrinsic pathway of the mammalian clotting cascade, such as kaolin, celite, glass, ellagic acid, sulfatide, amylose sulfate and dextran sulfate, or sulfatides, do not activate factor C at all. The activation of factor C by LPS is dependent on the concentration of LPS; however, too high concentrations of LPS inhibit the activation: there is an optimum concentration of LPS to activate factor C. The optimum concentration of LPS shifts depending on the concentration of factor C, indicating that the molar ratio of factor C and LPS is important for the activation (Nakamura et al. 1988b). This fact strongly suggests that the activation occurs through intermolecular interactions between factor C molecules bound to LPS.

The entire amino acid sequence of factor C has been revealed by its cDNA cloning (Muta et al. 1991). Factor C is a novel mosaic protein with five "sushi" domains, an EGF-like domain, a C-type lectin like-domain and a serine protease domain. In addition to these domains, a Cys-rich region and a Pro-rich sequence have also been found in the NH_2-terminal and the COOH-terminal portions of the H chain. The B chain contains a typical structure of serine proteases. Human α-thrombin shows the highest sequence identity with the factor $\overline{C}$-derived B chain (36.7%), being consistent with the substrate specificity mentioned above. The "sushi" domain, which is also called the SCR (short consensus repeat) or β_2–GP–I (β_2–glycoprotein–I) like domain, was named from the schematically illustrated domain structure based on its two disulfides, which looks like sushi, a Japanese traditional cuisine (Ichinose et al. 1990). "Sushi" domains are now found in more than 20 proteins. Among them, factor C is only the protein isolated from invertebrates. Five internal "sushi" domains are present: four in the H-chain and one in the A-chain. This type of domain has been mainly found in proteins participating in the mammalian complement system. Thus, factor C is the first example in invertebrates of a protein that has a complement-like structure. Most of the complement factors containing this domain interact with C3b or C4b (Müller-Eberhard 1988). An EGF-like domain is present in the H-chain. This type of domain has been found in several blood coagulation factors, complement factors, low density lipoprotein (LDL) receptor and other proteins. The EGF-like domain of factor C is most homologous to that in the laminin B2 chain. A lectin-like domain is found between the third and fourth "sushi" domains. This region shows similarity to so-called C-type lectin or a CRD (carbohydrate recognition domain) (Drickamer 1988). No serine protease has, until now, been known to include this type of lectin-like domain. Coagulation factor X activating enzyme from Russell's viper venom (RVV-X), which is a metalloprotease, is known to contain this type of lectin-like domain (Takeya et al. 1992). However, the lectin-like domain and the protease domain are present as distinct subunits held by a disulfide bond. At present, we do not know the function of this domain in factor C. This type of domain is also found in proteins whose functions are not considered to only bind carbohydrates, such as IgE receptor (Kikutani et al. 1986), pulmonary surfactant-associated protein (White et al. 1985) and tetranectin, which has been identified as a plasminogen kringle 4 binding protein (Fuhlendorff et al. 1987). Moreover, this domain has some sequence similarity with cartilage proteoglycan core protein (Doege et al. 1987). Two other characteristic sequences are found in the H-chain. One is a Cys-rich region (11 out of 76 residues), which is located in the NH_2-terminal portion of the H chain. The other is in the COOH-terminal portion of the H-chain. This second region contains 13 proline residues out of 48 residues and has homology with the connecting region of coagulation factor XII. The significance, however, is unknown.

Serine proteases containing "sushi" domains, except for factor C, have been found only in mammalian complement systems. Proteases shown to have both "sushi" and "EGF" domains include horseshoe crab factor C and mammalian Clr and Cls, each of which is associated with the initiation of cascade reactions. "Sushi" domains in this molecule may be sites for interaction with other proteins such as horseshoe crab factor B or other components. In 1989, three membrane proteins, E-selectin (Bevilacqua et al. 1989), L-selectin (Siegelman et al. 1989), and P-selectin (Johnston et al. 1989), have been reported to contain this type of domain. These proteins also contain an EGF-like domain and several "sushi"-like domains, although the orders of these domains along the polypeptide chains are different from those of factor C. It is also of interest that the functions of these three proteins are associated with biological defense systems.

Factor C, the initiator of the horseshoe crab coagulation cascade, is structurally related to mammalian complement factors. The initiator of this cascade, factor C, is a newly discovered type of serine protease zymogen, a "coagulation-complement factor", which may play an important role both in hemostasis and defense mechanisms. In considering the structural similarity of initiators of two systems, the horseshoe crab coagulation and mammalian complement systems may have evolved from a common origin. Concerning the evolution of host defense systems in the animal kingdom, this is consistent with a possible common evolutionary origin of the coagulation and complement cascades.

5.5 Factor G and Factor $\overline{G}$

During the diagnostic application of the limulus test, it was pointed out that positive reactions are observed with plasma of some patients even in the absence of LPS (Pearson et al. 1984). Since some of these patients suffered from fungus infection or were undergoing hemodialysis with cellulosic dialyzers, this pseudo-positive reaction had been suggested to be at least in part caused by β-(1,3)-glucans. In 1981, we and others found the presence of β-(1,3)-glucan-sensitive protease in the hemocyte lysate (Morita et al. 1981; Kakinuma et al. 1981). We named the zymogen factor G and recently succeeded in purification and cDNA cloning of factor G (Seki et al. 1994; Muta et al. 1995). The presence of β-(1,3)-glucan binding proteins, which are responsible for the activation of prophe-noloxidase, have also been reported in crayfish (Duvic and Söderhäll 1990), cockroaches (Söderhäll et al. 1988), and silkworms (Ochiai and Ashida 1988). The molecular entity of these proteins, however, still remains to be analyzed. Horseshoe crab factor G described here is the first β-(1,3)-glucan-responsive protein whose characteristics were biochemically analyzed at molecular level.

The zymogen factor G is a heterodimeric protein composed of two non-covalently-associated subunits, α (72 kDa) and β (37 kDa) (Muta et al. 1995; Fig. 8). The two subunits are derived from separate mRNA species and thus are

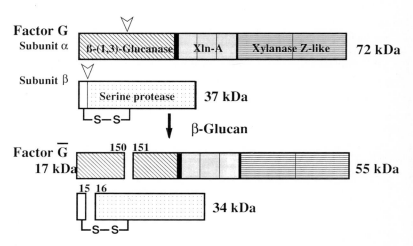

Fig. 8. Gross structure of zymogen factor G. Upon incubation with β-(1,3)-glucan, the subunits α and β are activated by limited proteolysis

encoded by different genes (Seki et al. 1994). Subunit β is a serine protease zymogen with a 31-amino-acid NH_2-terminal preprosequence preceding the 278 amino acids of a mature protein with a calculated molecular mass of 30 846 daltons; it exhibits sequence similarity to serine proteases with the catalytic triad, $His^{58}-Asp^{107}-Ser^{211}$. The substrate binding site (S_1) corresponding to Ser^{189} in chymotrypsin is Asp, suggesting a trypsin-type enzyme. Subunit β shows the strongest sequence similarity with horseshoe crab factor B (40.5% identity). This is consistent with the fact that factor G shares a similar substrate specificity toward synthetic fluorogenic peptides and that it also catalyzes the activation of proclotting enzyme (Nakamura et al. 1986b). Subunit β also shows strong sequence similarity to proclotting enzyme (37.7% identity), suggesting that the horseshoe crab clotting factors, proclotting enzyme, factor B, and factor G-subunit β, may have evolved from a common origin.

In the sequence of subunit β, there are 31 amino acid residues preceding the NH_2-terminus of the mature protein. According to the algorithm, of von Heijne (von Heijne 1983), a signal peptidase would cleave between Ala^{-18} and Leu^{-17}, resulting in a predicted propeptide of 17 amino acids. In addition, the COOH-terminal region has the sequence Arg–Val–Arg–Arg (in amino acids from −4 to −1), similar to those in proclotting enzyme and factor B, a sequence which is often found in the propeptide regions of the mammalian vitamin K-dependent clotting zymogens and other protein precursors. The presence of this sequence suggests that the mature protein of factor G subunit β may be generated in the hemocytes through a processing mechanism similar to those of proclotting enzyme and factor B.

The cDNAs of subunit α encode a 19-amino-acid signal peptide and the mature protein consisting of 654 residues with a calculated molecular mass of 73 916 daltons. Being distinct from subunit β, subunit α is a new type of mosaic protein with intriguing features. The NH_2-terminal portion of subunit α (Pro^4–Ala^{236}) shows homology to the COOH-terminal portion of β-(1,3)-glucanase Al (EC 3.2.1.39) from *Bacillus circulans* (Yahata et al. 1990), some bacterial lichenases (Murphy et al. 1984; Hofemeister et al. 1986; Borriss et al. 1990; Lloberas et al. 1991; Zverlov et al. 1991; Gosalbes et al. 1991), and the gene product of meri-5 from *Arabidopsis thaliana* (Medford et al. 1991), which contains a glucanase catalytic domain. A dot matrix plot indicates that subunit α contains two types of internal repeats. In the central portion of the subunit, there are three tandem repeats of 47 amino acids. Each repeat shows a sequence similarity with the repeats found in xylanase A (EC 3.2.1.8) from *Streptomyces lividans* (Shareck et al. 1991), *Rarobacter faecitabidus* protease I (Shimoi et al. 1992), β-(1,3)-glucanase from *Oerskovia xanthineolytica* (Shen et al. 1991), and ricin B chain (Lamb et al. 1985), as shown in Fig. 9. Because each of these three repeating units serves to recognize or bind to polysaccharides, one or all of these repeating units in factor G-subunit α may be the binding domain for β-(1,3)-glucan. The COOH-terminal portion of subunit α has two long repeats (126 amino acids in length) with 91.3% identity and is homologous to the NH_2-terminal portion of xylanase Z (EC 3.2.1.8) from *Clostridium thermocellum* (Grépinet et al. 1988) and

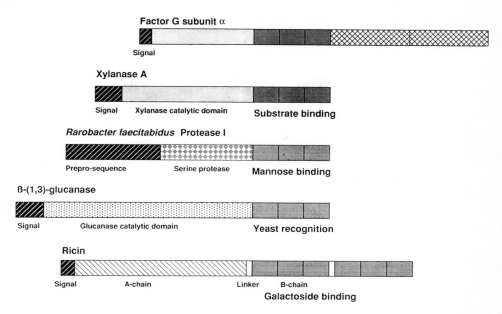

Fig. 9. Structure of the carbohydrate binding proteins. The structures of the proteins containing the three tandem repeats *(shaded)* in the middle of factor G-subunit α are shown. The tandem repeat structures are responsible for carbohydrate binding in xylanase A from *Streptomyces lividans* (Shareck et al. 1991), *Rarobacter faecitabidus* protease I (Shimoi et al. 1992), β-(1,3)-glucanase from *Oerskovia xanthineolytica* (Shen et al. 1991), and ricin B chain. (Lamb et al. 1985)

endo-1,4-β-xylanase from *B. polymyxa* (Gosalbes et al. 1991). The function of this NH_2-terminal part of xylanase Z is so far unknown. This portion of subunit α contains two cysteine residues, Cys^{421} and Cys^{559}, and their positions along the sequences are conserved in these two repeats but not in xylanase Z.

The purified factor G is easily activated by other various glucans containing β-(1,3) linkages from different origins, but it is not activated by LPS, sulfatides, or cholesterol sulfates (Muta et al. 1995). The most effective activators which have been examined are linear β-(1,3)-glucans, such as curdlan and paramylon (Table 2). As little as 1 ng of curdlan significantly activates the zymogen factor G. Branching of the linear chain with (1,4) or (1,6)-β-linkages appears to reduce the factor G activating activity. Shorter oligosaccharides containing two to six glucose residues do not activate factor G at all. There is an optimum concentration of curdlan to activate factor G (Muta et al. 1995). When the concentration of factor G is ten fold higher or lower, the optimum concentration for the activation shifts approximately 10-fold, depending on the concentration of factor G, implying that the molar ratio of factor G and β-(1,3)-glucan is important in the zymogen activation. These kinetics indicate that the activation of factor G occurs through an intermolecular interaction with each factor G molecule bound to β-glucan. Similar phenomena are observed in the activation of horseshoe crab factor C by LPS (Nakamura et al. 1988b) and mammalian

Table 2. Activation of horseshoe crab factor G by various glucans

Saccharides	Type of linkages	Acti-vation	ED_{50}[b] (g/ml)
D-Glucans			
Laminaran oligosac-charides (d.p.[a] 2–7)	$(1{\to}3)$-β-D	–	
Curdlan	$(1{\to}3)$-β-D	+	4×10^{-9}
Carboxymethylcurdlan	$(1{\to}3)$-β-D	+	$(10^{-5}, 35\%)$[c]
Paramylon	$(1{\to}3)$-β-D	+	4×10^{-8}
Laminaria digitata laminarin	$(1{\to}6), (1{\to}3)$-β-D	+	$(10^{-9}, {<}10\%)$[c]
Schizophyllan	$(1{\to}6), (1{\to}3)$-β-D	+	$(10^{-6}, 25\%)$[c]
Lentinan	$(1{\to}6), (1{\to}3)$-β-D	+	3×10^{-6}
Baker's yeast β-D-glucan	$(1{\to}6), (1{\to}3)$-β-D	+	4×10^{-8}
Cetraria islandica lichenan	$(1{\to}3), (1{\to}4)$-β-D	+	3×10^{-7}
Usnea barbata lichenan	$(1{\to}3), (1{\to}4)$-β-D	+	1×10^{-7}
Barley β-glucan	$(1{\to}3), (1{\to}4)$-β-D	+	5×10^{-8}
Nigeran	$(1{\to}3), (1{\to}4)$-β-D	+	$(10^{-4}, {<}10\%)$[c]
Krestin	$(1{\to}4)$-α,$(1{\to}4)$-β-D	+	4×10^{-8}
Yeast α-D-mannan	$(1{\to}2),(1{\to}3),(1{\to}6)$-$\alpha$-D	+	4×10^{-4}
$(1{\to}3)$-β-D-galactan	$(1{\to}3)$-β-D	+	2×10^{-4}
Oat-Spelt $(1{\to}4)$-β-D-xylan	$(1{\to}4)$-β-D	+	4×10^{-5}

[a]Degree of polymerization.
[b]Minimal concentration (ED_{50}) of each saccharide required to activate 50% of the purified factor G (0.32 μg/ml) is shown.
[c]Less than 50% of factor G was activated in the range from 10^{-9} to 10^{-4} g/ml. The optimum concentration and the amount of activated factor G at the optimum are shown.

coagulation factor XII by a negatively-charged surface (Sugo et al. 1985). Upon activation of factor G, both subunits α and β exhibit structural changes by limited proteolyses (Muta et al. 1995) (Fig. 8). The 72 kDa subunit α is converted to a 55 kDa and a 17 kDa fragment, and the 37 kDa subunit β is converted to a 34 kDa fragment. In both cases, the COOH-terminal sides of arginine residues are cleaved, being consistent with the substrate specificity of factor $\bar{G}$. The cleavage site in subunit α is between Arg^{150} and Glu^{151}, in the middle of the glucanase Al-like domain. An Arg^{15}–Ile^{16} bond in subunit β is cleaved to form an active serine protease with a 15-amino-acid lig ht chain linked through a disulfide bond. A longer incubation with β-(1,3)-glucan causes a fragmentation of the 55 kDa fragment to a 46 kDa fragment, being concomitant with a loss of the amidase activity. Although the activation of factor G is associated with limited proteolyses of the carboxyl side of the arginine residue, neither factor $\bar{C}$, factor $\bar{B}$ nor clotting enzyme activate the zymogen factor G. In addition to these coagulation factors, digestive proteases, such as trypsin or chymotrypsin, also do not activate factor G. This fact implies that the activation site in subunit β is sterically hindered, possibly by subunit α. β-Glucan-binding to subunit α may open that site in the subunit β, resulting in autocatalytic activation through intermolecular interaction between the subunit βs. The active subunit β then quickly hydrolyzes the Arg^{150}–Glu^{151} bond in subunit α. Then, another site in the same subunit is cleaved, which leads to the inactivation of the protease activity. Since this cleavage in the subunit α appears to reduce the amidolytic activity, subunit α is likely to regulate the catalytic activity of the subunit β.

Reconstitution experiments using purified proteins demonstrated that factor $\overline{G}$ is capable of activating proclotting enzyme directly, without the presence of any other proteins (Muta et al. 1995). In another set of experiments, the β-(1,3)-glucan-mediated gelation is reconstituted with factor G, proclotting enzyme, and coagulogen, thus indicating that factor G, proclotting enzyme and coagulogen are the minimum requirements for the coagulation, and that factor $\overline{G}$ induces clot formation through the activation of proclotting enzyme followed by the activation of coagulogen. Horseshoe crab hemocytes are known to contain two types of serpin-type serine protease inhibitors, LICI-1 and -2 (see below). Whereas LICI-1 has no effect on factor $\overline{G}$, LICI-2 strongly inhibits the factor $\overline{G}$ amidase activity (Muta et al. 1995).

6 Protease Inhibitors

6.1 Anticoagulants

6.1.1 Limulus Intracellular Coagulation Inhibitor, Types 1 and 2 (LICI-1 and -2)

At least three types of factor $\overline{C}$ inhibitors are present in the hemocyte. Two of them have been, so far, purified and characterized and named Limulus intracellular coagulation inhibitor (LICI) types 1 and 2 (Miura et al. 1994, 1995). Both LICI-1 and -2 belong, structurally and functionally, to the serpin family of serine protease inhibitors. They inhibit amidase activity of factor $\overline{C}$ by forming a covalent 1:1 complex. The second-order rate constants of LICIs-1 and -2 for factor $\overline{C}$ are $2.5 \times 10^6 \ M^{-1}s^{-1}$ and $7.1 \times 10^4 \ M^{-1}s^{-1}$, respectively. LICI-2 inhibits clotting enzyme and factor $\overline{G}$ as well as factor $\overline{C}$, whereas LICI-1 is specific to factor $\overline{C}$. Glycosaminoglycans such as heparin and heparan sulfate have no effect on the inhibitory activity, being clearly different from the case of antithrombin III.

LICI-1 is a single-chain glycoprotein consisting of 394 amino acids and has an apparent molecular mass of 48 kDa. LICI-2 is also a single-chain glycoprotein composed of 386 amino acids. Its apparent molecular mass is 42 kDa. In the tissues which have been examined the mRNAs of LICIs are expressed only in hemocytes: the mRNAs are not detected in heart, hepatopancreas, brain, stomach, intestine, skeletal muscle or coxal gland. LICIs are stored specifically in the large granules of the hemocytes, and are released in response to external stimuli. LICIs show significant sequence identities to members of the intracellular serpin superfamily, such as human plasminogen activator inhibitor type 2 and human monocyte/neutrophil elastase inhibitor. LICI-1 contains a putative reactive site, Arg^{359}–Ser^{360}, at the corresponding position present in several inhibitors of the serpin superfamily. While LICI-2 shows a significant sequence similarity to the previous horseshoe crab serpin, LICI-1 (42% identity), LICI-2 contains a different putative reactive site, Lys^{350}–Ser^{351}. The phylogenic tree for 12 serpins is shown in Fig. 10. Phylogenically LICIs could well be placed near the position of ovalbumin and an intracellular serpin branch rather than insect serpins or other mammalian serpins.

Of intracellular serpins, the entire amino acid sequences of human leukocyte/monocyte elastase inhibitor (Remold-O'Donnell et al. 1992), equine elastase inhibitor (Dubin et al. 1992), porcine leukocyte neutral protease inhibitor (Teschauer et al. 1993), placental thrombin inhibitor (Coughlin et al. 1993; Morgenstern et al. 1993), and maspin (Zou et al. 1994) have been reported. The key features of these intracellular serpins are the lack of signal sequences destined for the endoplasmic reticulum and the presence of an oxidation-sensitive residue, methionine, in proximity to their reactive sites (Coughlin et al. 1993). LICI-2 containing a methionine residue at the position of the P_3 site also has a higher sensitivity to N-cholorosuccinimide, and the oxidation inactivates the LICI-2 inhibitory activity. Although the physiological significance of the labile methionine residue remains to be defined, these intracellular inhibitors may be involved in the regulation of intracellular protein degradation, cell apoptosis and tumour suppression. In despite of the structural similarity between LICIs and these intracellular serpins, both LICI-1 and -2 have typical NH_2-terminal signal peptides and are stored in large granules. Thus, they may be coreleased with several coagulation factors and bactericidal peptides, in response to an external LPS stimulation. By contrast, mammalian intracellular serpins are localized in

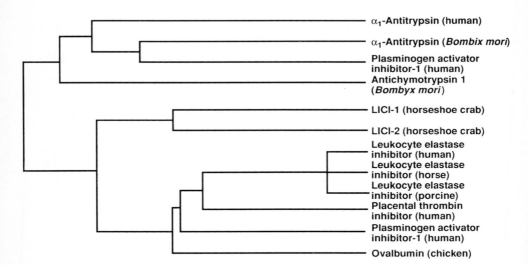

Fig. 10. Phylogenic tree of serpins

the cytosolic fraction and are not secreted into plasma or culture media (Dubin et al. 1992; Morgenstern et al. 1993), except for PAI-2 (Belin et al. 1989). A part of PAI-2 enters the endoplasmic reticulum through the NH_2-terminal hydrophobic region of the molecule and is consitutively secreted into the blood circulation (von Heijne et al. 1991). Therefore, in terms of subcellular localization, LICIs could be classified into a new subfamily of intracellular serpins, a regulated secretory serpin.

6.2 Trypsin Inhibitor in the Hemocytes

A Kunitz-type protease inhibitor with 61 amino acids is present in the hemocytes (Nakamura et al. 1987). It inhibits trypsin ($Ki = 4.60 \times 10^{-10}$M), α-chymotrypsin ($Ki = 5.54 \times 10^{-9}$M), and elastase ($Ki = 7.20 \times 10^{-8}$M). It also inhibits plasmin and plasma kallikrein, but shows no effect on horseshoe crab factor $\overline{C}$ or clotting enzyme. The amino acid sequence of this inhibitor has a high degree of identity with those isolated from sea anemone inhibitor 5-II (47%) and bovine pancreatic trypsin inhibitor (43%).

Two other proteins with trypsin-inhibitory activity have been found in American horseshoe crabs (*L. polyphemus*). One is called *Limulus* endotoxin-binding protein-protease inhibitor (LEBP-PI) since it has ability to bind LPS (Minetti et al. 1991). LEBP-PI, purified by an LPS-affinity column, binds LPS with a *Kd* of the order of 10^{-7} M. The binding is competed by polymyxin B, lipid A and D-glucosamine. It inhibits trypsin with an IC_{50} value of approximately 1×10^{-7} M. This protein is composed of 105 amino acids with a calculated molecular mass of 12 289 daltons, although it shows an apparent molecular mass of 8000 daltons on SDS-PAGE. The sequence of LEBP-PI exhibits no homology with any other known proteins including mammalian and invertebrate LPS-binding proteins and trypsin inhibitors. It localizes in the larger granules of the hemocytes and is released after the stimulation of LPS.

The other protein, called *Limulus* trypsin inhibitor (LTI), has been copurified with coagulogen and inhibits trypsin with a *Ki* value of 3.3×10^{-7}M (Donovan and Laue 1991). LTI also inhibits chymotrypsin, but does not affect the horseshoe crab clotting factors. LTI has an apparent molecular mass of 15 kDa on SDS-PAGE. Although its NH_2-terminal sequence has been determined, there is no sequence similarity with other known trypsin inhibitors.

6.3 Inhibitor in Plasma

6.3.1 α_2-Macroglobulin

Although the hemolymph plasma of horseshoe crab is presumed to contain several proteinase inhibitors, only one of them is well characterized. The inhibitor purified from the American horseshoe crab is a homologue of vertebrate

α_2-macroglobulin (α_2M) (Quigley and Armstrong 1994). Its concentration in plasma is approximately 2 mg/ml. Since it is present in the hemocytes, that may be where it is synthesized. The molecular mass of the monomer is 185 kDa, similar to that of human α_2M (Armstrong et al. 1991). In contrast to human α_2M, which is a homotetramer, it is composed of homodimer with a native molecular mass of 360 kDa. It inhibits a wide range of proteolytic enzymes including trypsin, chymotrypsin, elastase, subtilisin, thermolysin, and papain. These broad specificities against protease are a characteristic of α_2M. The structural analysis of the *Limulus* α_2M is now undergoing cDNA cloning in our laboratory. The overall structure of *Limulus* α_2M is very similar to that of human α_2M. *Limulus* α_2M, indeed, contains a thiol ester bond detected in human α_2M and complement factors such as C3 and C4. The thiol ester is sensitive to primary amines such as methylamine. The exposure of *Limulus* α_2M to methylamine results in the loss of the inhibitory activity.

Hemocyanin-depleted horseshoe crab plasma exhibits cytolytic activity on exogenous erythrocytes (Enghild et al. 1990). The molecular mechanism for this reaction remains to be elucidated. Studies on such a defense system, however, should be very interesting, since the reaction might involve recognition process of self and nonself materials. *Limulus* α_2M appears to be essential in this reaction, since methylamine treatment of the hemolymph leads to inactivation of this cytolytic activity (Enghild et al. 1990). The activity is restored by the addition of purified *Limulus* α_2M, but not by methylamine-treated *Limulus* α_2M. This cytolytic reaction seems to require C-reactive protein in the plasma (Armstrong et al. 1993). *Limulus* α_2M may bind to exogenous cell surfaces through its internal thiol ester, just like vertebrates C3 and C4 which have indeed sequence similarities to those of *Limulus* and human α_2M. Thus, the complement factors may have evolved from a common origin with α_2Ms.

7 Transglutaminase and Its Substrates

7.1 Transglutaminase

It is well known that in the mammalian clotting system, the fibrin clot generated via the cascade reaction is finally cross-linked to form a huge fibrin network in addition to being cross-linked to other plasma proteins, such as soluble fibronectin, α_2-macroglobulin and α_2-plasmin inhibitor. This final step is catalyzed by plasma transglutaminase (TGase), factor XIIIa, and the cross-linking of fibrin with itself and with other proteins is essential for normal hemostasis and wound healing (Lorand et al. 1981). It is, therefore, expected that in the horseshoe crab clotting system a TGase might participate in the cross-linking of the coagulin gel and/or in the immobilization against invading microorganisms. Indeed, in 1973 and later, two groups reported the existence of Ca^{2+}-dependent TGase in *L. polyphemus* hemocytes (Campbell-Wilkes 1973; Chung et al. 1977).

However, Roth et al. (1989) reported that TGase activity could not be detected in the *Limulus* lysate or gelled lysate, and that the coagulin gel formed after coagulation of the lysate mediated by LPS was not covalently cross-linked. This controversy was settled in 1993 by the purification and cDNA cloning of a TGase from the horseshoe crab hemocytes (Tokunaga et al. 1993a,b). The discrepancy was possibly caused by instability of the enzyme. Horseshoe crab TGase is sensitive to NaCl and KCl, probably to chloride ions, although no such effect has been reported in any other TGase from mammals or microorganisms. This instability of the TGase in the lysate was dealt with by a dialysis procedure applied to freshly prepared extract from the hemocytes. The procedure made it possible to purify the horseshoe crab TGase from the lysate. In invertebrates, hemolymphs of horseshoe crab, lobster (Fuller and Doolittle 1971a,b), sand crab (Madaras et al. 1979), and sponge (Brozen et al. 1987) have been reported to contain TGase activity. However, to our knowledge, this is the first example of the purification and characterization of an invertebrate TGase.

The purified TGase shows an apparent molecular mass of 86 kDa and has mammalian-type II (cytosolic) TGase-like enzymatic properties (Tokunaga et al. 1993a). The activity is Ca^{2+}-dependent and is inhibited by primary amines, EDTA and SH-reagents. Furthermore, the TGase is not inhibited by GTP, ATP and CTP, unlike some of mammalian TGases. In a similar fashion to guinea pig liver TGase, horseshoe crab TGase does not exist as a zymogen form, in contrast to plasma factor XIII.

Two cDNAs for horseshoe crab TGase have been cloned: these differ only in three amino acid residues suggesting the existence of isoforms (Tokunaga et al. 1993b). Both cDNAs encode 764 residues of mature TGase. The mRNA of TGase is observed in hemocyte, hepatopancreas, and stomach, but not in heart, intestine and muscle. Horseshoe crab TGase shows significant sequence identity with members of mammalian TGase family, in the catalytic region as follows: guinea pig liver TGase (32.7%), human factor XIII a subunit (36.9%), human keratinocyte TGase (39.9%), and human erythrocyte band 4.2 (23.7%). In addition to the catalytic domain, horseshoe crab TGase has a unique NH_2-terminal cationic 60-residues extension with no homology to other mammalian TGases or any other known proteins. The genealogical tree of known TGases suggests that horseshoe crab (invertebrate) TGases and mammalian (vertebrate) TGases are divergent in the oldest root.

Although two potential TGase substrates have been found in hemocyte lysate (see below), the physiological function of these substrates is not clear (Tokunaga et al. 1993a). No dansylcadaverine (DCA) is incorporated into hemocyte membrane proteins, hemolymph plasma proteins, or coagulogen. It should be noted, however, that if large amounts of LPS are added to the hemocyte lysate leading to clot formation, the protein band of coagulogen disappears and new bands of 6 and 40 kDa appear on SDS-PAGE; these are labeled with DCA. Coagulin derived from coagulogen after clot formation could be cross-linked by TGase with some cell-wall components of the invading microorganisms.

7.2 Intracellular Substrates of Transglutaminase

When the dialyzed hemocyte lysate is incubated in the presence of Ca^{2+}, dithio-threitol, and DCA, two major proteins are found to incorporate DCA by the endogenous TGase activity in the lysate (Tokunaga et al. 1993a). A rapid (<5 min) incorporation of DCA is observed with 10 and 80 kDa bands, and a slow incorporation (<60 min) is observed in a 20-kDa protein. Of these protein substrates, an 80 kDa protein contains a large number of proline residues, amounting to about 22% of the total amino acids. It is, thus, called proline-rich protein and its cDNA cloning is now the subject of research. According to the partial sequence, this protein contains many internal repeat structures each of which is rich in proline. On the other hand, the 10 kDa protein, which is also identified as an intracellular substrate for TGase, consists of 81 amino acid residues with a calculated molecular weight of 8671, thus it has been named 8.6 kDa protein (Tokunaga et al. 1993a). It is a Cys-rich protein consisting of 14 half-cystine residues. The 8.6 kDa protein is readily intermolecularly cross-linked by TGase, forming multimers as large as pentamers. A significant sequence similarity is observed between the 8.6 kDa protein and fulvocin C, a bacteriocin isolated from *Myxococcus fulvus,* which consists of a total of 45 amino acid residues (Tsai and Hirsch 1981). However, the 8.6 kDa protein does not show any bactericidal activities (*Escherichia coli* K12).

Both proline-rich protein and 8.6 kDa protein are major proteins in the large granules of the hemocyte, whereas horseshoe crab TGase is cytosolic type. Therefore, there is no possibility that the TGase and those potential substrates could interact with each other. If such TGase is secreted as is the case of factor XIII, they would come in contact in degranulation of the hemocyte. It is known that a cross-linking of melittin by TGase increases its hemolytic activity three fold (Perez-Paya et al. 1991) and that cross-linking of phospholipase A_2 also results in an enhancement of enzyme activity (Cordella-Miele et al. 1990). Therefore, a multimer of the 8.6 kDa protein cross-linked by TGase may have a bactericidal activity like fulvocin. Furthermore, the possibility could not be ruled out that TGase may catalyze cross-linking of coagulin or microbial cell wall with other proteins.

8 Antibacterial Substances

In the course of the purification of LPS-sensitive coagulation factors, two types of LPS binding substances have been identified in the horseshoe crab hemocytes (Iwanaga et al. 1994b). The subsequent studies have also demonstrated that the hemocyte granules contain an unexpectedly large number of antibactericidal peptides and proteins. These include anti-LPS factor, the tachyplesin family, big defensin, factor D and L6 (Table 1).

8.1 Antilipopolysaccharide Factor (ALF)

Antilipopolysaccharide factor (ALF) has been purified from Japanese and American horseshoe crab hemocytes, and their primary structures have been determined (Tanaka et al. 1982; Morita et al. 1985; Aketagawa et al. 1986; Muta et al. 1987). It is a single-chain polypeptide of 101 or 102 amino acids with a molecular mass of 12 kDa. There is a single disulfide linkage along the chain. ALF shows strong inhibitory activities on the growth of Gram-negative bacterial (Morita et al. 1985). This activity is expressed probably through the interaction with the lipid A portion of the LPS molecule, since it has no activity on Gram-positive bacteria. The inhibitory activity against Gram-negative bacteria is dependent on the length of O-antigen polysaccharide chains; the shorter chain shows stronger inhibitory activity. Indeed, ALF inhibits the activation of factor C by synthetic lipid A. Two cationic amino acid clusters in the ALF molecule may be important for the interaction with phosphate groups in lipid A (Aketagawa et al. 1986; Muta et al. 1987). Since ALF shows cytolytic activity on mammalian cells presensitized with LPS (Ohashi et al. 1984), the growth inhibitory activity on Gram-negative bacteria is very probably caused by membrane perturbation. The NH_2-terminal region of ALF is hydrophobic and may be inserted into the hydrophobic lipid bilayer of the target cell. Recently, the 3D structure of ALF has been determined by X-ray crystallography (Hoess et al. 1993); the structure is consistent with the hypotheses mentioned above. ALF is localized in the large granules of hemocytes and would be released with coagulation factors in the degranulation caused by LPS (Toh et al. 1991).

8.2 Tachyplesin Family

Tachyplesin is a cationic peptide composed of 17 amino acids and localized in the membrane fraction from hemocyte debris (Nakamura et al. 1988c). Recently, we have found that tachyplesin and its analogues are located exclusively in the smaller granules (Shigenaga et al. 1990; Shigenaga et al. 1993). Several isoforms have also been identified in tachyplesin; tachyplesins I and II from Japanese horseshoe crabs, tachyplesins I and III from Southeast Asian species, and polyphemusins I and II from American species (Nakamura et al. 1988c; Miyata et al. 1989; Muta et al. 1990a). All tachyplesins consist of 17 amino acids and all polyphemusins consist of 18 amino acids. They contain five to six lysines and arginines but no acidic residues. Its COOH-terminal arginine is α-amidated. cDNA sequences for tachyplesins I and II indicate that tachyplesin is synthesized as a larger precursor (Shigenaga et al. 1990; Fig. 11). The precursor contains an NH_2-terminal hydrophobic signal sequence and a possible processing sequence, –Glu–Ala–Glu–Ala–, followed by the mature tachyplesin, COOH-terminal processing and amidation signal, –Gly–Lys–Arg–, and a COOH-terminal 34 extension peptide with an unknown function. The COOH-terminal extension peptide

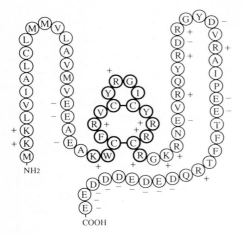

Fig. 11. Structure of *T. tridentaus* tachyplesin precursor

is highly acidic, in contrast to the mature form. In particular, the COOH-terminal 9 residues are all acidic. This extension peptide may be neutralizing cationic charges of the mature tachyplesin in order to protect the granule membrane. The 3D structure of the mature form of tachyplesin I has been determined by ^{1}H NMR study (Kawano et al. 1990). Tachyplesin has an antiparallel β-sheet structure with a β-turn. Positive charges are clustered at a side of the plane formed by the β-sheet. The other side of the plane is composed of hydrophobic side chains of amino acids. Thus, it has an amphiphilic nature, which seems important for antimicrobial activities.

Tachyplesin binds LPS and has strong antimicrobial activity against Gram-negative bacteria as ALF does (Nakamura et al. 1988c). However, in contrast to ALF, the spectrum of antimicrobial activity is much broader and also shows growth inhibitory activity against Gram-positive bacteria and fungi. The tachyplesin content in the hemocytes is extremely high. More than 80% of proteins in the smaller granule are tachyplesin (Shigenaga et al. 1993); a single horseshoe crab contains approximately 10 mg of tachyplesin (Muta et al. 1990a). Thus, tachyplesin appears to be a major antimicrobial substance for this animal. In the presence of such a high concentration of tachyplesin, the LPS-mediated factor C activation could be inhibited. However, the electron microscopic study indicates that the smaller granules are released a little later than the large granules (Toh et al. 1991). Therefore, the exocytosis of small granules containing tachyplesin occurs after the activation of the LPS-sensitive coagulation factors in the larger granules, and so would not affect the activation of the cascade reaction.

8.3 Big Defensin

During the characterization of protein components in separated large and small granules of the hemocyte, we recently found an additional antibactericidal protein in both large and small granules (Shigenaga et al. 1993). The structural analysis of this protein indicates that the COOH-terminal half of a 79 amino-acid mature protein has a sequence similar to those of defensins found both in mammals and insects. Since it contains an NH_2-terminal hydrophobic extension having no homology with any other proteins, it was named big defensin (Saito et al. 1955b). It inhibits not only the growth of Gram-negative and Gram-positive bacteria, but also fungi, as do tachyplesins. The NH_2-terminal hydrophobic extension can be separated by trypsin digestion from the COOH-terminal cationic defensin region. Surprisingly, each of them shows a different spectrum of antimicrobial activities; the NH_2-terminal region shows growth inhibitory activity against Gram-positive bacteria and the COOH-terminal portion against Gram-negative bacteria. Big defensin, therefore, may prove to be a new class of defensin family, a chimeric or hybrid antimicrobial peptide, possessing two functional domains with different antimicrobial activities. cDNA sequence analysis indicates that a precursor of big defensin contains a signal sequence followed by the mature protein sequence, giving no information on the sorting mechanism of the two granules in biosynthesis (unpublished results). It is of interest to study whether the proteolytic processing of big defensin into the two NH_2-terminal and COOH-terminal fragments is mediated by horseshoe crab proteases or by proteases from infective microorganisms, after degranulation.

9 Agglutinins/Lectins

9.1 Plasma Agglutinins

The presence of hemagglutinin activity in the horseshoe crab hemolymph was first reported by Noguchi at the beginning of this century (Noguchi 1903). Thereafter, several lectins have been purified from the hemolymph. More recently, several agglutinins have been isolated as single components from the hemocytes, which are stored in the large granules. These lectins, after degranulation of the hemocytes, are likely to play roles in the aggregation of invading microorganisms and possibly in opsonization of foreign cells. In mammalian lectins, two major families are well-known: Ca^{2+}-dependent (C-type) lectins and SH-dependent (S-type) lectins (Drickamer 1988). However, none of the horseshoe crab agglutinins, whose primary structures have been determined, fall into these two categories. The structural analyses of horseshoe crab lectins are of interest because they may form a new lectin family.

9.1.1 Limulin/C-Reactive Protein/Carcinoscorpin

Since Noguchi's first report, several groups have studied the hemagglutinin activity in the hemolymph. Limulin isolated from American horseshoe crab is a Ca^{2+}-dependent lectin that binds sialic acid (N-acetylneuraminic acid) and 2-keto-3-deoxyoctonate (KDO) (Cohen 1968; Marchalonis and Edelman 1968;

Finstad et al. 1974; Vaith et al. 1979). It has a high affinity for homopolymers of
N-acetylneuraminic acid. Since its partial NH_2-terminal amino acid sequence
(Kaplan et al. 1977) has extensive identity (90.5%) to that of C-reactive protein
(CRP) subsequently isolated from the same species, they are likely to be polymorphic forms of the same protein family. As is well known, mammalian CRP is
an acute phase protein that binds phosphorylcholine in the C-polysaccharide of
the pneumococcal cell wall. It is induced by interleukin-6 in the liver. On the other
hand, horseshoe crab CRP is constitutively present as a major glycoprotein in the
hemolymph plasma at a concentration of 1–5 mg/ml (Robey and Liu 1981). The
native limulin/CRP has a molecular mass of more than 500 kDa and is composed
of a homohexamer of approximately 25-kDa subunits (Fernández-Morán et al.
1968, Osmand et al. 1977). They have polymorphic forms; three closely related
genes for CRP have been cloned, which are distinct from the limulin (Nguyen et
al. 1986a,b). The genes for horseshoe crab CRP encodes a 24-amino-acid signal
peptide and a mature protein of 218 amino acids. The sequence identity with
human CRP is about 25%. A sialic-acid-binding lectin, which has similar
characteristics to limulin/CRP but is antigenically distinct, has also been isolated
from Southeast Asian horseshoe crab and named carcinoscorpin (Bishayee and
Dorai 1980). This lectin binds O-(N-acetylneuraminyl) (2→6)2-acetamido-2-
deoxy-D-galacitol with a Ka of 1.82×10^3 M^{-1}. Its native form has a molecular
mass of 420 kDa composed of two different subunits of 27 and 28 kDa.

9.1.2 Polyphemin

A *Staphylococcus aureus* agglutinating protein, which is distinct from the sialic-
acid-binding lectin described above, has been isolated by an affinity chromatography method using a teichoic acid-immobilized column (Brandin and Pistole
1983). The lectin was termed polyphemin, because it was purified from the
hemolymph plasma of American horseshoe crab, *L. polyphemus*. It requires N-
acetylglucosamine for binding to teichoic acid. However, N-acetylglucosamine
monomers do not inhibit the binding. The covalent structural analysis and
further precise characterization remain to be performed. Polyphemin may play a
role in the defense mechanism of horseshoe crabs against Gram-positive bacteria, which has teichoic acid on the surface.

9.1.3 Lectins from Japanese Horseshoe Crab
(Tachypleus tridentatus)

The hemolymph plasma of the Japanese horseshoe crab contains several different
kinds of lectins (Shimizu et al. 1977). The hemolymph plasma shows hemagglutinating activity on human O-erythrocytes, which is inhibited by N-
acetylneuraminic acid, N-acetyl-D-glucosamine or N-acetyl-D-galactosamine,
but not by glucose, galactose, mannose, fucose, glucosamine or galactosamine.
The bovine submaxillary gland mucin-affinity column chromatography of the

plasma results in fractionation of a major lectin and three minor lectins, M1, M2, M3, and M4. M1 and M2 are eluted by 0.1 M NaCl in 0.1 M sodium borate (pH 8.5); M3, the major lectin, by 50 mM N-acetyl-D-glucosamine in Tris-buffered saline; and M4 by 1 M NaCl in 50 mM Tris-HCl (pH 7). M3, the major component showing the strongest lectin activity was further purified by N-acetyl-D-galactosamine-Sepharose, which is eluted by 50 mM N-acetyl-D-glucosamine. On SDS-PAGE analysis under reducing conditions, M2 and M3 give a single band with 20 and 23 kDa, respectively, although their native molecular masses are much larger. These lectins have optimal activities at about neutral pH and are heat-sensitive; over 80% of the activity is lost when treated at 60 °C for 10 min. Although $CaCl_2$ enhances the lectin activities of M1, M2, and M4, M3 is not affected by the metal ions. The hemagglutinating activities of these four lectins are inhibited by N-acetylneuraminic acid, N-acetyl-D-glucosamine, N-acetyl-D-galactosamine, N-acetyl-D-mannosamine, N-acetyl-D-muramic acid, fetuin, and mucins from various species. Glucosamine, galactosamine, glucose, galactose, mannose, fucose, and glucuronic acid show no effect as in the case of crude plasma. Thus, these lectins specifically recognize N-acetylamino sugar.

9.2 Intracellular Agglutinins

9.2.1 LPS-Binding Protein (L6)

A new type of lectin has been purified from hemocyte by using an agarose-matrix column (Saito et al. 1995a). This novel lectin contains Zn^{2+} and requires metal ions for carbohydrate binding, since it is eluted from the column with EDTA or O-phenanthroline as well as with monosaccharides, such as glucose, mannose, and galactose. It agglutinates sheep erythrocytes coated with Re-LPS, but shows no hemagglutinin activity for sheep, rabbit, or human red blood cells. However, it does not inhibit the LPS-mediated activation of factor C, suggesting that it binds an external polysaccharide containing an O-specific chain and/or a core region such as 2-keto-3-deoxy-D-manno-octanate (KDO). It has significant antibacterial activity against Gram-negative bacteria but not against Gram-positive bacteria and fungi. The NH_2-terminal sequence of this protein indicates that it is identical to one of the major proteins found in the larger granule previously designated L6 (Shigenaga et al. 1993; Saito et al. 1995a). The entire amino acid sequence of L6 reveals that it consists of 221 residues with no N-linked sugar and is composed of six tandem repeats, each of which consists of 33–38 amino acid residues with 32–61% internal sequence identities. It contains three intrachain disulfide bonds and one free cysteine. The sequence of L6 does not show any similarity to other proteins, including various animal lectins and LPS-binding proteins, thus indicating a novel protein with a lectin-activity. It is very likely that it also participates in the defense system of the animal, since it could be released from the large granule in response to LPS, together with other coagulation factors and antimicrobial substances.

9.2.2 Limunectin

A protein of 54 kDa, designated limunectin, that binds phosphorylcholine has been purified from the hemocytes (Liu et al. 1991). It is distinct from limulin/CRP in that its binding ability is fully active even in the absence of Ca^{2+}. The amino acid sequence deduced from its gene structure reveals that it contains 10 internal repeats consisting of 45 amino acids. Limunectin shows partial sequence similarity to those of vitronectin, gelatinase and collagenase. It binds extracellular matrix molecules like collagen, heparin, and fibronectin as well as fixed cells including E. coli and S. aureus, and the horseshoe crab hemocyte itself. It localizes in the larger granule and may act as an adhesion molecule mediating cell-to-cell and/or cell-to-matrix interaction after exocytosis induced by LPS.

9.2.3 Limulus 18K Agglutination-Aggregation Factor (18K-LAF)

An 18-kDa hemagglutinin has been isolated from American horseshoe crabs, L. polyphemus (Fujii et al. 1992). It shows agglutinating activity not only for horse erythrocyte but also for the horseshoe crab hemocyte itself, hence it has been named Limulus 18K agglutination-aggregation factor (18K-LAF). The hemagglutinating activity is not inhibited by glucosamine, N-acetylglucosamine, N-acetylneuraminic acid, glucuronic, or galacturonic acids, thus indicating a different protein from limulin/CRP. The whole sequence of 153 amino acids of 18K-LAF shows three internal repeats, each of which contains two highly conserved stretches of Val–Asn–Asp/Ser–Trp–Asp and Glu–Asp–Arg–Arg–Trp. Ten half-cystines in the molecule form five intrachain disulfide bonds. This protein shows overall sequence similarity to that of a 22-kDa extracellular matrix protein of mammalians (Neame et al. 1989). The sequence identity between the two proteins is 37%, although net charges are different: pI of 18K-LAF is 8.3, whereas that of the 22-kDa protein is 4.6. The 18K-LAF is present in the large granules of the hemocytes and may promote cell aggregation after degranulation.

10 Summary

The blue blood of the horseshoe crab contains a sophisticated defense system very sensitive to pathogens or foreign materials. The hemocytes circulating in the hemolymph detect trace amounts of LPS molecules on the invading microorganisms and respond quickly to release the granular components into the external milieu. The coagulation system composed of three serine protease zymogens, factor C, factor B, and proclotting enzyme, and a clottable protein, coagulogen, is activated by LPS to form insoluble coagulin gel. The coagulation system also responds to β-(1,3) glucan through the activation of unique heterodimeric serine protease zymogen, factor G. The pathogens are, thus, engulfed in the gel and subsequently killed by antimicrobial substances with various specificities, which are also released from the cells. The horseshoe crab has developed two kinds of serine protease zymogens as biological sensors, factor C and factor G, which are responsive to LPS and β-(1,3) glucan on the surface of Gram-negative bacteria and fungi, respectively. These are possible invaders for horseshoe crabs and also for most animals including humans. This novel heterodimeric serine protease zymogen, factor G, may open a new way to develop an innovative assay system to quantitate β-(1,3) glucans. Furthermore, these LPS and β-(1,3) glucan sensitive factors could be utilized as a unique tool to analyze other biological reactions caused by LPS or the glucan. Although the coagulation reaction in horseshoe crabs is famous, it is not the only defense mechanism of this animal. Many agglutinins are present either in hemolymph plasma or in the cell. The hemolymph plasma also has cytolytic activity against foreign cells. These cellular and humoral defense systems, in concert, defend themselves from invading foreign organisms. Such a sophisticated defense system has allowed the horseshoe crab to survive for more than 200 million years on the earth. Horseshoe crabs are often called 'living fossils.' However, they are not fossils. They are living.

Acknowledgments. The authors wish to express our thanks to Drs. T. Morita, T. Nakamura, F. Tokunaga, T. Miyata, S. Kawabata, T. Shigenaga, Y. Miura, M. Niwa, Y. Toh, and M. Hirata, and Messrs. N. Seki, R. Hashimoto, T. Saito, N. Okino, T. Oda, A. Iwanaga, D. Iwaki, and Ms. Y. Takaki for collaboration in this work. The authors are also grateful to Ms. M. Fujita for secretarial assistance. This work was supported by Grants-in-Aid for Scientific Research from the Ministry of Education, Science and Culture of Japan.

References

Aketagawa J, Miyata T, Ohtsubo S, Nakamura T, Morita T, Hayashida H, Miyata T, Iwanaga S, Takao T, Shimonishi Y (1986) Primary structure of *Limulus* anticoagulant anti-lipopolysaccharide factor. J Biol Chem 261: 7357–7365

Armstrong PB, Mangel WF, Wall JS, Hainfield JF, Van HK, Ikai A, Quigley JP (1991) Structure of alpha 2-macroglobulin from the arthropod *Limulus polyphemus*. J Biol Chem 266: 2526–2530

Armstrong PB, Armstrong MT, Quigley JP (1993) Involvement of alpha 2-macroglobulin and C-reactive protein in a complement-like hemolytic system in the arthropod *Limulus polyphemus*. Mol Immunol 30: 929–934

Bang FB (1956) A bacterial disease of *Limulus polyphemus*. Bull Johns Hopkins Hosp 98: 325–351

Barr PJ (1991) Mammalian subtilisins: The long-sought dibasic processing endoprotease. Cell 66: 1–3

Belin D, Wohlwend A, Schleuning W-D, Kruithof EKO, Vassalli J-D (1989) Facultative polypeptide translocation allows a single mRNA to encode the secreted and cytosolic forms of plasminogen activators inhibitor 2. EMBO J 8: 3287–3294

Bevilacqua MP, Stengelin S, Gimbrone MA Jr, Seed B (1989) Endothelial leukocyte adhesion molecule 1: an inducible receptor for neutrophils related to complement regulatory proteins and lectins. Science 243: 1160–1165

Bishayee SH, Dorai DT (1980) Isolation and characterization of a sialic acid-binding lectin (carcinoscorpin) from Indian horseshoe crab *Carcinoscorpius rotunda cauda*. Biochim Biophys Acta 623: 89–97

Bohn H (1986) Hemolymph clotting in insects. In: Brehélin M (ed) Immunity in invertebrates. Springer, Berlin Heidelberg New York, 188–207

Boman HG (1994) Antimicrobial peptides, Ciba Foundation Symp 186. Wiley, Chichester

Borriss R, Buettner K, Maentsaelae P (1990) Structure of the beta-1,3-1,4-glucanase gene of *Bacillus macerans*: Homologies to other beta-glucanases. Mol Gen Genet 222: 278–283

Brandin ER, Pistole TG (1983) Polyphemin: A teichoic acid-binding lectin from the horseshoe crab, *Limulus polyphemus*. Biochem Biophys Res Commun 113: 611–617

Brozen R, Sands P, Riesen W, Weissmann G, Lorand L (1987) The antiquity of transglutaminase: an intracellular enzyme from marine sponge cells enhances clotting of lobster plasma. Biol Bull 173: 423

Campbell-Wilkes LK (1973) Thesis, Northwestern University. Univ Microfilms, Ann Arbor, MI

Chasan R, Anderson KV (1989) The role of *easter*, an apparent serine protease, in organizing the dorsal-ventral pattern of the *Drosophila* embryo. Cell 56: 391–400

Cheng SM, Suzuki A, Zon G, Liu TY (1986) Characterization of a complementary deoxyribonucleic acid for the coagulogen of *Limulus polyphemus*. Biochim Biophys Acta 868: 1–8

Chung SI, Seid RC Jr, Liu T-Y (1977) Demonstration of transglutaminase activity in *Limulus* lysate. Thromb Haemostasis (Abstr) 38: 182

Cohen E (1968) Immunologic observations of the agglutinin of the hemolymph of *Limulus polyphemus* and *Birgus latro*. Trans NY Acad Sci 30: 427–443

Cordella-Miele E, Miele L, Mukherjee AB (1990) A novel transglutaminase-mediated post-translational modification of phospholipase A2 dramatically increases its catalytic activity. J Biol Chem 265: 17180–17188

Coughlin P, Sun J, Cerruti L, Salem HH, Bird P (1993) Cloning and molecular characterization of a human intracellular serine proteinase inhibitor. Proc Natl Acad Sci USA 90: 9417–9421

Day NKB, Gewurz H, Johannsen R, Finstad J, Good RA (1970) Complement and complement-like activity in lower vertebrates and invertebrates. J Exp Med 132: 941–950

DeLotto R, Spierer P (1986) A gene required for the specification of dorsal-ventral pattern in *Drosophila* appears to encode a serine protease. Nature 323: 688–692

Doege K, Sasaki M, Horigan E, Hassell JR, Yamada Y (1987) Complete primary structure of the rat cartilage proteoglycan core protein deduced from cDNA clones. J Biol Chem 262: 17757–17767

Donovan MA, Laue TM (1991) A novel trypsin inhibitor from the hemolymph of the horseshoe crab *Limulus polyphemus*. J Biol Chem 266: 2121–2125

Drickamer K (1988) Two distinct classes of carbohydrate-recognition domains in animal lectins. J Biol Chem 263: 9557–9560

Dubin A, Travis J, Enghild JJ, Potempa J (1992) Equine leukocyte elastase inhibitor. Primary structure and identification as a thymosin-binding protein. J Biol Chem 267: 6576–6583

Duvic B, Söderhäll K (1990) Purification and characterization of a β-1,3-glucan binding protein from plasma of the crayfish *Pacifastacus leniusculus*. J Biol Chem 265: 9327–9332

Enghild JJ, Thogersen IB, Salvesen G, Fey GH, Figler NL, Gonias SL, Pizzo SV (1990) Alpha-macroglobulin from *Limulus polyphemus* exhibits proteinase inhibitory activity and participates in a hemolytic system. Biochemistry 29: 10070–10080

Fernández-Morán H, Marchalonis JJ, Edelman GM (1968) Electron microscopy of a hemagglutinin from *Limulus polyphemus*. J Mol Biol 32: 467–469

Finstad CL, Good RA, Litman GW (1974) The erythrocyte agglutinin from *Limulus polyphemus* hemolymph. Molecular structure and biological function. Ann NY Acad Sci 234: 170–182

Fortes-Dias CL, Minetti CA, Lin Y, Liu TY (1993) Agglutination activity of *Limulus polyphemus* coagulogen following limited proteolysis. Comp Biochem Physiol [B] 105: 79–85

Fuhlendorff J, Clemmensen I, Magnusson S (1987) Primary structure of tetranectin, a plasminogen kringle 4 binding plasma protein: homology with asialoglycoprotein receptors and cartilage proteoglycan core protein. Biochemistry 26: 6757–6764

Fujii N, Minetti CA, Nakhasi HL, Chen SW, Barbehenn E, Nunes PH, Nguyen NY (1992) Isolation, cDNA cloning, and characterization of an 18-kDa hemagglutinin and amebocyte aggregation factor from *Limulus polyphemus*. J Biol Chem 267: 22452–22459

Fuller GM, Doolittle RF (1971a) Studies of invertebrate fibrinogen. I. Purification and characterization of fibrinogen from the spiny lobster. Biochemistry 10: 1305–1311

Fuller GM, Doolittle RF (1971b) Studies of invertebrate fibrinogen. II. Transformation of lobster fibrinogen into fibrin. Biochemistry 10: 1311–1315

Gosalbes MJ, Pérez-González JA, González R, Navarro A (1991) Two beta-glycanase genes are clustered in *Bacillus polymyxa:* molecular cloning, expression, and sequence analysis of genes encoding a xylanase and an endo-beta-(1,3)-(1,4)-glucanase. J Bacteriol 173: 7705–7710

Grépinet O, Chebrou MC, Béguin P (1988) Nucleotide sequence and deletion analysis of the xylanase gene (*xyn Z*) of *Clostridium thermocellum.* J Bacteriol 170: 4582–4588

Hoess A, Watson S, Siber GR, Liddington R (1993) Crystal structure of an endotoxin-neutralizing protein from the horseshoe crab, *Limulus* anti-LPS factor, at 1.5 Å resolution. EMBO J 12: 3351–3356

Hofemeister J, Kurtz A, Borriss R, Knowles J (1986) β-Glucanase gene from *Bacillus amyloliquefaciens* shows extensive homology with that of *Bacillus subtilis.* Gene 49: 177–187

Hosaka M, Nagahama M, Kim WS, Watanabe T, Hatsuzawa K, Ikemizu J, Murakami K, Nakayama K (1991) Arg-X-Lys/Arg-Arg motif as a signal for precursor cleavage catalyzed by furin within the constitutive secretory pathway. J Biol Chem 266: 12127–12130

Ichinose A, Bottenus RE, Davie EW (1990) Structure of transglutaminase. J Biol Chem 265: 13411–13414

Iwanaga S (1993) The *Limulus* clotting reaction. Curr Opin Immunol 5: 74–82

Iwanaga S, Morita T, Harada T, Nakamura S, Niwa M, Takada K, Kimura T, Sakakibara S (1978) Chromogenic substrates for horseshoe crab clotting enzyme. Its application for the assay of bacterial endotoxin. Haemostasis 7: 183–188

Iwanaga S, Miyata T, Tokunaga F, Muta T (1992) Molecular mechanism of hemolymph clotting system in *Limulus.* Thromb Res 68: 1–32

Iwanaga S, Muta T, Shigenaga T, Miura Y, Seki N, Saito T, Kawabata S (1994a) Role of hemocyte-derived granular components in invertebrate defense. Ann NY Acad Sci 712: 102–116

Iwanaga S, Muta T, Shigenaga T, Seki N, Kawano K, Katsu T, Kawabata S (1994b) Structure-function relationships of tachyplesins and their analogues. In: Ciba Foundation Symposium 186 Antimicrobial peptides. Wiley, Chichester, pp 160–175

Johnston GI, Cook RG, McEver RP (1989) Cloning of GMP-140, a granule membrane protein of platelets and endothelium: sequence similarity to proteins involved in cell adhesion and inflammation. Cell 56: 1033–1044

Kakinuma A, Asano T, Torii H, Sugino Y (1981) Gelation of *Limulus* amebocyte lysate by an antitumor (1→3)-β-D-glucan. Biochem Biophys Res Commun 101: 434–439

Kaplan R, Li SS, Kehoe JM (1977) Molecular characterization of limulin, a sialic acid binding lectin from the hemolymph of the horseshoe crab, *Limulus polyphemus.* Biochemistry 16: 4297–4303

Kawano K, Yoneya T, Miyata T, Yoshikawa K, Tokunaga F, Terada Y, Iwanaga S (1990) Antimicrobial peptide, tachyplesin I, isolated from hemocytes of the horseshoe crab (*Tachypleus tridentatus*). NMR determination of the beta-sheet structure. J Biol Chem 265: 15365–15367

Kikutani H, Inui S, Sato R, Barsumian EL, Owaki H, Yamasaki K, Kaisho T, Uchibayashi N, Hardy RR, Hirano T, Tsunasawa S, Sakiyama F, Suemura M, Kishimoto T (1986) Molecular structure of human lymphocyte receptor for immunoglobulin E. Cell 47: 657–665

Lamb FI, Roberts LM, Lord JM (1985) Nucleotide sequence of cloned cDNA coding for preproricin. Eur J Biochem 148: 265–270

Levin J, Bang FB (1964) The role of endotoxin in the extracellular coagulation of *Limulus* blood. Bull Johns Hopkins Hosp 115: 265–274

Liu T, Lin Y, Cislo T, Minetti CA, Baba JM, Liu TY (1991) Limunectin. A phosphocholine-binding protein from *Limulus* amebocytes with adhesion-promoting properties. J Biol Chem 266: 14813–14821

Lloberas J, Perez-Pons JA, Querol E (1991) Molecular cloning, expression and nucleotide sequence of the endo-β-1,3-1,4-D-glucanase gene from *Bacillus licheniformis.* Predictive structural analyses of the encoded polypeptide. Eur J Biochem 197: 337–343

Lorand L, Credo RB, Janus TJ (1981) Factor XIII (fibrin-stabilizing factor). Methods Enzymol 80: 333–341

Madaras F, Parkin JD, Castaldi PA (1979) Coagulation in the sand crab (*Ovalipes bipustulatus*). Thromb Haemostasis 42: 734–742

Marchalonis JJ, Edelman GM (1968) Isolation and characterization of a hemaggulutinin from *Limulus polyphemus*. J Mol Biol 32: 453–465

Medford JI, Elmer JS, Klee HJ (1991) Molecular cloning and characterization of genes expressed in shoot apical meristems. Plant Cell 3: 359–370

Minetti CA, Lin YA, Cislo T, Liu TY (1991) Purification and characterization of an endotoxin-binding protein with protease inhibitory activity from *Limulus* amebocytes J Biol Chem 266: 20773–20780

Miura Y, Kawabata S, Iwanaga S (1994) A *Limulus* intracellular coagulation inhibitor with characteristics of the serpin superfamily. Purification, characterization, and cDNA cloning. J Biol Chem 269: 542–547

Miura Y, Kawabata S, Wakamiya Y, Nakamura T, Iwanaga S (1995) A *Limulus* intracellular coagulation inhibitor type 2. Purification, characterization, cDNA cloning, and tissue localization. J Biol Chem 270: 558–565

Miyata T, Hiranaga M, Umezu M, Iwanaga S (1984a) Amino acid sequence of the coagulogen from *Limulus polyphemus* hemocytes. J Biol Chem 259: 8924–8933

Miyata T, Usui K, Iwanaga S (1984b) The amino acid sequence of coagulogen isolated from Southeast Asian horseshoe crab, *Tachypleus gigas*. J Biochem (Tokyo) 95: 1793–1801

Miyata T, Matsumoto H, Hattori M, Sakaki Y, Iwanaga S (1986) Two types of coagulogen mRNAs found in horseshoe crab (*Tachypleus tridentatus*) hemocytes: Molecular cloning and nucleotide sequences. J Biochem (Tokyo) 100: 213–220

Miyata T, Tokunaga F, Yoneya T, Yoshikawa K, Iwanaga S, Niwa M, Takao T, Shimonishi Y (1989) Antimicrobial peptides, isolated from horseshoe crab hemocytes, tachyplesin II, and polyphemusins I and II. Chemical structures and biological activity. J Biochem (Tokyo) 106: 663–668

Morgenstern KA, Henzel WJ, Baker JB, Wong S, Pastuszyn A, Kisiel W (1993) Isolation and characterization of an intracellular serine proteinase inhibitor from a monkey kidney epithelial cell line. J Biol Chem 268: 21560–21568

Morita T, Tanaka S, Nakamura T, Iwanaga S (1981) A new (1→3)-β-D-glucan-mediated coagulation pathway found in *Limulus* amebocytes. FEBS Lett 129: 318–321

Morita T, Ohtsubo S, Nakamura T, Tanaka S, Iwanaga S, Ohashi K, Niwa M (1985) Isolation and biological activities of *Limulus* anticoagulant (anti-LPS factor) which interact with lipopolysaccharide (LPS). J Biochem (Tokyo) 97: 1611–1620

Murphy N, McConnell DJ, Cantwell BA (1984) The DNA sequence of the gene and genetic control sites for the excreted *B. subtilis* enzyme β-glucanase. Nucleic Acid Res 12: 5355–5367

Muta T, Miyata T, Tokunaga F, Nakamura T, Iwanaga S (1987) Primary structure of anti-lipopolysaccharide factor from American horseshoe crab, *Limulus polyphemus*. J Biochem (Tokyo) 101: 1321–1330

Muta T, Fujimoto T, Nakajima H, Iwanaga S (1990a) Tachyplesins isolated from hemocytes of Southeast Asian horseshoe crabs (*Carcinoscorpius rotundicauda* and *Tachypleus gigas*): Identification of a new tachyplesin, tachyplesin III, and a processing intermediate of its precursor. J Biochem (Tokyo) 108: 261–266

Muta T, Hashimoto R, Miyata T, Nishimura H, Toh Y, Iwanaga S (1990b) Proclotting enzyme from horseshoe crab hemocytes. cDNA cloning, disulfide locations, and subcellular localization. J Biol Chem 265: 22426–22433

Muta T, Miyata T, Misumi Y, Tokunaga F, Nakamura T, Toh Y, Ikehara Y, Iwanaga S (1991) *Limulus* factor C. An endotoxin-sensitive serine protease zymogen with a mosaic structure of complement-like, epidermal growth factor-like, and lectin-like domains. J Biol Chem 266: 6554–6561

Muta T, Oda T, Iwanaga S (1993) Horseshoe crab coagulation factor B. A unique serine protease zymogen activated by cleavage of an Ile–Ile bond. J Biol Chem 268: 21384–21388

Muta T, Seki N, Takaki Y, Hashimoto R, Oda T, Iwanaga A, Tokunaga F, Iwanaga S (1995) Purified horseshoe crab factor G: Reconstitution and characterization of the (1→3)-β-D-glucan-sensitive serine protease cascade. J Biol Chem 270: 892–897

Müller-Eberhard HJ (1988) Molecular organization and function of the complement system. Annu Rev Biochem 57: 321–347

Mürer EH, Levin J, Holme R (1975) Isolation and studies of the granules of the amebocytes of *Limulus polyphemus*, the horseshoe crab. J Cell Physiol 86: 533–542

Nakamura S, Iwanaga S, Harada T, Niwa M (1976a) A clottable protein (coagulogen) from amoebocyte lysate of Japanese horseshoe crab (*Tachypleus tridentatus*). Its isolation and biochemical properties. J Biochem (Tokyo) 80: 1011–1021

Nakamura S, Takagi T, Iwanaga S, Niwa M, Takahashi K (1976b) Amino acid sequence studies on the fragments produced from horseshoe crab coagulogen during gel formation: Homologies with primate fibrinopeptide B. Biochem Biophys Res Commun 72: 902–908

Nakamura S, Morita T, Harada-Suzuki T, Iwanaga S, Takahashi K, Niwa M (1982) A clottable enzyme associated with the hemolymph coagulation system of horseshoe crab (*Tachypleus tridentatus*): Its purification and characterization. J Biochem (Tokyo) 92: 781–792

Nakamura T, Morita T, Iwanaga S (1985) Intracellular proclotting enzyme in *Limulus* (*Tachypleus tridentatus*) hemocytes: its purification and properties. J Biochem (Tokyo) 97: 1561–1574

Nakamura T, Morita T, Iwanaga S (1986a) Lipopolysaccharide-sensitive serine-protease zymogen (factor C) found in *Limulus* hemocytes. Isolation and characterization. Eur J Biochem 154: 511–521

Nakamura T, Horiuchi T, Morita T, Iwanaga S (1986b) Purification and properties of intracellular clotting factor, factor B, from horseshoe crab (*Tachypleus tridentatus*) hemocytes. J Biochem (Tokyo) 99: 847–857

Nakamura T, Hirai T, Tokunaga F, Kawabata S, Iwanaga S (1987) Purification and amino acid sequence of Kunitz-type protease inhibitor found in the hemocytes of horseshoe crab (*Tachypleus tridentatus*). J Biochem (Tokyo) 101: 1297–1306

Nakamura T, Tokunaga F, Morita T, Iwanaga S (1988a) Interaction between lipopolysaccharide and intracellular serine protease zymogen, factor C, from horseshoe crab (*Tachypleus tridentatus*) hemocytes. J Biochem (Tokyo) 103: 370–374

Nakamura T, Tokunaga F, Morita T, Iwanaga S, Kusumoto S, Shiba T, Kobayashi T, Inoue K (1988b) Intracellular serine-protease zymogen, factor C, from horseshoe crab hemocytes. Its activation by synthetic lipid A analogues and acidic phospholipids. Eur J Biochem 176: 89–94

Nakamura T, Furunaka H, Miyata T, Tokunaga F, Muta T, Iwanaga S, Niwa M, Takao T, Shimonishi Y (1988c) Tachyplesin, a class of antimicrobial peptide from the hemocytes of the horseshoe crab (*Tachypleus tridentatus*). Isolation and chemical structure. J Biol Chem 263: 16709–16713

Neame PJ, Choi HU, Rosenberg LC (1989) The isolation and primary structure of a 22-kDa extracellular matrix protein from bovine skin. J Biol Chem 264: 5474–5479

Nguyen NY, Suzuki A, Boykins RA, Liu TY (1986a) The amino acid sequence of *Limulus* C-reactive protein. Evidence of polymorphism. J Biol Chem 261: 10456–10465

Nguyen NY, Suzuki A, Cheng SM, Zon G, Liu TY (1986b) Isolation and characterization of *Limulus* C-reactive protein genes. J Biol Chem 261: 10450–10455

Noguchi H (1903) A study of immunization – haemolysins, agglutinins, precipitins and coagulins in cold-blooded animals. Zentralbl Bakteriol Abt Orig 33: 353–362

Ochiai M, Ashida M (1988) Purification of a β-1,3-glucan recognition protein in the prophenoloxidase activating system from hemolymph of the silkworm, *Bombyx mori*. J Biol Chem 263: 12056–12062

Ohashi K, Niwa M, Nakamura T, Morita T, Iwanaga S (1984) Anti-LPS factor in the horseshoe crab, *Tachypleus tridentatus*. Its hemolytic activity on the red blood cell sensitized with lipopolysaccharide. FEBS Lett 176: 207–210

Olafsen JA (1986) Invertebrate lectins-Biochemical heterogeneity as a possible key to their biological function. In: Brehélin M (ed) Immunity in invertebrates. Springer, Berlin Heidelberg New York, pp 94–111

Ornberg RL, Reese TS (1979) Secretion in *Limulus* amebocytes is by exocytosis. Prog Clin Biol Res 29: 125–130

Osmand AP, Friedenson B, Gerwurz H, Painter RH, Hofmann T, Shelton E (1977) Characterization of C-reactive protein and the complement subcomponent Clt as homologous proteins displaying cyclic pentameric symmetry (pentraxins). Proc Natl Acad Sci USA 74: 739–743

Pearson FC, Bohon J, Lee W, Bruszer G, Sagona M, Dawe R, Jakubowski G, Morrison D, Dinarello
    C (1984) Comparison of chemical analyses of hollow-fiber dialyzer extracts. Artif Organs 8:
    291–298
Perez-Paya E, Thiaudiere E, Abad C, Dufourcq J (1991) Selective labelling of melittin with a
    fluorescent dansylcadaverine probe using guinea-pig liver transglutaminase. FEBS Lett 278: 51–54
Quigley JP, Armstrong PB (1994) Invertebrate α_2-macroglobulin: Structure-function and the ancient
    thiol ester bond. Ann NY Acad Sci 712: 131–145
Remold-O'Donnell E, Chin J, Alberts M (1992) Sequence and molecular characterization of human
    monocyte/neutrophil elastase inhibitor. Proc Natl Acad Sci USA 89: 5635–5639
Robey FA, Liu T-Y (1981) Limulin: A C-reactive protein from Limulus polyphemus J Biol Chem 256:
    969–975
Roth RI, Chen JC-R, Levin J (1989) Stability of gels formed following coagulation of Limulus
    amebocyte lysate. Lack of covalent crosslinking of coagulin. Thromb Res 55: 25–36
Saito T, Kawabata S, Hirata M, Iwanaga S (1995a) A novel type of Limulus lectin-L6: Purification,
    covalent structure and antibacterial activity. J Biol Chem 270: 14493–14499
Saito T, Kawabata S, Shigenaga T, Takayenoki Y, Cho J, Nakajima H, Hirata, M, Iwanaga S (1995b)
    A novel big defensin identified in horseshoe crab hemocytes. Isolation, amino acid sequence and
    antibacterial activity. J Biochem (Tokyo) 117: 1131–1137
Seki N, Muta T, Oda T, Iwaki D, Kuma K, Miyata T, Iwanaga S (1994) Horseshoe crab (1,3)-β-D-
    glucan-sensitive coagulation factor G. A serine protease zymogen heterodimer with similarities to
    β-glucan-binding proteins. J Biol Chem 269: 1370–1374
Shareck F, Roy C, Yaguchi M, Morosoli R, Kluepfel D (1991) Sequences of three genes specifying
    xylanases in Streptomyces lividans. Gene 107: 75–82
Shen SH, Chrétien P, Bastien L, Slilaty SN (1991) Primary sequence of the glucanase gene from
    Oerskovia xanthineolytica. Expression and purification of the enzyme from Escherichia coli. J Biol
    Chem 266: 1058–1063
Shigenaga T, Muta T, Toh Y, Tokunaga F, Iwanaga S (1990) Antimicrobial tachyplesin peptide
    precursor. cDNA cloning and cellular localization in the horseshoe crab (Tachypleus tridentatus).
    J Biol Chem 265: 21350–21354
Shigenaga T, Takayenoki Y, Kawasaki S, Seki N, Muta T, Toh Y, Ito A, Iwanaga S (1993) Separation
    of large and small granules from horseshoe crab (Tachypleus tridentatus) hemocytes and charac-
    terization of their components. J Biochem (Tokyo) 114: 307–316
Shimizu S, Ito M, Niwa M (1977) Lectins in the hemolymph of Japanese horseshoe crab, Tachypleus
    tridentatus. Biochim Biophys Acta 500: 71–79
Shimoi H, Iimura Y, Obata T, Tadenuma M (1992) Molecular structure of Rarobacter faecitabidus
    protease I. A yeast-lytic serine protease having meannose-binding activity. J Biol Chem 267:
    25189–25195
Siegelman MH, van de Rijn M, Weissman IL (1989) Mouse lymph node homing receptor cDNA clone
    encodes a glycoprotein revealing tandem interaction domains. Science 243: 1165–1172
Söderhäll K, Smith VJ (1986) The prophenoloxidase system: The biochemistry of its activation and
    role in arthropod cellular immunity with special references to crustaceans. In: Brehélin M (ed)
    Immunity in Invertebrates. Springer, Berlin Heidelberg New York, pp 208–223
Söderhäll K, Rögener W, Söderhäll I, Newton RP, Ratcliff NA, (1988) The properties and purification
    of a Blaberus cranifer plasma protein which enhances the activation of haemocyte prophenoloxi-
    dase by a β-1,3-glucan. Insect Biochem 18: 323–330
Sottrup JL, Borth W, Hall M, Quigley JP, Armstrong PB (1990) Sequence similarity between alpha 2-
    macroglobulin from the horseshoe crab, Limulus polyphemus, and proteins of the alpha 2-macro-
    globulin family from mammals. Comp Biochem Physiol B 96: 621–625
Srimal S, Miyata T, Kawabata S, Morita T, Iwanaga S (1985) The complete amino acid sequence of
    coagulogen isolated from Southeast Asian horseshoe crab, Carcinoscorpius rotundicauda. J
    Biochem (Tokyo) 98: 305–318
Sugo T, Kato H, Iwanaga S, Takada K, Sakakibara S (1985) Kinetic studies on surface-mediated
    activation of bovine factor XII and prekallikrein. Effect of kaolin and high-Mr kininogen on the
    activation reactions. Eur J Biochem 146: 43–50

Takagi T, Hokama Y, Miyata T, Morita T, Iwanaga S (1984) Amino acid sequence of Japanese horseshoe crab (*Tachypleus tridentatus*) coagulogen B chain: completion of the coagulogen sequence. J Biochem (Tokyo) 95: 1445–1457

Takeya H, Nishida S, Miyata T, Kawada S, Saisaka Y, Morita T, Iwanaga S (1992) Coagulation factor X activating enzyme from Russell's viper venom (RVV-X). A novel metalloproteinase with disintegrin (platelet aggregation inhibitor)-like and C-type lectin-like domains. J Biol Chem 267: 14109–14117

Tanaka S, Iwanaga S (1993) *Limulus* test for detecting bacterial endotoxins. Methods Enzymol 223: 358–364

Tanaka S, Nakamura T, Morita T, Iwanaga S (1982) *Limulus* anti-LPS factor: An anticoagulant which inhibits the endotoxin-mediated activation of *Limulus* coagulation system. Biochem Biophys Res Commun 105: 717–723

Teschauer WF, Mentele R, Sommerhoff CP (1993) Primary structure of a porcine leukocyte serpin. Eur J Biochem 217: 519–526

Toh Y, Mizutani A, Tokunaga F, Muta T, Iwanaga S (1991) Morphology of the granular hemocytes of the Japanese horseshoe crab *Tachypleus tridentatus* and immunocytochemical localization of clotting factors and antimicrobial substances. Cell Tissue Res 266: 137–147

Tokunaga F, Miyata T, Nakamura T, Morita T, Kuma K, Miyata T, Iwanaga S (1987) Lipopolysaccharide-sensitive serine-protease zymogen (factor C) of horseshoe crab hemocytes. Identification and alignment of proteolytic fragments produced during the activation show that it is a novel type of serine protease. Eur J Biochem 167: 405–416

Tokunaga F, Nakajima H, Iwanaga S (1991) Further studies on lipopolysaccharide-sensitive protease zymogen (factor C). Its isolation from *Limulus polyphemus* hemocytes and identification as an intracellular zymogen activated by alpha-chymotropsin, not by trypsin. J Biochem (Tokyo) 109: 150–157

Tokunaga F, Yamada M, Miyata T, Ding YL, Hiranaga KM, Muta T, Iwanaga S, Ichinose A, Davie EW (1993a) *Limulus* hemocyte transglutaminase. Its purification and characterization, and identification of the intracellular substrates. J Biol Chem 268: 252–261

Tokunaga F, Muta T, Iwanaga S, Ichinose A, Davie EW, Kuma K, Miyata T (1993b) *Limulus* hemocyte transglutaminase. cDNA cloning, amino acid sequence, and tissue localization. J Biol Chem 268: 262–268

Tsai H, Hirsch HJ (1981) The primary structure of fulvocin C from *Myxococcus fulvus*. Biochim Biophys Acta 667: 213–217

Vaith P, Uhlenbruck G, Müller WEG, Cohen E (1979) Reactivity of *Limulus polyphemus* hemolymph with D-glucuronic acid containing glycosubstances. Prog Clin Biol Res 29: 579–587

Von Heijne G (1983) Patterns of amino acids near signal-sequence cleavage sites. Eur J Biochem 133: 17–21

Von Heijine G, Liljeström P, Mikus P, Andersson H, Ny T (1991) The efficiency of the uncleaved secretion signal in the plasminogen activator inhibitor type 2 protein can be enhanced by point mutations that increase its hydrophobicity. J Biol Chem 266: 15240–15243

White RT, Damm D, Miller J, Spratt K, Schilling J, Hawgood S, Benson B, Cordell B (1985) Isolation and characterization of the human pulmonary surfactant apoprotein gene. Nature 317: 361–363

Yahata N, Watanabe T, Nakamura Y, Yamamoto Y, Kamimiya S, Tanaka H (1990) Structure of the gene encoding β-1,3-glucanase A1 of *Bacillus circulans* WL-12. Gene 86: 113–117

Zou Z, Anisowicz A, Hendrix MJC, Thor A, Neveu M, Sheng S, Rafidi K, Sefter E, Sager R (1994) Maspin, a serpin with tumor-suppressing activity in human mammary epithelial cells. Science 263: 526–529

Zverlov VV, Laptev DA, Tishkov VI, Velikodvorskaja GA (1991) Nucleotide sequence of the *Clostridium thermocellum* laminarinase gene. Biochem Biophys Res Commun 181: 507–512

Cytotoxic Activity of Tunicate Hemocytes

N. Parrinello

1 Introduction

Tunicates (protochordates) are filter-feeding marine invertebrates with a worldwide distribution. In their larval form, they exhibit many of the features characteristic of the vertebrates. The larva, with a tail, notochord, and dorsal neural tube, upon settlement undergoes a remarkable metamorphosis in which it loses most of its chordate characteristics and becomes a sessile invertebrate adult. Thus, due to these characteristics, tunicates are considered to be the most primitive members of the phylum Chordata. Owing to their position in the phylogenetic line leading to the vertebrates, they have attained importance as experimental organisms and have been examined by researchers from a variety of disciplines (developmental biology, ecology, physiology, and immunology).

Studies on the immune system have revealed that tunicates are reliant upon the nonspecific inflammatory reactions of the blood cells observed in invertebrates and vertebrates, and have furthermore revealed a vertebrate-like histocompatibility system resembling the MHC (Scofield et al. 1982a).

Natural cell killing capability, independent of phagocytosis, probably appeared early during the evolution of the immune system (Cooper 1980; Parrinello and Arizza 1992). In vertebrates various kinds of cell-mediated cytotoxicity are expressed by the following effector cells (Evans and Mckinney 1991): (1) macrophages, (2) natural killer (NK) cells, (3) NK-like nongranular lymphoid-type nonspecific cytotoxic cells (NCC) (4) activated macrophages, (5) cytotoxic T-lymphocytes, and (6) antibody dependent cell-mediated cytotoxicity (ADCC) by Fc receptor bearing cells, and K-cells. The first two types of cytotoxic cells may be found throughout all the animal groups, and may be considered the most primitive cytotoxic cells; NCC was found in teleost fishes. The last three classes of cells require induction to activate their cytotoxic mechanisms, so they probably developed late in evolution. These cells are components of acquired cytotoxicity and they may be considered as highly specialized.

Hemocyte-mediated cytotoxic responses in vitro have been reported in different invertebrate phyla: Sipunculidae, Annelida, Mollusca, Arthropoda, and

Institute of Zoology, Laboratory of Marine Immunobiology, University of Palermo, Via Archirafi, 18 # 90123 Palermo, Italy

Echinodermata (Parrinello and Arizza 1992). Multiple modes of allorecogniton and hemocyte cytotoxicity have been reported in tunicates, as revealed by various experimental procedures: (1) tunic graft unilateral rejection in solitary ascidians; (2) tunic reaction to erythrocyte subcuticular injection; (3) nonfusion reaction (NFR) in colonial ascidians; (4) in vitro bilateral "contact reaction" among allo- or xenogeneic hemocytes; (5) in vitro hemocyte cytotoxic activity against mammalian targets. These cytotoxic reactions include both spontaneous (NK- or macrophage-like) and induced levels of cytotoxicity.

In this review an attempt has been made to identify the hemocytes that have been described as involved in effector cytotoxic reactions.

2 Ascidian Hemocytes

It has been proposed that ascidian hemocytes can be divided, by TEM studies, into two main categories (De Leo 1992): stem cells and granulocytes. The former show features of undifferentiated elements including hemoblasts and lymphocyte-like cells (LLCs); they are morphologically similar to each other but LLCs lack a prominent nucleolus (Ermak 1976). Typically, granulocytes are cells containing vesicles, vacuoles, and granules classified on the basis of their variously electrondense contents (De Leo 1992). These cell types include: clear granulocytes (microgranular/hyaline amoebocyte), clear vesicular granulocytes (vacuolar hyaline amoebocyte/phagocyte), microgranulocytes (granular amoebocyte/macrophage), and vacuolar granulocytes. Vacuolar granulocytes are distinguished in univacuolar (signet ring cells) and globular (compartment cells and morula cells) granulocytes. Among the granulocyte types, transitional forms are probably included. The names in brackets are based exclusively on the apparent shape of the cells when viewed under light microscopy.

In this paper, the hemocytes are usually named according to De Leo (1992), although the alternative terminology is also reported as expressed by the authors.

3 Multiple Modes of Self/Nonself Recognition and Cytotoxicity

3.1 Graft Rejection in Solitary Ascidians. Specific Immunorecognition and Hemocyte Cytocidal Responses to Foreignness

Hemocyte cytotoxic reactions restricted by cellular polymorphic histocompatibility antigens, have been implicated with tunic graft rejection in solitary ascidians (Table 1).

Analyses of histocompatibility in *Styela plicata* have identified cell-mediated responses controlled by a single gene locus that may encode discrete, cellular histocompatibility antigens (Raftos et al. 1987a,b; Raftos and Briscoe 1990). After the successful fusion of the tunic graft to their hosts, the effect or processes become evident and culminate in the destruction of allogeneic vascular elements

Table 1. Cytotoxic reactions in several ascidian species. Cytotoxic reactions have been attributed to the ascidian species on the basis of available data

Ascidians	Cytotoxic reaction								References
	Specific Multiallelic locus(i)				Nonspecific reaction				
	Allograft rejection	Contact reaction		Nonfusion reaction	Tunic reaction	NK-like			
		Xenogeneic	Allogenic			Anti-RBC	Anti-tumor		
Ciona intestinalis	+	+			+	+	+		Fuke (1980); Parrinello et al. (1977); Reddy et al. 1975
Ciona robusta		+							Fuke (1980)
Styela plicata	+					+			Raftos et al. (1987a,b); Raftos and Briscoe (1990); Parrinello et al. (in Prep.)
Styela clava		+	+						Fuke (1980); Kelly et al. (1992)
Molgula manhattensis					+				Anerson (1971)
Halocynthia roretzi		+	+						Fuke (1980); Fuke and Nakamura (1985)
Halocynthia aurantium		+							Fuke (1980)
Pyura mirabilis		+							Fuke (1980)
Botryllus primigenus				+					Tanaka and Watanabe (1973); Taneda and Watanabe (1982a,b)
Botryllus schlosseri				+					Sabbadin (1962)
Botrylloides simodensis				+					Saito and Watanabe (1984); Hirose et al. (1990)
Botrylloides violaceus				+					Mukai and Watanabe (1975); Hirose et al. (1988)

RBC = red blood cells.

that degenerate within 30 days. At the end, all that remains is fine cellular debris (Raftos 1990). This reaction is provided with immunologic memory, as shows by the results on second-set and third-party allografts.

Raftos et al. (1988) determined the mode of allorecognition in *S. plicata* by establishing reciprocal first-set tunic allografts within pairs of individuals. Unilateral response was exhibited in 27% of pairs, i.e., one animal retained reciprocal grafts while the other rejected them. The appearance of unilateral rejection is supported by second-set grafting and is consistent with an allorecognition mechanism involving specific identification of foreignness.

Immunological specificity of the reaction was confirmed by preimmunization of individuals with allogeneic hemocytes (Raftos 1991). Hosts primed with allogeneic hemocytes acquired the ability to rapidly (second-set type) eliminate tunic allograft from the original donor of preimmunizing hemocytes. In contrast, most (71%) third-party transplants obtained from individuals other than the original preimmunizing donors, did not undergo rapid rejection. Genetic analyses of a population from Balmoral Beach (Sidney Harbor) identified five major tissue variants encoded by a single genetic locus, and supported a codominant expression of discrete haplotypes.

The genetic patterns of graft rejection, the unilateral response in allorecognition, as well as the results of the in vitro allogeneic cytotoxic hemocyte reactions in *Styela clava* (Kelly et al. 1992) and the requirement for cell contact, are consistent with a regime in which a discrete immunocompetent cell population recognizes foreign histocompatibility antigens. A unique genetic pattern of multiple alleles at a single histocompatibility locus, appears to be implicated in the transplantation of *S. plicata* and the cytotoxic hemocyte reactions of *S. clava*.

The allorecognition mode is similar to those of vertebrates and annelid worms (Du Pasquier 1979) whereby the difference between individuals of a single-tissue haplotype is sufficient to initiate allograft rejection. On the other hand, this mode differs from those of *Halocynthia roretzi* contact reactions (Fuke and Nakamura 1985) and the nonfusion reactions of colonial tunicates (see below; Oka and Watanabe 1957) in which hemihomologous combinations yield compatibility).

3.2 Tunic Reaction to Erythrocyte Subcuticular Injection

The tunic reaction of *Ciona intestinalis* is an inflammatory dose dependent response typical of invertebrates. Small quantities of erythrocytes injected into tunic are cleared by phagocytosis, whereas larger amounts induce encapsulation and in some specimens also cause tissue injury (Parrinello 1981; Parrinello and Patricolo 1984; Parrinello et al. 1977, 1984). The injury process, as well as the encapsualtion, may be related to the massive infiltration of cells coming from the hematogeneic sites of the body wall. Studies of fine structures revealed crypts or nodules where numerous detaching stem cells were observed (paper in prep.). Tunic tissue disintegration is a nonspecific reaction probably associated with the degranulation of the microgranulocytes (granular amoebocytes) which are very numerous in the inflammed tunic (Parrinello et al. 1990). Tunic disintegration is not related to encapsulation, as confirmed by statistically significant differences between some groups of results. Among the inflammatory cells, numerous stem cells were found which probably come from the hematogeneic nodules. Histological observations suggest that granulocytes and degranulation may be directed to erythrocyte disintegration. Tunic damage may be a lesser effect of the reaction, the biochemical nature of which is unknown. Cytotoxic mechanisms, such as those proposed for mammalian neutrophil (Hirschhorn 1974; Gleisner 1979), monocytes, and NK-like cells (Roder et al. 1982) could participate in killing and inflammation. Numerous morula cells at different developmental stages are present (paper in prep.) in the injured tissue. Since this cell type contains prophenoloxidase (Smith and Peddie 1992), products obtained by its activation could be involved both in lysis and encapsulation processes.

A high frequency (92%) of suffering specimens was also found after injection of bovine serum albumin solution. In this case, tissue damage was very fast (24 h), and the same features previously described for erythrocyte injection characterized the injured tunics. BSA was an irritant that activated mechanisms for stem – cell proliferation, granulocyte infiltration, and degranulation with release of lytic molecules (Parrinello et al. 1984).

The timing sequence of the tunic reaction after second-set erythrocyte injection is characterized by a heightened nonspecific response (Parrinello et al. 1977). This second response is probably not anamnestic, although more rapid proliferation and infiltration of stem cells from hematogenic nodules may be assumed.

3.3 Nonfusion Reaction in Colonial Ascidians

Botryllus colonies are established by the surface settlement of sexually produced swimming tadpole larvae and their subsequent metamorphosis to an oozoid form (founder individual). The oozoid undergoes cyclic blastogenesis to produce new individuals (zooid) in a large star-shaped clonal cluster. More clusters can form a single colony. All zooids, as well as colony clusters, are connected to each other within a common tunic matrix via common blood vessels which have spherical to elongate termini (ampullae) near the surface of the tunic. When a colony is split, naturally or experimentally, into two or more fragments (subclones, ramets), each fragment usually continues to grow independently and forms a colony.

In the natural environment colonies undergo natural transplantation when they come into contact. On space-limited surfaces, colony growth by budding may result in contact between neighboring animals. This is followed by the reciprocal transplantation of blood vessels and limited mixing of blood and tissue elements (fusion). The same phenomenon is observed when two colonies (or fragments) meet with their extending ampullae.

Different types of assays have revealed several aspects of allorecognition and colony specificity. Allogeneic outcomes were analyzed at the morphological and cellular level. The most common method used was the colony-allorecognition assay. Two subclones were put on glass slides in pairs and placed so that they came into contact with each other with their extending ampullae. Complementary to this technique was the cut-surface assay.

A rapid and sensitive tool was the oozooid-allorecognition assay in which the thin oozoid preparation allows easy visualization of ampullae under the microscope: within 24 paired oozooids undergo vascular fusion or rejection similar to that between adult colonies (Scofield et al. 1982a,b).

Colony specificity is the capacity for self/nonself distinction leading to fusion by completion of the vascular anastomosis of compatible colonies or rejection by cytotoxic reaction and the ampullar amputation or disintegration of non-compatible colonies. Nonfusion reaction (NFR) stages took place in 1–2 days depending on temperature (Tanaka and Watanabe 1973).

Natural allorecognition phenomena in botryllid ascidians manifested specific features within a given species or between populations (Taneda et al. 1985; Hirose et al. 1988, 1990; Rinkevich and Weissman 1988; Boyd et al. 1990). Histocompatibility discrimination is controlled by a single gene locus named Fu/HC (Weissman et al. 1990) which contains approximately 100 codominantly expressed alleles. Genetic analyses revealed that, unlike the MHC vertebrate locus, a *Botryllus* colony heterozygotic (AB) at the Fu/Hc locus has the capacity to fuse with any other colony carrying at least one of these two alleles, even though it may differ at the other. Rejecting colonies share no alleles (AB/CD).

According to Rinkevich (1992) and Sabbadin et al. (1992), various rejection types, based on extent of ampullae movement, ampullar epithelium contact, and tunic function in allogeneic processes, have been distinguished.

In the most common processes, hemocyte (morula cell) infiltration from the ampullar lumen into the tunic forms pads or corona-like dark black hemorrhages which is followed by the formation of a necrotic region. Subsequently, no more hemocytes appear to infiltrate the tunic from the bleeding ampullae (Brancoft 1903; Sabbadin 1962; Milanesi and Burighel 1978; Scofield et al. 1982a,b; Scofield and Nagashima 1983; Rinkevich and Weissman 1988; Boyd et al. 1990). Less common types of rejection were: (1) the complete destruction of the ampullar tips, and large hemorrhages resulting from ampullar monolayered epithelium disappearance; (2) cellular reaction developed within the ampullar lumen without bleeding; (3) alterations associated with vessels frequently plugged by clotted masses of hemocytes which subsequently degenerated, while the tunic rapidly disintegrated. This latter process resulted in separation of the two allogeneic partners.

Secondary, tertiary, and additional sets of assays done with the same pairs of subclones of several different Monterey vs. Woods Hole *B. schlosseri* did not reveal accelerated or enhanced NFR (Rinkevich and Weissman 1991).

3.4 Natural Cytocidal Response. In Vitro "Contact Reaction"

The identification of fast (few minutes) in vitro reactions between allogeneic or xenogeneic combinations of hemocytes is evidence for the existence of direct cellular cytotoxic responses (Fuke 1980). Reactivity is initiated by cell contact which causes cellular degranulation. Degranulation results either in the mutual death of opposing cells or in prolonged incapacitation.

Shortly after two xenogeneic hemocytes (congeneric species pairs as well as different genera) came into contact, a "tight adherence" stage was followed by mutual lysis of the cells. The ascidian species examined are listed in Table 1. Cell–cell interaction at the membrane level may result in a forced suicide for both parties, whereas an effector–target – cell relationship does not seem to exist between two cells in a contact reaction. Cytotoxicity also occurs in vivo, as shown by injecting hemocytes marked with a vital stain into a xenogeneic host (Fuke 1980) in which all the donor cells were surrounded by one or more host derived cells.

The xenogeneic bilateral cytotoxicity is a property of the hemocytes from all the examined solitary ascidians (Table 1). The lack of self markers at species level may trigger the reaction. Allogeneic reactions have been observed between hemocytes for some donor combinations.

In *Halocynthia roretzi*, contact reaction appears to be mediated by two histocompatibility (H) genes with considerable allogeneic polymorphism (Fuke and Numakunai 1982; Fuke and Nakamura 1985). Genetic analysis carried out employing several individuals (Fuke and Nakamura 1985) in all the possible combinations, showed that allogeneic reactions occurred between hemocytes from some, but not all, donor combinations. None of the mutually nonreactive specimens have the same pattern of reactivity against the remainder of the animals, therefore it has been suggested that the reaction takes place only when two individuals do not share an allele of an H gene or genes. This genetic model is identical to that of *Botryllus primigenus* colony fusibility: a nonreactive pair must share at least one allele. In *H. roretzi* there are two H loci, and it has been assumed that the maximum number of alleles would be 73.

The existence of a cytotoxicity dependent on allogeneic recognition was also shown by the in vitro response of *Styela clava* hemocytes, which effect rapid (from 15 min to 4 h) cytotoxic reactions following cell contact in allogeneic cultures. Significantly higher frequencies of nonviable cells (13.2% differences with regard to autogeneic combinations) were observed by dye exclusion assays (microscopic observation or spectrophotometric quantification of eosin-y stained hemocytes), as a result of cytocidal activity (Kelly et al. 1992). Allogeneic combination analysis showed three discrete levels of cytotoxicity (low, inter-mediate, high) which are consistent with the hierarchical nature of the cytotoxic responses, in accordance with the presence of polymorphic cellular histocom-patibility antigens similar to those detected by tissue transplantation in *S. plycata* (Raftos et al. 1988).

3.5 In Vitro Spontaneous Hemocyte Cytotoxic Activity Against Mammalian Target Cells

The ability to lyse mammalian target cells (such as tumor line, virally infected cells, and erythrocytes) without prior sensitization, is primarily a function of vertebrate natural killer cells (NK cells) and macrophages (Klein and Mantovani 1993). Hemocyte-mediated cytotoxic activity against mammalian tumoral cells or erythrocytes was examined in solitary ascidians.

3.5.1 Erythrocyte Targets

Parrinello and Arizza (1992) were the first to report that hemocytes of *Ciona intestinalis* showed spontaneous cytotoxic capacity (HCA) when assayed in vitro against several erythrocyte targets. The membrane structure and lipid distribution of erythrocytes are well known (Van Deenen 1981), and represent useful targets for outlining a model for cytotoxic activity.

Human ABO, guinea pig, rabbit, and sheep erythrocytes (SE) were non-specifically lysed. The assays were carried out using isoosmotic media with targets (450 mOsm) or effector cells (1090 mOsm). Parrinello et al. (1995) showed that hemocytes effected cytotoxicity in isosomotic media including artificial sea water (SW), tunicate solution (according to Raftos and Cooper 1991), and marine solution (MS; according to Peddie and Smith 1993), as well as in a medium (TBS) isoosmotic for erythrocytes. TBS promotes erythrolysis, even if the hemocytes are adapted to the low osmolarity. In in vitro experiments, medium composition and osmolarity may be crucial in activating and/or increasing the hemocyte reactivity and/or the target susceptibility.

Experiments carried out in the TBS medium revealed that HCA, at an effector/target ration of about 1:3, is a calcium-dependent reaction that is evident at 20–25 °C within 15 min incubation time, reaching its highest value at 37 °C. Assays carried out using (1) the medium (TBS) in which hemocytes were maintained for 1 h at 37 °C, and (2) debris prepared from hemocyte lysates, suggest that HCA, assayed against SE, is a cell-mediated process that requires effector–target cell contact (Parrinello et al. 1993). Although such spontaneous cellular cytotoxicity was also a property of *S. plicata* hemocytes (Parrinello et al., submitted), further results show that a soluble lysin against rabbit erythrocytes, sphingomyelin-inhibitable, is contained in the lysate of *C. intestinalis* hemocytes. Probably, cytoplasmic lysins can be released as a result of the cell contact. An explanation of the cytotoxic mechanism could derive from the molecular characterization of the lysin (current research).

3.5.2 Tumor Cells

Additional results on the natural cytotoxic activity of *C. intestinalis* were obtained by Peddie and Smith (1993). They separated populations of hemocytes which were cytotoxic towards mammalian tumor cell lines, labeled them with carboxyfluorescein diacetate, and investigated them in vitro using fluorochromasia. The cytotoxic mechanism involves at least two stages distinguishable by their divalent cation requirements. In particular, selective removal of Ca^{2+} inhibits target cell lysis but does not reduce the formation of effector-to-target-cell conjugates. The response involves energy metabolism, requires an intact cytoskeleton, and entails active secretory processes in the effector cells (Peddie and Smith 1994).

4 The Hemocytes Involved in Cytotoxic Reactions (Table 2)

4.1 Are Immunocompetent Lymphocyte-Like Cells Cytotoxic?

The lymphocyte-like cell (5–7 μm diameter), presents a large nucleus which lacks a nucleolus, and a small amount of basophilic cytoplasm containing a few small vesicles (Fig. 1a). The cytoplasm of larger LLCs was similar in appearance to that of small amoebocytes. Cells, apparently intermediate to lymphocyte-like cells and small amoebocytes, containing more mitochondria, lysosome-like granules, and pseudopodia, were observed by TEM (Warr et al. 1977; De Leo 1992; Fuke and Fukumoto 1993). The resemblance of such cells to the typical lymphocytes of mammals was much greater in smears than in corresponding thin sections for TEM.

There is no direct evidence that LLCs are responsible for cytotoxic reaction, although, it has been postulated that they are responsible for the recognition and destruction of allogeneic tissue. In *S. plicata,* the association between LLC infiltration of incompatible transplants and accumulation around the allogeneic

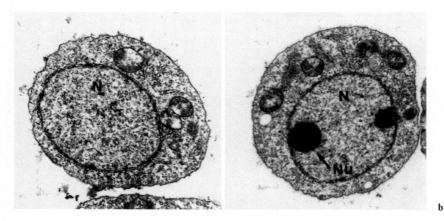

Fig. 1. a Lymphocyte-like cells of *Ciona intestinalis,* × 8000 **b** Hemoblast, × 8 000. (Electron micrograph kindly provided by Dr. G. De Leo)

Table 2. Ascidian hemocytes involved in cytotoxic reactions. Cytotoxic hemocytes have been defined on the basis of available data

Hemocytes	Cytotoxic reaction						References
	Specific Multiallelic locus/i-dependent			Nonspecific reaction			
					NK-like		
	Allograft rejection	Contact reaction	Nonfusion reaction	Tunic reaction	Anti-RBC	Anti-tumor	
Stem cells	?		?	-	-		Raftos et al. (1987a,b); Taneda and Watanabe (1982c)
Clear granulocytes (hyaline amoebocytes)						+	Peddie and Smith (1993)
Clear visicular granulocytes (visicular hyaline amoebocytes)						+	Peddie and Smith (1993)
Microgranulocytes (granular amoebocytes)		+		+		+	Fuke (1980); Parrinello et al. (1984, 1990); Peddie and Smith (1993)
Vacuolar granulocytes:							
Univacuolar (Signet-ring cells)		+	+	?	+		Fuke (1980); Parrinello and Patricolo (1984); Cammarata et al. (in prep.)
Globular - compartment cells			+				Raftos et al. (1987a,b); Fuke (1980)
- morula cells	+	+	+	?			Sabbadin et al. (1992); Parrinello et al. (in prep.)

RBC = red blood cells.

tunic blood vessels (Raftos et al. 1987a,b; Raftos and Cooper 1991) with subsequent cytocidal activity leading to graft degeneration, could reflect the capacity of this cell type to specifically recognize allogeneic cell determinants (Raftos et al. 1987b; Raftos 1991). In *Ciona intestinalis*, eight weeks after transplantation, tunic first-set allografts contained numerous LLCs (Reddy et al. 1975).

LLCs exhibited a specific adaptive response to allogeneic tissue: they accumulated within second-set allografts more rapidly than within first set allografts; however, they failed to rapidly penetrate third party grafts from animals other than the original first-set donor (Raftos 1991). In addition, adaptive proliferation of LLCs in response to allogeneic stimuli has been shown in *S. clava* preimmunized against allogeneic tissue by injecting hemocytes obtained from separate individuals (Raftos and Cooper 1991). Immunization protocols included first-set, second-set, and third-party challenge. Proliferative activity in host ascidians, monitored by bromodeoxyuridine (analog of thymidine) incorporation, was identified by immunohistochemical procedures (mAb anti-BrDu) in all tissues sampled (gut, mantle, pharynx) with the exception of circulating hemocytes. The incorporation of the BrDu was restricted to proliferative hemocytes involved in DNA replication as indicated by thymidine competition and gamma irradiation which inhibited BrDu uptake. Discrete crypts within the body wall of the allogeneic recipient were sites of significantly greater proliferative activity after first-set and second-set immunizations. At these sites, lymphocyte-like cells and microgranulocytes (granular amoebocytes) were BrDu positive. Differential counts of these two cell types revealed that variations observed between autogeneically and allogeneically primed animals were due largely to an increased frequency of LLCs: three times more LLCs were observed within 5 days of first-set immunization and a more rapid and powerful response was detected in second-set recipients, whereas in third-party hosts the LLC proportion did not differ significantly from those of first-set hosts. In contrast to the responses of LLCs, the frequency of BrDu-positive granular amoebocytes within the body wall increased after first-set and second-set allogeneic immunizations in both autogeneic and allogeneic grafts. The proliferative activity of these hemocytes was also evident in in vitro cultures of pharyngeal explants (Raftos et al. 1990). In that study, details of the lymphocyte-like cell morphology were not shown, and they were defined as small (5–7 μm diameter) spherical cells characterized immunohistochemically by their large (about 4μm) nuclei and limited cytoplasm. The possibility exists that the above cells, although named LLCs, were actually hemoblasts. In this respect, Sawada et al. (1993) have shown that the small cells contained in the hemolymph of this ascidian can be considered hemoblasts when examined by light and electron microscopy.

The response of microgranulocytes has been regarded as nonspecific, probably stimulated by wounding or inflammation rather than allogeneic recognition (Raftos and Cooper 1991). These hemocytes were described as large cells (8–12 μm diameter) with small nuclei and a distinct, granular cytoplasm that was often intensely stained following BrDu treatment.

4.2 Are Lymphocyte-Like Cells Involved in NFR?

The presence of lymphocyte-like cells was observed in the nonfusion reaction of *Perophora sagamiensis* (Koyama and Watanabe 1982). Histological observations showed that amoebocytes and lymphocyte-like cells were found 4 h after contact, near or on the epidermal cells of the contact area.

Studies of X-ray effects were performed on *Botryllus primigenus* colony specificity (Taneda and Watanabe 1982). When the cut surfaces of two previously irradiated colony pieces were brought into contact, only weak NFR took place in the contact area. Infiltrated cells were rarely observed, although fusion never occurred. In the circulating hemocytes, the mean percentage of LLCs was reduced (from 11.4 to 5.8% after 5000 R irradiation), amoebocytes decreased while phagocytes and morula cells increased greatly.

The X-ray results from *Botryllus* and the histological observations in *Perophora* suggest that lymphocyte-like cells can be involved in rejection, although direct correlation between these cells and cytotoxic processes remains to be determined.

A recent report demonstrated that *B. schlosseri* lymphocyte-like cells may proliferate in primary cultures of hemocytes from circulating hemolymph (Rinkevich and Rabinowitz 1993). In vitro differentiation of these cells to various hemocytes has not been achieved, although they were responsive to several mitogenic factors (*Botryllus* hemolymph, mixed IL-1Ó, IL-1β, and IL-2, and TNF). The surfaces of these cells present sugars, as shown by Con A and WGA-specific (but nonmitogenic) binding. Sugars are also present on the surface of signet-ring cells, compartment cells, and morula cells.

This evidence of LLCs involvement in NFR is predictable in view of their stem-cell role.

4.3 Stem-Cell Function of Lymphocyte-Like Cells

Morphological and functional distinctions between hemoblasts and lymphocyte-like cells have not been made in reports on ascidian cellular immunology. Hemoblasts within body-wall nodules appear to respond to immunologic stimulation by hemopoietic activity. These structures represent a major hemopoietic organ in *S. clava* (Ermak 1975, 1976, 1982), and both autogeneic and allogeneic immunizations substantially induce hemocyte proliferation in the body wall close to the site of injections (Raftos and Cooper 1991). Hemoblasts were also found in the circulating hemolymph Sawada et al. 1991, 1993; De Leo 1992; Fuke and Fukumoto 1993).

The hemoblast has the nucleus/cytoplasm ratio and ultrastructural features of undifferentiated elements (44–74 μm diameter; Fig. 1b). The round nucleus contains abundant chromatin which is attached to the inner nuclear envelope, and a prominent often granular nucleolus. Homogenous cytoplasm presents a large number of free ribosomes often arranged in polyribosomes, roundish or rod-like mitochondria, occasional rough endoplasmic reticulum, and a few lipid droplets. Frequently, differentiated structures characteristic of other cell types can be recognized in the hemoblasts. This finding agrees with the role of this cell as a source of other hemocyte types (Milanesi and Burighel 1978; De Leo 1992).

As far as we know, the small cells, named lymphocyte-like cells or hemoblasts (Fig. 1) may represent multipotent stem cells that differentiate the diverse hemocyte types (De Leo 1992; Ermak 1976, 1982; Wright 1981). Ermak (1976) suggested that hemoblasts in hemopoietic nodules produce lymphocyte-like cells which may give rise to all other hemocyte types.

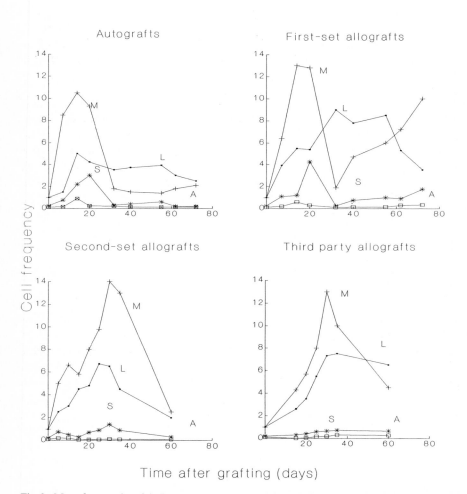

Fig. 2. Mean frequencies of the hemocytes surrounding *Styela plicata* autografts, first-set, second-set, and third-party allografts. (Graphs have been obtained by plotting data reported by Raftos et al. 1987b)

Also, the differential hemocyte counts within the allograft beds and tissues of *S. plicata*, partially support the stem cell function of LLCs. The graphs presented in Fig. 2 were obtained by plotting the data reported by Raftos et al. (1987b). The frequency of morula cells, signet-ring cells, and amoebocytes decreased before first-set allograft rejection (at 30 days), whereas LLC frequency drastically increased. However the results also showed that the frequency of the hemocytes, especially globular granulocytes (morula-cells), rose after 50 days when the LLC frequency started to decrease. Second-set allografts at 30 days showed a low influx of unilocular granulocytes (signet-ring cells) and amoebocytes, whereas morula cells were very frequent. In both cases, the frequency values of morula-cells appeared higher than those of LLCs. Moreover, a rapid onset of morula cells can be observed in second-set allografts. Thus, it may be postulated that LLCs could be the stem cells which differentiate into the hemocytes in allografts. On the other hand, postrecognition processes which need effector hemocytes, morula cells in particular, appear to be immunologically nonspecific (see also Sect. 6).

Similarly, among the inflammatory cells, stem cells have been found to infiltrate the tunic tissue (void of blood vessels) of *C. intestinalis* near the mantle epithelium as a result of the subcuticular injections of particulate or soluble irritants (Parrinello 1981; Parrinello et al. 1984; Parrinello and Particolo 1984; Parrinello et al. 1990). Stem cells (hemoblasts and LLCs) originate from the hemopoietic nodules situated in the pharyngeal and body walls under the tunic adjacent to the injection site. Studies of fine structures show crypts of hemopoietic tissue composed of hemoblasts, LLCs, and prevalently transitional cells, some of which pass through the epythelium layer into the tunic (paper prep.).

4.4 Hemocytes Involved in Botryllid Cytotoxic Reactions

Histological studies of NFR in *Botryllus* pairs have revealed that rejection implies infiltration of blood cells, especially globular granulocytes (morula cells), from the ampullae into the tunic where they disintegrate. Cell infiltration is mediated by increased permeability of the ampullar epithelium (Taneda and Watanabe 1982a), and induced by chemotactic factors (Hirose et al. 1990) which may congregate the cells at the ampullar tips before release into the tunic.

In *B. primigenus*, test cells and cells near the cut surface underwent a degenerative process, darkening completely and disintegrating. TEM observations (Tanaka and Watanabe 1973) showed that at first the outer membrane of the test cells disappeared, thus releasing mitochondria and vesicles into the test matrix. Electron-dense material in the vesicles, the vesicles themselves, and mitochondria gradually disappeared; swellings of the outer nuclear envelope membrane often preceded its complete disappearance.

NFR, at cellular level, directly appears to initiate in the ampullar lumen. Hemocytes from the ampullae, especially morula cells, infiltrate into the tunic where they disintegrate (reviewed in Sabbadin et al. 1992). Upon destruction of these cells, the electron-dense bodies were observed to have disintegrated into minute granules dispersing into the test matrix.

The roles of the hemocytes and plasma in nonself recognition and effector mechanisms were examined in *B. simodensis* (Saito and Watanabe 1984) and *B. primigenus* (Taneda and Watanabe 1982b) using the technique of allogeneic/syngeneic injection of whole hemolymph or its fractions into the ampullae of the recipient. In *B. primigenus*, allogeneic injection of whole hemolymph always induced inflammatory responses similar to NFR: formation of clusters within the vessels, contraction of ampullae, regression of zooids. The allogeneic challenge of separated hemocytes induced strong reactions, whereas only weak positive responses were induced by cell-free allogeneic plasma injection (Taneda and Watanabe 1982b). The rejection process is irreversible. When one partner (off-colony) in a colony allorecognition assay is taken away from the juxtaposed allogeneic partner (on-colony), after interaction has been established but before rejection was identifiable, rejection proceeds to completion in the remaining partner at the same rate as in the undisturbed allogeneic pairs (Tanaka 1973). This result is evidence for the role of immigrated humoral factors and infiltrated hemocytes from the removed partner in continuing the reactions in spite of the artificial removal. It is postulated that NFR is initiated by interactions between hemocytes and possibly a humoral (degranulating) factor(s) in the nonself plasma. Destruction of on-colony test cells appears to be brought about, in the short term, by a factor(s) released off-colony. The subsequent destruction of test cells may in turn release additional inflammatory factors.

The sequence of cellular events that follows allorecognition and leads to complete rejection was examined in vitro in paired oozooids of *B. schlosseri* (Scofield and Nagashima 1983). Besides morula cells, signet-ring cells, compartment cells (both considered morula-cell precursors) and granular amoebocytes were spread into the tunic as the ampullae retreated from the contact point. An allogeneic plasmic factor(s) and hemocytes could have a role in initiating a definite cascade of events: (1) granular amoebocytes in the blood or test cells in the tunic ground substance respond to a humoral factor(s) introduced reciprocally, and/or to cellular cytotoxicity, and are destroyed; (2) inflammatory factors generated by hemocyte destruction may elicit the migration of morula cells out of the ampullae and then cause their destruction. As observed in the *C. intestinalis* tunic reaction (Parrinello et al. 1977, 1984) the cytotoxic reactions following allogeneic contact are extremely destructive to surrounding tissues.

Alloimmune memory in botryllid NFR was not found (Rinkevich and Weissman 1991). However, Neigel (1988) reported that significant in vitro cytotoxicity between incompatible *B. schlosseri* colonies was only observed with combinations of hemocytes from colonies that had engaged in a previous in vivo colony rejection reaction. There was no cytotoxicity if the effector cells (unidentified) were from a colony primed against another unrelated to the source of the target cells. These results suggest a unilateral specific reaction.

5 Hemocytes Involved in the Contact Reaction

A number of morphologically distinct hemocytes are capable of performing xeno- or allogeneic reactions. Contact reaction was observed between cells of the same type, as well as between cells of different types. Vacuolated cells, granular amoebocytes of differently sized (large, small, minute, fine) granules, and large basophilic cells, were susceptible to this reaction. Because of their low frequency (6.9%) it has not been determined whether lymphocyte-like cells are involved.

Vacuolar granulocytes (vacuolated cells) are the most conspicous and abundant cells in the hemolymph of *H. roretzi* (Fuke and Fukumoto 1993). They are characterized by fluid-filled vacuoles or vesicles. Viable cells show active amoeboid movement and phagocytosis. The vacuolated cells have been distinguished into at least three subtypes which show the following main features: (1) one large vacuole containing granular material, an eccentrically located nucleus, free ribosomes and a few mitochondria in the cytoplasm (univacuolar granulocyte/signet-ring like cell); (2) several vacuoles containing electron-dense spherical bodies, many small vesicles, multivesicular bodies and residual bodies (globular granulocytes/compartment cells); (3) a large number of vacuoles similar in appearance to those of other vacuolated cells (probably, globular granulocytes/morula cells). Granular amoebocytes, are characterized by the presence of minute, electron-dense, membrane bordered granules usually not distributed throughout the cytoplasm. Macrogranular cells are spherical (8 μm in diameter), with indistinct spherules or vacuoles (10–20 per cell) containing very electron-dense material (possible globular granulocytes/morula cells); they differ from vacuolated cells because they lack amoeboid movement and show different stainability (vital stains) of the vacuoles. Large basophilic (15–20 μm in diameter) ovoid cells have numerous vacuoles which occupy the cytoplasm and contain characteristic microfilaments. The vacuolated cells contain vanadium (Michibata et al. 1987) and protein-bound iron (Michibata et al. 1986). They are reported to play some defensive roles, i.e., phagocytosis (Fuke 1979), encapsulation (Anderson 1971) and self/nonself recognition (Fuke 1980, 1990).

The exact nature of lysis is unclear. It was unpredictable which of the two cells would lyse first; in any case, lysis of one cell was always followed immediately by that of the other cell (bidirectional lysis). Microscope observations suggested that the reaction was triggered by direct cell contact, although the effector mechanisms may include both cellular and diffusible mediators. When contact between the vesicular cells of *H. roretzi* and *P. mirabilis* took place, some of the nearby *P. mirabilis* vesicular cells were found to be lysed without prior contact with the *H. roretzi* cells. Although the supernatants from cultures in which contact reactions had taken place did not show lytic activity, diffusible factor(s) released from one lysed cell may be able to trigger a chain reaction among nearby vesicular cells in a short-range cytotoxic reaction. Fuke (1980) suggested that the lysis of the first cell in a cell pair is effected by direct cellular interactions, and the lysis of the second cell may be a direct consequence of the lysis of the first. This reaction differs from cell-mediated cytotoxicity in higher animals.

Bilateral contact reactions was not found when hemocytes from incompatible *B. schlosseri* colonies were mixed. Experiments performed by a chromium release assay did not show specific lysis (>10%) (Neigel 1988).

6 Globular Granulocytes (Morula Cells): Are They Involved in Cytotoxic Reactions?

This hemocyte is usually a round cell (berry-like under the light microscope; 5.5–11.0 μm diamater) with large membrane-bound vacuoles, containing homogenous round masses of very electron-dense material (Fig. 3). Often, a single large mass is present, partially filling the vacuole and a few highly electron-dense granular inclusions, with microgranular or floccular features. Globular contents of various densities and aggregation characteristics ("loose fibrogranular and moderately dense material", De Leo 1992) are present (Fig. 3). In young *B. schlosseri* morula cells, the nucleus resembles that of a hemoblast; there are RER and free ribosomes, and near the nucleus a well-developed Golgi complex gives rise to many small opaque vesicles. Membrane-bound granules, of the same electron density as the dark masses of the vacuoles, seem to originate in the Golgi field, and then to approach the vacuoles and fuse with them. This granular product could be responsible for the formation of the large dark masses (Milanesi and Burighel 1978).

The morula cells were conspicuous participants in the rejection lesions of *S. plycata* tunic allografts. Where large numbers of these cells were evident within the contact zone graft/host (Raftos 1990), they were mainly responsible for NFR colonial ascidians, displayed (*S. plicata, C. intestinalis*) in vitro cytotoxic activity

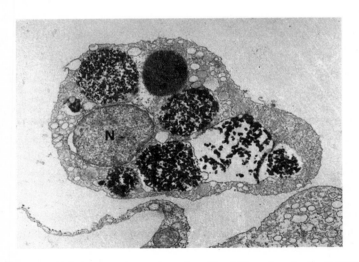

Fig. 3. Globular granulocyte (morula cell) *of Ciona intestinalis*, × 8000 *N* Nucleus. (Electron micrograph kindly provided by Dr. G. De Leo)

against mammalian erythrocytes, and were capable of contact reaction in solitary ascidians. Finally, the results of tunic grafting experiments in the solitary ascidian *Molgula manhattensis* (Anderson 1971) support the role of morula cells as inflammatory effector cells. Neither autografts nor allografts (tissue fragments of the branchial wall) were accepted by this ascidian, probably because they were isolated by mucus production within the tissue bed. The grafts were destroyed by a marked hemocyte response characterized by the infiltration of a great number of morula cells. Moreover, in the same ascidian, signet-ring cells and morula cells were present during the early response to glass fragments implanted in the branchial wall, whereas after 2 days, exclusively morula cells and their released substances encapsulated the foreign material.

Numerous morula cells were observed in *Ciona* inflammatory reactions produced in the tunic tissue by erythrocyte subcuticular injection. Various transitional stages, until cell break-up, suggested that the vacuolar material underwent changes and was released into the surrounding tissue which was in degeneration (paper in preparation). The morula cells have also been regarded as being involved in wound repair and tunic formation (Smith 1970; De Leo et al. 1981; Zaniolo 1981).

Two cytotoxic mechanisms may be effected by morula cells.

(1) As shown by electron-probe X-ray microanalysis, these cells contain reduced iron and sulfuric acid (Milanesi and Burighel 1978). The color reaction observed after they are released into the tunic during the NFR, may reflect a change in the oxidation state of the iron that in some way may participates in allogeneic effector reactions. It is known that hydrogen peroxide, ferrous sulfate, and potassium iodide can together generate cytotoxic iodide and hydroxyl radicals (Klebanoff 1982). All tunicates have amounts of bound iodine in the blood and tunic matrix (Barrington 1975), therefore, in a free radical-generating reaction such intermediaries could participate in killing mechanisms, as in vertebrate neutrophils, monocytes, and NK cells (Roder et al. 1982).

(2) Another cytotoxic mechanism may be related to the phenoloxidase (PO) activity showed by morula cells, and probably involved in inflammatory responses. In crustacean hemocytes, the proPO system, when activated to PO, exhibits biological functions similar to vertebrate complement, such as the production of fungal, opsonic, lytic, and degranulating factors (reviewed in Johansson and Söderhäll 1992). Upon activation, "sticky proteins" appeared to lyse semigranular crayfish hemocytes in vitro (Smith and Söderhäll 1983), while semigranular and granular cells which contain the proPO system are able to be cytotoxic towards a murine tumor cell line (Söderhäll et al. 1985).

The morula cells of several ascidian species (Smith and Söderhäll 1991), including *S. plicata* (Arizza et al. 1995) and *B. schlosseri* (Ballarin et al. 1993), contain proPO. In the solitary ascidian *C. intestinalis*, morula cells (and compartment cells) showed serine protease and PO activity, and played and indirect role in phagocytosis by contributing opsonic factors which mediate cell cooperation with phagocytic amoebocytes (vacuolar and granular) as shown by in vitro assays (Smith and Peddie 1992). It has been proposed that the opsonic effect is associated with the activation of prophenoloxidase (proPO) or protease by foreign agents (i.e. bacterial LPS). The extent to which these enzymes are linked to immune responses in ascidians is unknown.

There is evidence that in allograft reaction and (probably) NFR, recognition events could trigger stem cells to proliferate and differentiate into morula cells and their precursors. The specificity imposed by cellular preimmunization suggests that allogeneic recognition in *Styela plicata* was restricted to the initial phase of graft rejection, in which LLCs were extremely common, whereas the second phase was strongly characterized by inflammatory cells and cell destruction within incompatible transplantation. The initial self/nonself recognition generally leads to the activation of an amplification system involving cytokinetic mediators. Interleukin-1-like activity was shown in plasma or tissue extracts from eight tunicate species including *B. schlosseri*, *S. plicata*, *C. intestinalis* and *H. pyriformis*. The chromatographically isolated tunicate IL-1 fraction showed a molecular weight of 20000 daltons, stimulated mice thymocyte proliferation assayed in the presence of submitogeneic concentrations of ConA, and caused an increase in vascular permeability (Beck et al. 1989). Therefore tunicate IL-1 may be related to similar inflammatory molecules in mammals. The IL-1 fraction of *S. clava*, isolated by gel filtration and chromatofocusing chromatography, also stimulated the in vitro proliferation of lymphocyte-like cells and granular amoebocytes (Raftos et al. 1991).

7 Hemocytes with NK-Like Activity

7.1 Antierythrocytes

C. intestinalis hemocyte separations on a discontinuous Percoll gradient, showed that univacuolar granulocytes (Fig. 4), are killer cells (Parrinello et al., submitted). Attempts are currently being made to examine the erythrolytic capability of other hemocytes. This cell type (5–8 μm diameter) presents a flattened or kidney-shaped highly eccentric nucleus, and a large vacuole (3–6 μm diameter) which reduces the cytoplasm to the form of a thin peripheral ring. The vacuole contains material of various electron densities, which can be apparently homogeneous, or finely granular, or arranged in flocculent structures and masses. These morphological features could be linked with functional variations. The lytic mechanism is unknown. However, erythro-lysins have been found in the lysate of the separated cytotoxic hemocytes (paper in prep.), and inhibition experiments with lipidic components of the erythrocyte membrane showed that, among the phospholipids tested (phosphorylethanolamine, phosphatydilserine, phosphatidylcholine) at various concentrations (0.03–250 μg/ml) in TBS or MS medium, sphingomyelin was active in inhibiting the hemolytic activity of the hemocytes (Parrinello and Arizza 1992; Parrinello et al. 1995). Cholesterol was inactive, whereas ceramid and phosphorylcholine (components of the sphingomyelin molecule) displayed a low inhibitory

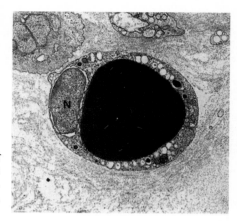

Fig. 4. Univacuolar refractile granulocyte of *Ciona intestinalis*, × 10 000 (Electron micrograph kindly provided by Dr. G. De Leo) *N* Nucleus. Dense, apparently homogeneous and compact, finely granular material occupies the vacuole

capability. Attempts to show the sphingomyelinase activity of the chlorophorm-methanol extracts from hemocyte membranous debris, failed. On the basis of a cytotoxic inhibition experiments, phospholipid involvement has been suggested in the regulation of the mammalian cytotoxic activities exerted by NK cells or cytotoxic T-lymphocytes (Yue et al. 1987; Tschop et al. 1989; Ojcius and Young 1990). Although further attempts must be made to look for hemocyte membrane-bound phospholipases, the results suggest that *Ciona* hemocyte lysin is a protein which is specifically active with sphingomyelin, causing changes in erythrocyte membrane permeability leading to lysis. Also the reactions of soluble hemolysins contained in the cell-free coelomic fluid of some echinoderms, may be mediated by interaction with membrane targets by means of lipids (sphingomyelin, cholesterol, phosphatidyl-inositol and phosphatidylethanolamine) (Canicattí et al.1987; Canicattí 1989,1991).

Hemocyte cytotoxicity appears to be triggered by an unrestricted recognition mechanism. Agglutinins are released from the hemocytes into the medium during in vitro experiments (Arizza et al. 1991, 1993; Cammarata et al. 1993; Parrinello et al. 1993). These factors may be involved in self-recognition. The absence of self markers on erythrocyte membranes may activate a mechanism which induces lipid-lysin interaction leading to cytotoxic reactions.

7.2 Antitumor

Comparison of the cytotoxic activity by each cell band separated from *C. intestinalis* hemolymph, showed that cytotoxic cells were enriched in bands containing clear granulocytes (nonvacuolar hyaline amoebocytes, non phagocytic) or clear vesicular granulocytes (vacuolar hyaline amoebocytes) and microgranulocyte (granular amoebocytes) (phagocytic) (Peddie and Smith 1993).

Clear granulocytes, are characterized by a cytoplasm containing numerous vesicles, a small amount of RER, profiles of the Golgi complex, and a few mitochondria; the peripheral region devoid of organelles contains a few micro-tubules and microfilaments. Morphologically, the vesicles resemble the primary lysosomes of mammalian monocytes/macrophages, apparently originating from the mature face and ends of the Golgi cisternae (Rowley 1982).

Clear vesicular granulocytes, have a cytoplasm containing several large and small vacuoles, which are apparently empty or contain fibrous material. In some cells vacuoles, an amorphous electron-dense material is enclosed associated with the limiting membrane; vanadium-bearing vacuoles are also contained. The cytoplasm contains small amounts of RER, occasional Golgi complexes and a few pinocytotic vesicles. Rowley (1982) suggested that this cell type may be part of the developmental series which leads to granular amoebocytes (phagocytic).

Microgranulocytes are readily identifiable since they contain: (1) large gran-ules which consist of an electron-dense area often located to one side of the granule and surrounded by microtubules; (2) vacuoles containing "droplets" of electron-dense unstructured material which consists primarily of vanadium.

Statistical analysis by paired t-test has revealed that clear granulocytes (non-phagocytic) cause a significantly larger lysis than the phagocytic hemocytes.

Experiments to characterize cytolysis have demonstrated that activity increases proportionally with the effector-to-target cell ratio, occurs within 15 min, and is maximal at 20 °C incubation temperature.

Experiments have not been performed to examine the possibility that this "spontaneous" activity can be increased by priming challenges.

8 Conclusions

Ascidians are marine benthic invertebrates which undergo transplantation reac-tions in the wild. Field experiments have shown that the sibling larvae of solitary and colonial ascidians settle in aggregations (Young and Braithwaite 1980; van Duyl et al. 1981; Schmidt 1982; Grosberg and Quinn 1986; Davis 1987). Cosettle-ment on space-limited surfaces may result in contact and fusion between neigh-boring genetically distinct but immunologically compatible specimens. In this way, heterozygosity is promoted, cross-fertilization in hermaphroditic animals is favoured, and advantages may consequently arise for the organisms involved (Rinkevich 1992; Rinkevich and Weissman 1992, and literature therein). Fusion is mediated by tunic contacts and is accomplished by commixture of vascular systems and limited mixing of the hemolymph and tissue elements. Rinkevich and Weissman (1992) suggest that, in *Botryllus*, a pleiotropic role for Fu/Hc locus may be used as a screening tool for heterozygosity/homozygosity of the whole genoma of both partners within a single chimera. Colony growth by budding may result in tunic contact between neighboring animals, or alter-natively, two colonies may meet with their extending ampullae.

The limits of fusibility are indicated by the appearance of rejection which, in turn, restricts the eventual substrate surface that each specimen may dominate. This may be a benefit because the substrate is a limited resource for sedentary marine organisms, and may be particularly significant for colonial species. Tunic allograft rejection in solitary ascidians, and nonfusion reactions in botryllids could be the result of incompatible transplantation in the wild when tunic contact is established. It has been suggested that the recognition site shifts from the ampullae towards the tunic surface during evolution. Thus, subcuticular rejection in *Botrylloides simodensis* and *Botrylloides violaceus* is the most advanced (see Rinkevich 1992, and literature therein). On the other hand, the hemocyte contact reactions may indicate that these cells developed xeno- and allospecific reactivity to selected hemolymph commixtures when tunic fusion occurred in solitary ascidians.

The development of a susceptibility to pathogenic infection could express another pleiotropic role for allorecognition loci. Weissman (1988) suggested that vertebrate MHC originated from an ancient polymorphic gene family which encoded interacting cell-surface molecules.

Tunicates have developed multiple modes of allorecognition and cytotoxicity. It is manifest that cytotoxicity in graft reaction, NFR, and bilateral allogeneic contact reaction is triggered by specific recognition (dependent on multiallelic locus/i), whereas NK-like (anti-erythrocyte, and anti-tumor cells) reaction is not specific. As shown in Table 1, specific and nonspecific cytotoxicity may be found in the same species.

There is an ongoing debate about recognition mechanisms. The simplest mechanism is based on a single superficial receptor which tests for the presence or absence of self-markers, rather than one based on a specific reaction to nonself markers. Such a recognition mechanism could account for the rejection of foreignness leading to cell-contact-dependent NK-like activity. It has been proposed that rapid constitutive NK cytotoxic activity in mammals is also governed by a regimen to detect the deleted or reduced expression of self-MHC (Karre et al. 1986).

The genetics of historecognition can be explained by a nonself recognition model: if a single allele difference does not produce a nonself signal above a necessary threshold, it does not induce a nonself response. This threshold could exist at the level of individual cells on whose surface a minimum number of receptors have to be engaged for activation. The nonself-recognition model was tested in *B. schlosseri* by application of a chromium release assay of cytotoxicity, and the specificity of in vitro cytotoxic reactions was examined (Neigel 1988).

Tunic graft rejection in solitary ascidians is characterized by specific immuno-recognition showing anamnestic potential. This response is mediated by lymphocyte-like cells which are restricted by polymorphic histocompatibility antigens. The genetic patterns of graft rejection and the results of allogeneic cytotoxic reactions are consistent with a regime in which a discrete immunocompetent cell population recognizes foreign histocompatibility antigens. The recognition of allogeneic cells should represent the inductive phase of transplantation rejection. It has been found that hemocytes can be primed to specific incompatible allogeneic colonies by prior in vivo colony rejection assay. Alloresponses might have a precursor-effector activation that results from allorecognition. In vitro hemocyte assays, which measured cytotoxic reactions, revealed a sensitization to the former allotypic combination when compared to new allogeneic partners. This ability to respond selectively to different types of nonself is in agreement with the detection of nonself attributes, as is observed in the vertebrate immune system.

Several hemocyte types appear to be capable of self/nonself recognition and cytotoxic activity, and several cytotoxic mechanisms may be activated by the recognition events. Spontaneous cytotoxic activity (named contact reaction) appears to be a property of vacuolar granulocytes signet-ring cells, globular-compartment cells and morula cells), microgranulocytes (granular amoebo-cytes), and large basophilic cells. Natural NK-like activity against erythrocytes and tumor cells is a characteristic of univacuolar granulocytes and clear vesicular granulocytes (nonphagocytic, nonvacuolar amoebocytes), respectively, and to a lesser extent clear granulocytes (phagocytic vacuolar granular amoebocytes).

Although morula cells had been considered responsible for tunic tissue formation and regeneration (Smith 1970; De Leo et al. 1981; Zaniolo 1981), their role in allograft tissue rejection and in nonfusion reaction, and their cytotoxic in bilateral contact reactions has also been demonstrated. Probably by way of transitional forms, globular amoebocytes (morula cells) develop from stem cells which have been stimulated to proliferate by allogeneic challenges in restricted responses. Similarly, stem cells proliferate following inflammatory stimuli (Parrinello et al. 1977, 1984) and differentiate effector cells causing disruptive reactions. Lymphocyte-like cells are stem cells.

As suggested by in vitro experiments (Parrinello et al., submitted) univacuolar refractile granulocytes may be involved in the cytotoxic reactions, causing inflammatory aspecific responses in which microgranulocytes (granular amoebocytes) and globular granulocytes (morula cells) also participate. The inflammatory reaction starts with cell death and necrosis which is extended (also by lysosomal factors) to the adjacent tunic matrix (allograft, NFR, "tunic reaction"). Tunic reaction in *Ciona intestinalis* showed that cellular responses caused the destruction of foreignness (including the surrounding tissue), isolated the damaged area, and reestablished body integrity. Different aspecific inflammatory treatments (heating, UV irradiation, gamma irradiation) initiate rapid necrosis of the tunic matrix, observed by graft degeneration (Raftos 1991).

The natural activity of hemocytes is a nonspecific response as shown by the modes observed in the contact reaction and antimammalian cell-target natural cytotoxicity. We do not know about effector molecular mechanisms. When compared to mammals, these immune responses show no relationship with cytotoxic effector lymphocytes (Tc, K and NK cells) but could be similar to macrophage activity.

A considerable role may be assigned to cytokines, but further research is needed in order to characterize them, identify the releasing cells, and examine the structural and functional aspects of hemocyte-activating molecules.

Aknowledgements. Supported in part by CNR grant CT 92.02739.04 and MURST 1993. Many thanks are also due to Dr. Matteo Cammarata for his assistance in preparing the manuscript.

References

Anderson RS (1971) Cellular responses to foreign bodies in the tunicate *Molgula manhattensis* (De Kay). Biol Bull 141: 91–98

Arizza V, Parrinello N, Cammarata M (1993) Immunocytochemical localization of cellular lectins in *Phallusia mamillata* hemocytes. Anim Biol 2: 15–22

Arizza V, Parrinello N, Schimmenti S (1991) In vitro release of lectins by *Phallusia mamillata* hemocytes. Dev Comp Immunol 15: 219–226

Arizza V, Cammarata M, Tomasino MC, Parrinello N (1995) Phenoloxidase characterisation in vacuolar hemocytes from the solitary ascidian *Styela plicata*. J Invertebr Path, in press

Ballarin L, Cima F, Sabbadin A (1993) Histoenzymatic staining and characterization of the colonial ascidian *Botryllus schlosseri* hemocytes. Boll Zool 60: 19–24

Ballarin L, Cima F, Sabbadin A (1994) Phenoloxidase in the colonial ascidian *Botryllus schlosseri* (Urochordata: Ascidiacea) Anim Biol 3: 41–48

Barrington EJW (1975) Problems of iodine binding in ascidians. In: Barrington EJW, Jefferies RPS (eds) Protochordates. Academic Press, London, pp 129–158

Beck G, Vasta GR, Marchlonis JJ (1989) Characterization of interleukin-1 activity in tunicates. Comp Biochem Physiol 92B: 93–98

Boyd HC, Weissman IL, Saito Y (1990) Morphologic and genetic verification that *Monterey Botryllus* and Woods Hole *Botryllus* are the same species. Biol Bull 178: 239–250

Brancoft FW (1903) Variation and fusion of colonies in compound ascidians. Proc Calif Acad Sci 8: 137–186

Cammarata M, Parrinello N, Arizza V (1993) Lectin release by density gradient separated and in vitro cultured hemocytes from *Phallusia mamillata*. J Exp Zool 266: 319–327

Canicatti C (1989) Evolution of lytic system in echinoderms. II Naturally occurring hemolytic activity in *Marthasterias glacialis* (Asteroida). Comp Biochem Physiol 93A: 587–591

Canicatti C (1991) Binding properties of *Paracentrotus lividus* (Echinoidea) hemolysin. Comp Biochem Physiol 98A: 463–468

Canicatti C, Parrinello N, Arizza V (1987) Inhibitory activity of sphingomyelin of hemolytic activity of coelomic fluid of *Holothuria polii* (Echinodermata). Dev Comp Immunol 11: 29–35

Cooper EL (1980) Phylogeny of cytotoxicity. Endeavour 4: 160–165

Davis AR (1987) Variation in recruitment of the subtidal colonial ascidian *Podoclavella cylindrica* (Quoy & Gaimard): the role of substratum choice and early survival. J Exp Mar Biol Ecol 106: 57–71

De Leo G (1992) Ascidian hemocytes and their involvement in defence reactions. Boll Zool 59: 195–213

De Leo G, Patricolo E, Frittitta G (1981) Fine structure of the tunic of *Ciona intestinalis* L. II. Tunic morphology, cell distribution and their functinal importance. Acta Zool 62: 259–271

Du Pasquier L, Blomber B, Bernard CCA (1979) Ontogeny of immunity in amphibians: changes in antibody repertoires and appearance of adult major histocompatibility antigens in *Xenopus*. Eur J Immunol 9: 900–906

Ermak TH (1975) An autoradiographic demonstration of blood cell renewal in *Styela clava* (Urochordata: Ascidiacea). Experientia 31: 837–838

Ermak TH (1976) The hematogenic tissues of tunicates. In: Wright RK, Cooper EL (eds) Phylogeny of thymus and bone marrow-bursa cells. Elsevier, Amsterdam, pp 45–55

Ermak TH (1982) The renewing cell populations of ascidians. Am Zool 22: 795–805

Evans DL, McKinney EC (1991) Phylogeny of cytotoxic cells. In: Warr GW, Cohen N (eds) Phylogenesis of immune functions. CRC Press, Boca Raton, pp 215–239

Fuke MT (1979) Studies on coelomic cells of some Japanese ascidians. Bull Mar Biol Stn Asamushi Tohoku Univ 16: 142–159

Fuke MT (1980) "Contact reaction" between xenogeneic or allogeneic celomic cells of solitary ascidians. Biol Bull 158: 304–315

Fuke MT (1990) Self and nonself recognition in the solitary ascidian *Halocynthia roretzi*. In: Marchalonis J, Reinish C (eds) Defense molecules. Alan R Liss, New York, pp 107–117

Fuke MT, Fukumoto M (1993) Correlative fine structural behavioral, and histochemical analysis of ascidian blood cells. Acta Zool 74: 61–71

Fuke MT, Nakamura I (1985) Patterns of cellular alloreactivity of the solitary ascidian, *Halocynthia roretzi*, in relation to genetic control. Biol Bull 178: 239–250

Fuke MT, Numakunai T (1982) Allogeneic cellular reactions between intraspecific types of a solitary ascidian, *Halocynthia roretzi*. Dev Comp Immunol 6: 253–261

Gleisner JM (1979) Lysosmal factors in inflammation. In: Houck JC (ed) Chemical messangers of the inflammatory process. Elsevier, Amsterdam, pp 229–260

Grosberg RK, Quinn JF (1986) The genetic control and consequences of kin recognition by the larvae of a colonial marine invertebrate. Nature 322: 456–459

Hirsochhron R (1974) Lysosomal mechanisms in the inflammatory process. In: Zweifach BW, Grant L, Mc Cluskey RT (eds) The inflammatory process, vol 1. Academic Press, New York, pp 259–285

Hirose E, Saito Y, Watanabe H (1988) A new type of the manifestation of colony specificity in the compound ascidian, *Botrylloides violaceus* Oka. Biol Bull 175: 240–245

Hirose E, Saito Y, Watanabe H (1990) Allogeneic rejection induced by cut surface contact in the compound scidian, *Botrylloides simodensis*. Invertebr Reprod Dev 17: 159–164

Jackson AD, Smith VJ, Peddie CM (1993) In vitro phenoloxidase activity in the blood of *Ciona intestinalis* and other ascidians. Dev Comp Immunol 17: 97–108

Johansson MW, Söderhäll K (1992) Cellular defence and cellular adhesion in crustacean. Parasitol Today 5: 171–176

Karre K, Ljunggren HG, Piontek G, Kiessling R (1986) Selective rejection of H-2-deficient lymphoma variants suggests alternative immune defence strategy. Nature 319: 675–678

Kelly KL, Cooper EL, Raftos DA (1992) In vitro allogeneic cytotoxicity in the solitary urochordate *Styela clava*. J Exp Zool 262: 202–208

Klebanoff SJ (1982) The iron-H_2O_2-iodide cytotoxic system. J Exp Med 256: 1262–1267

Klein E, Mantovani A (1993) Action of natural killer cells and macrophages in cancer. Curr Opin Immunol 5: 714–718

Koyama H, Watanabe H (1982) Colony specificity in the ascidian *Perophora sagamiensis*. Biol Bull 162: 171–186

Michibata H, Tareda T, Anada N, Yamakawa K, Numakunai T (1986) The accumulation and distribution of vanadium, iron, and manganese in some solitary ascidians. Biol Bull 171: 672–681

Michibata H, Hirata J, Uesaka M, Numakunai T, Sakurai H (1987) Separation of vanadocytes: determination and characterization of vanadium ion in the separated blood cells of ascidian, *Ascidia ahodori*. J Exp Zool 244: 33–38

Milanesi C, Burighel P (1978) Blood cell ultrastructure of the ascidian *Botryllus schlosseri* I. Hemoblast, granulocytes, macrophage, morula cell and nephrocyte. Acta Zool 59: 135–147

Mukai H, Watanabe H (1974) On the occurrence of colony specificity in some compound ascidians. Biol Bull 147: 411–421

Mukai H, Watanabe H (1975) Fusibility of colonies in natural populations of the compound ascidian *Botrylloides violamus*. Proc Jpn Acad 51: 48–50

Neigel JE (1988) Recognition of self or nonself? Theoretical implications and an empirical test. In: Grossberg RK, Hedgecock D, Nelson K (eds) Invertebrate historecognition. Plenum Press, New York, pp 127–142

Ojcius DM, Young JDE (1990) Characterization of the inhibitory effect of lysolipds on perforin-mediated hemolysis. Mol Immunol 27: 257–261

Oka H, Watanabe H (1957) Colony specificity in compound ascidians as tested by fusion experiments. Proc Jpn Acad 33: 657–659

Parrinello N (1981) The reaction of *Ciona intestinalis L.* (Tunicata) to subcuticular erythrocyte and protein injection. Dev Comp Immunol 5 (Suppl) 1: 105–110

Parrinello N, Arizza V (1992) Cytotoxic activity of the invertebrate hemocytes with preliminary findings on the tunicate *Ciona intestinalis*. Boll Zool 59: 183–189

Parinello N, Patricolo E (1984) Inflammatory-like reaction in the tunic of *Ciona intestinalis* (Tunicata). II. Capsule components Bio Bull 167: 238–250

Parrinello N, Cammarata M, Arizza V (1995) Univacuolar hemocytes from the tunicate *Ciona intestinalis* are cytotoxic for mammalian erythrocytes in vitro. Biol Bull, submitted

Parrinello N, Patricolo E, Canicatti C (1977) Tunicate immunobiology. Tunic reaction of *Ciona intestinalis L.* to erythrocyte injection. Boll Zool 44: 373–381

Parrinello N, Patricolo E, Canicatti C (1984) Inflammatory-like reaction in the tunic of *Ciona intestinalis* (Tunicata). I. Encapsulation and tissue injury. Biol Bull 167: 229–237

Parrinello N, De Leo G, Di Bella MA (1990) Fine structural observation of the granulocytes involved in the tunic inflammatory-like reaction of *Ciona intestinalis* (Tunicata). J Invertebr Pathol 56: 181–189

Parrinello N, Arizza V, Cammarata M, Parrinello DM (1993) Cytotoxic activity of the *Ciona intestinalis* (Tunicata) hemocytes: properties of the in vitro reaction against erythrocyte targets. Dev Comp Immunol 17: 19–27

Parrinello N, Cammarata M, Lipari L, Arizza V (1995) Sphingomyelin inhibition of *Ciona intestinalis* (Tunicata) cytotoxic hemocytes assayed against sheep erythrocytes. Dev Comp Immunol 19: 31–42

Peddie CM, Smith VJ (1993) In vitro spontaneous cytotoxic activity against mammalian target cells by the hemocytes of the solitary ascidian, *Ciona intestinalis*. J Exp Zool 267: 616–623

Peddie CM, Smith VJ (1994) Mechanism of cytotoxic activity by hemocytes of solitary ascidian, *Ciona intestinalis*. J Exp Zool 270: 335–342

Raftos DA (1990) The morphology of allograft rejection in *Styela plicata* (Urochordate: Ascidiacea). Cell Tissue Res 261: 389–396

Raftos DA (1991) Cellular restriction of histocompatibility responses in the solitary urochordate, *Styela plicata*. Dev Comp Immunol 15: 93–98

Raftos DA, Briscoe DA (1990) Genetic basis of histocompatibility in the solitary urochordate *Styela plicata*. J Hered 81: 96–100

Raftos DA, Cooper EL (1991) Proliferation of lymphocyte-like cells from the solitary tunicate, *Styela clava*, in response to allogeneic stimuli. J Exp Zool 260: 391–400

Raftos DA, Tait NN, Briscoe DA (1987a) Allograft rejection and alloimmune memory in the solitary urochordate, *Styela plicata*. Dev Comp Immunol 11: 343–351

Raftos DA, Tait NN, Briscoe DA (1987b) Cellular basis of allograft rejection in the solitary urochordate, *Styela plicata*. Dev Comp Immunol 11: 713–725

Raftos DA, Briscoe DA, Tait NN (1988) The mode of recognition of allogeneic tissue in the solitary urochordate *Styela plicata*. Transplantation 45: 1123–1126

Raftos DA, Stillmann DL, Cooper EL (1990) In vitro culture of tissue from the tunicate *Styela clava*. In Vitro Cell Dev Biol 26: 962–970

Raftos DA, Cooper EL, Habicht GS, Beck G (1991) Invertebrate cytokines: tunicate cell proliferation stimulated by an interleukin 1-like molecule. Proc Natl Acad Sci USA 88: 9518–9522

Reddey AL, Bryan B, Hidelmann WH (1975) Integumentary allograft versus autograft reactions in *Ciona intestinalis*: a protochordate species of solitary tunicate. Immunogenetics 1: 584–590

Rinkevich B (1992) Aspects of the incompatibility nature in botryllid ascidians. Anim Biol 1: 17–28

Rinkevich B, Rabinowitz C (1993) In vitro culture of blood cells from the colonial protochordate *Botryllus schlosseri*. In Vitro Cell Dev Biol 29A: 79–85

Rinkevich B, Weissman IL (1988) Retreat growth in the ascidian *Botryllus schlosseri*: a consequence of nonself recognition. In: Grosberg RK, Hedgecock D, Nelson K (eds) Invertebrate historecognition. Plenum Press, New York, pp 93–109

Rinkevich B, Weissman IL (1991) Interpopulational allogeneic reactions in the colonial protochordate *Botryllus schlosseri*. Inter Immun 3: 1265–1272

Rinkevich B, Weissman IL (1992) Chimeras vs genetically homogeneous individuals: potential fitness costs and benefits. OIKOS 63: 119–124

Roder JC, Helfand SJ, Werkemeister J, McGarry R, Beaumont TJ, Duwe A (1982) Oxygen intermediates are triggered early in the cytolytic pathway of human NK cells. Nature 298: 569–572

Rowley AF (1982) Ultrastructural and cytochemical studies on the blood cells of the sea squirt, *Ciona intestinalis* I. Stem cells and amoebocytes. Cell Tissue Res 223: 403–414

Sabbadin A (1962) le basi genetiche della capacita' di fusione fra colonie in *Botryllus schlosseri* (Ascidiacea). Rend Accad Naz Lincei 32: 1031–1035

Sabbadin A, Zaniolo G, Ballarin L (1992) Genetic and cytological aspects of histocompatibility in ascidians. Boll Zool 59: 167–173

Saito Y, Watanabe H (1984) Partial biochemical characterization of humoral factors involved in the nonfusion reaction of the botryllid ascidian *Botrylloides simodensis*. Zool Sci 1: 229–235

Sawada T, Fujikura Y, Tomonaga S, Fukumoto T (1991) Classification and characterization of ten types of hemocytes in tunicate *Halocynthia roretzi*. Zool Sci 8: 939–950

Sawada T, Zhang J, Cooper E (1993) Classification and characterization of hemocytes in *Styela clava*. Biol Bull 184: 87–96

Schmidt CH (1982) Aggregation and fusion between conspecifies of a solitary ascidian. Biol Bull 162: 195–201

Scofield VL, Nagashima LS (1983) Morphology and genetics of rejection reactions between oozoids from the tunicate *Botryllus schlosseri*. Biol Bull 165: 733–744

Scofield VL, Schlumpberger JM, Weissman IL (1982a) Colony specificity in the colonial tunicate *Botryllus* and the origin of vertebrate immunity. Am Zool 22: 783–794

Scofield VL, Schlumpberger JM, West LA, Weissman IL (1982b) Protochordate allorecognition is controlled by an MHC-like gene system. Nature 295: 499–502

Smith MJ (1970) The blood cells and tunic of the ascidian *Halocynthia aurantum* (Pallas). I. Hematology, tunic morphology, and partition of cells between blood and tunic. Biol Bull 138: 345–378

Smith MJ, Peddie CM (1992) Cell cooperation during host defence in the solitary turnicate *Ciona intestinalis* (L). Biol Bull 183: 211–219

Smith VJ, Söderhäll K (1983) Induction of degranulation and lysis of haematocytes in the freshwater crayfish, *Asctacus astacus* by components of the prophenoloxidase activating system in vitro. Cell Tissue Res 233: 295–303

Smith MJ, Söderhäll K (1991) A comparison of phenoloxidase activity in the blood of marine invertebrates. Dev Comp Immunol 15: 251–262

Söderhäll K, Wingren A, Johansson MW, Bertheussen K (1985) The cytotoxic reaction of hemocytes from the freshwater crayfish, *Asatacus astacus*. Cell Immunol 94: 326–332

Tanaka K (1973) Allogeneic inhibition in a compound ascidian, *Botryllus primigenus Oka*. II Cellular and humoral responses in "nonfusion" reaction. Cell Immunol 7: 427–443

Tanaka K, Watanabe H (1973) Allogeneic inhibition in a compound ascidian *Botryllus primigenus Oka*. I. Processes and features of "nonfusion" reaction. Cell Immunol 7: 410–426

Taneda Y, Watanabe H (1982a) Studies in colony specificity in the compound ascidian *Botryllus primigenus Oka*. I. Initiation of "nonfusion" reaction with special reference to blood cells infiltration. Dev Comp Immunol 6: 43–52

Taneda Y, Watanabe H (1982b) Studies in colony specificity in the compound ascidian *Botryllus primigenus Oka*. II. In vivo bioassay for analyzing the mechanism of "nonfusion" reaction. Dev Comp Immunol 6: 243–252

Taneda Y, Watanabe H (1982c) Effect of X-irradiation on colony specificity in the compound ascidian *Botryllus primigenus Oka*. Dev Comp Immunol 6: 665–673

Taneda Y, Saito Y, Watanabe H (1985) Self or nonself discrimination in ascidians. Zool Sci 2: 433–442

Tschopp J, Schafer S, Masson D, Peitsch M, Heusser C (1989) Phosphorylcholine acts as a Ca-Dependent receptor molecule for lymphocyte perforin. Nature 337: 272–274

Van Deenen LLM (1981) Topology and dynamics of phospholipids in membranes. FEBS Lett 123: 3–15

Van Duyl FC, Bak RPM, Sybesma J (1981) The ecology of the tropical compound ascidian *Trididemnum solidum*. I. Reproductive strategy and larval behaviour. Mar Ecol Prog Ser 6: 35–42

Warr GW, Decker JM, Mandel TE, De Luca D, Hudson R, Marchalonis JJ (1977) Lymphocyte-like cells of the tunicate, *Pyura stolonifera:* binding of lectins, morphological and functional studies. Aust J Exp Biol Med Sci 55: 151–164

Watanabe H, Taneda Y (1982) Self or nonself recognition in compound ascidians. Am Zool 22: 775–782

Weissman IL (1988) Was the MHC made for the immune system or did immunity take advantage of an ancient polymorphic gene family encoding cell surface interaction molecules? A speculative essay. Int Rev Immunol 3: 397–416

Weissman IL, Scofield V, Saito Y, Boyd H, Rinkevich B (1988) Speculations on the relationships of two *Botryllus* allorecognition reactions-colony specificity and resorption-to vertebrate histocompatibility. In: Grosberg RK, Hedgecock D, Nelson K (eds) Invertebrate hitorecognition. Plenum Press, New York, pp 67–78

Weissman IL, Saito Y, Rinkevich B (1990) Allorecognition histocompatibility in a protochordate species: Is the relationship to MHC semantic or structural? Immunol Rev 113: 227–241

Wright RK (1981) Unrochordates. In: Ratcliff NA, Rowley AF (eds) Invertebrate blood cells, vol 2. Academic Press, New York, pp 565–626

Young CM, Braithwaite LF (1980) Larval behavior and post-setting morphology in the ascidian, *Chelyosoma productum* Stimpson. J Exp Mar Biol Ecol 42: 157–169

Yue CC, Reynolds CW, Henkart PA (1987) Inhibition of cytolysin activity in large granular lymphocyte granules by lipids: evidence for a membrane insertion mechanism of lysis. Mol Immunol 24: 647–653

Zaniolo G (1981) Histology of the ascidian *Botryllus schlosseri* tunic: in particular, the test cells. Boll Zool 48: 169–178

Humoral Factors in Tunicates

Y. Saito

1 Introduction

It is well known that vertebrates have a sophisticated recognition system known as the immune system. Using this immune system, animals can eliminate invasive microorganisms, such as viruses, bacteria, and parasites, and can also clear denatured cells and metabolic wastes from their bodies. Furthermore, they can recognize allogeneic tissues and organs transplanted from other individuals and reject them as nonself. In the complex responses of the immune system, humoral factors (e.g., antibodies, complement, hemagglutinins, lectins, and cytokines) and cellular components (e.g., lymphocytes, macrophages, and natural killer cells) are involved in mutual relationships. Recognition of self vs. nonself should be important not only for vertebrates, but also for all other living things, so as to maintain the individuality of the organism and species levels. Therefore, it is interesting to study the evolution of the immune system from the lower invertebrates to the vertebrates.

Invertebrates do not have the ability to synthesize antigen-specific antibodies and do not have complement in their body fluid. They also do not have B or T lymphocytes, which are the main components involved in the immune responses of vertebrates. However, invertebrates should have some humoral and cellular factors which could be used for the elimination of invasive microorganisms from their bodies, and some invertebrates can recognize allogeneic tissues as nonself, as do vertebrates. Tunicates belong to the same phylum (Chordata) as vertebrates, and vertebrates are generally considered to have evolved directly from these animals (Berill 1955). Therefore, tunicates probably share immunological characteristics with both vertebrates and invertebrates. Although tunicates, like other invertebrates, do not have antibody and complement in their hemolymph, several humoral factors have been found in this body fluid, such as hemagglutinins, lectins, antimicrobial substances, opsonins, and some substances that act as "signals" between hemocytes. Thus, the possibility cannot be ruled out that tunicates have some ancestral molecules of vertebrate humoral factors in their hemolymph. However, the function of most humoral factors in tunicate bodies still remains obscure. The focus of this paper is some of the humoral factors that

Shimoda Marine Research Center, University of Tsukuba, Shimoda 5-10-1, Shizuoka 415, Japan have been relatively well studied, especially with respect to their functions in the host tunicates.

2 Hemagglutinins (Lectins)

Naturally occurring hemagglutinating proteins (lectins) are widely distributed in nature and are finding numerous applications in immunologic and cancer research. In particular, those derived from plants have been studied thoroughly and have contributed much to such research; examples include phytohemagglutinin (PHA; from the red kidney bean), concanavalin A (Con A; from the Jack bean), wheat germ agglutinin (WHA), and soybean agglutinin. These naturally occurring hemagglutinins agglutinate mammalian or other vertebrate erythrocytes, bacteria, and cancer cells. It is commonly known that there are some hemagglutinins that react with specific components of the ABO and MN blood groups. Each hemagglutinin (lectin) has a carbohydrate-binding specificity that reacts with cell surface glycoproteins or glycolipids. Furthermore, some lectins promote proliferation of lymphocytes and are cytotoxic.

Many hemagglutinins or lectins have been reported from several families of tunicates (Table 1). Tyler (1946) showed that agglutinins to spermatozoa from some invertebrates existed in the hemolymph of *Ciona intestinalis* and *Styela barnharti*. Fuke and Sugai (1972) found hemagglutinins to erythrocytes of some rodents in the sera of *Styela plicata* and *Halocynthia hilgendorfi*. They suggested that the hemagglutinin from *S. plicata* might be a polysaccharide or mucopolysaccharide because it was very heat stable and was destroyed by periodate, which is an agent known to destroy saccharide by oxidation. Furthermore, the agglutinin had no apparent opsonic effect, but it seemed to play a role in cell-to-cell and cell-to-glass adherence. These characteristics are similar to those of plant hemagglutinins, rather than to agglutinins of animals. The hemolymph of many tunicates has been tested and found to have agglutinating activity for vertebrate erythrocytes: most of these agglutinins are heat labile and are digested by proteolytic enzymes; their major components are protein.

Parrinello and his coworkers have undertaken extensive studies on the hemagglutinins of two species of the family Ascidiidae. From the serum of *Ascidia malaca*, an agglutinin was purified that had D-galactose binding specificity, a molecular weight of about 58 kDa, and intramolecular disulfide linkages. Furthermore, the lectin with the same carbohydrate-binding specificity as the serum lectin was found on the surface of about 34% of hemocytes (Parrinello and Arizza 1988). On the other hand, they found two types of hemagglutinins from the hemolymph of *Phallusia mammillata*. These have molecular weights of 61 to 65 kDa and lactose binding specificity (Parrinello and Canicatti 1983). In addition to these agglutinins, two agglutinins were found from a hemocyte extract of this species. These cellular agglutinins are composed of two subunits that are similar in molecular weight (36 and 35 kDa), and have the same binding specificity for lactose as the humoral agglutinins (Parrinello and Arizza 1989).

Table 1. Hemagglutinins (lectins) in some ascidians

Species	Form[a]	Source[b]	Reference
Order Enterogona			
Suborder Aplousobranchia			
Family Polyclinidae			
Aplidium australiensis	C	PL	Coombe et al. (1984)
Family Didemnidae			
Didemnum patulum	C	PL	Coombe et al. (1984)
D. candidum	C	PL	Vasta and Marchalonis (1987)
Family Polycitoridae			
Atapozoa fantsiana	C	PL	Coombe et al. (1984)
Suborder Phlebobranchia			
Family Cionidae			
Ciona intestinalis	S	PL	Wright (1974); Wright and Cooper (1975); Parrinello and Patricolo (1975)
		HC	Parrinello et al. (1993)
Family Ascidiidae			
Ascidia ceratodes	S	PL	Tyler (1946)
A. malaca	S	PL	Parrinello and Patricolo (1975); Parrinello and Canicatti (1982)
		HC	Parrinello and Arizza (1988)
A. thompsoni	S	PL	Coombe et al. (1984)
Phallusia mammillata	S	PL	Parrinello and Patricolo (1975); Parrinello and Canicatti (1983)
		HC	Parrinello and Arizza (1989); Cammarata et al. (1993)
P. despressiuscula	S	PL	Coombe et al. (1984)
Order Pleurogona			
Suborder Stolidobranchia			
Family Botryllidae			
Botrylloides leachii	C	PL	Coombe et al. (1982); Coombe et al. (1984)
B. mabnicoecus	C	PL	Coombe et al. (1984)
Family Styelidae			
Polyandrocarpa misakiensis	C	EP	Suzuki et al. (1990); Kawamura et al. (1991)
Stolonica australis	C	PL	Coombe et al. (1984)
Cemidocarpa etheridgii	S	PL	Coombe et al. (1984)
Polycarpa obtecta	S	PL	Coombe et al. (1984)
P. papillata	S	PL	Coombe et al. (1984)
Styela plicata	S	PL	Fuke and Sugai (1972)
S. clava	S	PL	Wright and Cooper (1984)
Family Pyuridae			
Halocynthia hilgendorfi	S	PL	Fuke and Sugai (1972)
H. pyriformis	S	PL	Anderson and Good (1975); Form et al. (1979); Vasta and Marchalonis (1983)
H. papillosa	S	PL	Bretting and Renwrantz (1973)
H. roretzi	S	PL	Yokosawa et al. (1982, 1986); Azumi et al. (1987)
		HC	Azumi et al. (1991a)
H. hispida	S	PL	Coombe et al. (1984)

Table 1. (*Contd.*)

Species	Form[a]	Source[b]	Reference
Herdmania momus	S	PL	Coombe et al. (1984)
Microcosmus sulcatus	S	PL	Bretting and Renwrantz (1973)
M. nichollsi	S	PL	Coombe et al. (1984)
Pyura praeputialis	S	PL	Coombe et al. (1984)
P. irregularis	S	PL	Coombe et al. (1984)

[a]S, solitary ascidian; C, colonial ascidian.
[b]PL, plasma; HC, hemocyte; EP, epithelium.

Cellular agglutinins are also located on the cell surface (Parrinello and Arizza 1989) and released from the in vitro cultured hemocytes (Arizza et al. 1991). Only one type of hemocyte, compartment cells, releases those hemagglutinins (Cammarata et al. 1993). Because there are few or no cellular agglutinins present in the serum, it seems likely that these agglutinins are not continuously released by hemocytes into the circulation, but that secretion is triggered when hemocytes come into contact with foreign substances. This assumption may be supported by the occurrence of agglutinin release immediately after contact with foreign materials, such as the wall of a syringe or centrifuge tube (Arizza et al. 1991). By mixing hemocytes and rabbit erythrocytes, three reactions demonstrating the release of agglutinins could be obtained in a few minutes. There were (1) clumps of aggregates composed of hemocytes and erythrocytes, which are probably secretory rosettes; (2) clumps of erythrocytes agglutinated by the released agglutinins; and (3) rosettes formed by hemocytes with not less than three erythrocytes attached. These results obtained from the two species *A. malaca* and *P. mammillata,* suggest that three types of agglutinins – humoral, cell surface, and released agglutinins – collaborate to recognize and eliminate foreign substances, as they have the same carbohydrate-binding specificity.

In *Halocynthia roretzi,* a hemagglutinin has also been isolated from hemocytes (Azumi et al. 1991a). Two hemagglutinins have been found in the serum of this species. They show different carbohydrate-binding specificity (galactose and N-acetyl-D-galactosamine) and have different molecular weights (41 and 28 kDA, respectively). The agglutinin from hemocytes has binding specificity for lipopolysaccharide (LPS), heparin, and chondroitin sulfate, and its molecular weight is 120 kDa. Like the cellular agglutinin of *P. mammillata,* this agglutinin is present in only one type of hemocyte, morula cells, which account for more than 50% of *H. roretzi* hemocytes and also contain antimicrobial substances (halocyamine A and B; Azumi et al. 1990a). As suggested by its binding specificity for LPS, which is a major component of the cell wall of Gram-negative bacteria, this agglutinin can readily agglutinate marine Gram-negative bacteria and, unexpectedly, it can also agglutinate most marine Gram-positive bacteria. However, it does not have activity as a bactericidal agent, and evidence of its release from hemocytes has not been observed yet. Another agglutinin having binding specificity for LPS was also reported from the hemolymph of *Styela clava* (Wright and Cooper 1984). It may be possible that these LPS-binding agglutinins take part in the tunicate immuno-defense system for bacterial infection, although more information is needed on this subject.

In *Ciona intestinalis,* the hemolymph has weak hemagglutinating activity (Wright 1974; Parrinello and Patricolo 1975; Wright and Cooper 1975). When mammalian erythrocytes (human and duck) were injected into the tunic or perivisceral cavity of *C. intestinalis,* these erythrocytes soon formed aggregated clumps and phagocytes rapidly adhered to the clumps to clear them. At this time, the agglutinin titer in the hemolymph decreased; it returned to the normal level after clearance of the injected erythrocytes (Wright 1974). This observation suggests that the humoral agglutinin is consumed during the formation of erythrocyte aggregates and during adherence of phagocytes to erythrocytes, that is, it is involved in the elimination of mammalian erythrocytes by phagocytosis. In addition, hemocytes of *C. intestinalis* showed a natural cytotoxic capacity when assayed in vitro against mammalian (rabbit, sheep, guinea pig, and human) erythrocytes (Parrinello et al. 1993). The hemocyte cytotoxicity is a cell-mediated process that requires effector-target cell contact. Anti-sheep erythrocyte (anti-SE; calcium-dependent) and anti-rabbit erythrocyte (anti-RE; calcium-inde-pendent) agglutinins were found in the reaction medium, probably released by hemocytes as a result of the in vitro experiments. The hemocytes did not release both agglutinins continuously, because no increase in agglutinating titer of anti-SE or anti-RE in the incubation medium was observed after prolonged incubation of hemocytes. Therefore, those agglutinins were released immediately after contact with the foreign substances, as occurs with the cellular agglutinin of *P. mammillata.* In contrast to the case of *P. mammillata,* the cellular anti-SE and anti-RE are similar to the naturally occurring anti-SE and anti-RE in the plasma, respectively. It is not known where these humoral agglutinins are produced.

An interesting agglutinin has been isolated from *Polyandrocarpa misakiensis,* that appears in the epithelial cells at only the earliest stage of morphogenesis in an asexual bud (Suzuki et al. 1990; Kawamura et al. 1991). This agglutinin shows the characteristics of C-type lectins (Drickamer 1988): calcium-dependent acti-vity, galactose-binding specificity, disulfide bonds in a molecule, and a molecular weight of 14 kDa. This agglutinin is secreted into the mesenchymal space, where the extracellular matrix containing the agglutinin developed. The agglutinin promotes aggregation of pluripotent stem cells with epithelial cells and the association of those cells with the extracellular matrix to begin morphogenesis of the bud. After the commencement of bud morphogenesis, the agglutinin and the extracellular matrix disappear from the developing bud. This agglutinin also has strong antibacterial activity (Suzuki et al. 1990). Kawamura et al. (1991) dis-cussed the possibility that this agglutinin is involved in the defense system against bacterial infection as well as functioning to promote morphogenesis. This may be the only tunicate agglutinin whose characteristics and functions are relatively clear, although its role in immune function is not clear.

Table 2. Molecular weight and carbohydrate specificity of isolated hemagglutinins from some tunicates

Species	Source[a]	Hemagglutinin	M.W. Subunit	Oligomer	Carbohydrate specificity	References
Didemnum candidum	PL	DCL-I	14000	56600	D-Galactose	Vasta and Marchalonis (1986); Vasta et al. (1986)
	PL	DCL-II	14500	57500	D-Galactose	Vasta and Marchalonis (1986); Vasta et al. (1986)
Ascidia malaca	PL	A-SL	58000		D-Galactose	Parrinello and Arizza (1988)
Phallusia mammillata	PL	Anti-RE	61–65000	>200000	α-Lactose	Parrinello and Canicatti (1983)
	PL	Anti-HE	61–65000	>200000	α-Lactose, Lactose	Parrinello and Canicatti (1983)
	HC		36900		α-Lactose	Parrinello and Arizza (1989)
	HC		35090		α-Lactose	Parrinello and Arizza (1989)
Botrylloides leachii	PL	HA-1	28200 29600	152000	Lactose, D-Galactose, Melibiose, D-Fucose	Coombe et al. (1984)
	PL	HA-2	33400	65000	Lactose	Coombe et al. (1984)
	PL	LPB-3	29600 28200 23600	140000 –160000	Lactose, D-Galactose	Coombe et al. (1984)
Polyandrocarpa misakiensis	EP		14000		D-Galactose	Suzuki et al. (1990); Kawamura et al. (1991)
Halocynthia pyriformis	PL	HPYL-I	40500		Sialic acid	Vasta and Marchalonis (1987)
	PL	HPYL-II	29500		Sialic acid	Vasta and Marchalonis (1987)
	PL	HPYL-III	19500		L-Fucose	Vasta and Marchalonis (1987)
H. roretzi	PL	I	41000	660000	D-Galactose	Yokosawa et al. (1982, 1986)
	PL	II	28000	500000	N-Acetyl-D-Galactoamine	Azumi et al. (1987)
	HC		120000		LPS, Heparin, Chondroitin sulfate	Azumi et al. (1991a)

[a]PL, plasma; HC, hemocyte; EP, epithelium.

At present, there are not many tunicate agglutinins whose chemical characteristics are defined, as seen in Table 2. However, in general, the humoral and/or cellular agglutinin population in a tunicate seems to be heterogeneous (Vasta and Marchalonis 1983; Wright and Cooper 1984). Furthermore, tunicate agglutinins are quite varied in molecular weight and carbohydrate-binding activity, which suggests that these agglutinins have a wide variety of functions.

3 Antimicrobial Substances

Because invertebrates lack humoral immunoglobulins, antimicrobial substances seem to be one of the important humoral factors involved in their defense mechanisms. Some antimicrobial or antiviral substances have been isolated from ascidians (Rinehart et al. 1981, 1984; Ireland et al. 1982; Kobayashi et al. 1984, 1988; Ishibashi et al. 1987). Didemnins isolated from the colonial tunicate *Trididemnum* sp. are cyclic peptides with a high level of antiviral and antitumor activity. They inhibit replication of various RNA and DNA viruses in vitro at low concentrations (0.05–1 μg/ml; Rinehart et al. 1981). Eudistomins isolated from the colonial tunicate *Eudistoma olivaceum* were found to be highly active against herpes simplex virus. The eudistomins are β-carboline derivatives containing bromine atoms (Rinehart et al. 1984). These substances were isolated from whole extracts of tunicate colonies, and thus it is not clear that they are involved in the immune system of these colonial tunicates because it is possible that they are produced by symbiotic microorganisms.

Azumi et al. (1990a) isolated two novel antimicrobial substances, halocyamine A and B, from the hemocytes of a solitary ascidian, *Halocynthia roretzi.* The structures of halocyamine A and B were L-histidyl-L-6,7-dihydroxyphenylalanyl-glycyl-6-bromo-8,9-didehydrotryptamine and L-threonyl-L-6,7,-dihydroxyphenylalanyl-L-histidyl-6-bromo-8,9-didehydrotryptamine, respectively. Both halocyamine A and B inhibited the growth of some Gram-positive bacteria (*Bacillus subtilis, B. megaterium,* and *B. cereus*) and a yeast (*Cryptococus neoformans*), but the inhibition was weak in the Gram-negative bacterium *Escherichia coli.* Halocyamine A also inhibited the growth of marine Gram-negative bacteria *Achromobacter aquamarinus* and *Pseudomonas perfectomarinus,* but it did not show inhibition of growth of two other marine Gram-negative bacteria, *Alteromonus putrefaciens* and *Vibrio anguillarum* (Azumi et al. 1990b). It is not clear why different intensities of growth inhibition were seen among Gram-negative bacteria. In addition to the antibacterial activity, halocyamine A inhibited the in vitro growth of fish RNA viruses (infectious hematopoietic necrosis virus and infectious pancreatic necrosis virus) and showed cytotoxic activity against some cultured mammalian cells. Both halocyamines were only found in the morula-like cells, which is the major hemocyte type in *H. roretzi* (Fuke 1979; Wright 1981).

Halocyamines show antiviral and antibacterial activity against marine viruses and bacteria. As mentioned before, an interesting hemagglutinin, which has binding specificity for LPS, was isolated from the same type of hemocyte as those containing halocyamines (Azumi et al. 1991a). LPS is the major component of the cell wall of Gram-negative bacteria, and this agglutinin can bind to various marine Gram-negative bacteria. These facts strongly suggest that *H. roretzi* may be using these factors in its defense against infection by marine viruses or bacteria. Halocyamines and the LPS-binding agglutinins are in the morula-like cells, not in the plasma. However, hemocytes of *H. roretzi* show a "contact reaction" in which reciprocal lysis occurs by self and nonself recognition between allogeneic or xenogeneic hemocytes (Fuke 1980). If lysis of hemocytes occurs by contact with bacteria as in the case of the contact reaction, the LPS-binding agglutinins could trap bacteria and then the halocyamines could kill them.

4 Other Humoral Factors

Cytokines are polypeptides released by activated vertebrate blood cells, which have profound effects on other blood cells and have hormone-like properties affecting other organ systems as well. Recently, a wide variety of these mediators has been isolated and characterized, such as interleukins (IL-1 to IL-10), interferons (IFNα, IFNβ and IFNγ), tumor necrosis factors (TNFα and TNFβ), colony-stimulating factors (CSFs), and transforming growth factor-β (TGFβ). These molecules are important to host immune and nonspecific defense systems, and are usually elicited during periods of inflammation or infection. Production of one cytokine may elicit the synthesis of others that help in the fight against infection. Many cytokines act synergistically with one another, and the elaboration of other factors (e.g. prostaglandins) can be influenced by cytokines (or combinations of cytokines). Both the cellular and humoral immune responses come under the control of cytokines (Beck et al. 1989a).

A cytokine-like substance found in tunicates has been described by Sigel et al. (1984). This substance (Ete) was extracted from the tunicate *Ecteinascidia turbinata* and studied for its effects on a variety of vertebrate and invertebrate cells. Studies with vertebrate cells showed it has effects on antibody production (inhibition), cell-mediated responses (inhibition), phagocytic activity (stimulation), and cell-mediated cytotoxicity (stimulation). In lower vertebrate and invertebrate systems, Ete stimulated eel (*Anguilla rostrata*) and prawn (*Macrobrachium rosenbergii*) hemocyte phagocytic activity, stimulated enhanced binding of sheep erythrocytes by eel and crayfish peripheral blood leukocytes, and stimulated the appearance of granulocyte-like cells in the peripheral blood of eels. The role of Ete in tunicate host defenses has not been described.

Eight North American species of tunicates, *E. turbinata*, *Botryllus schlosseri*, *Styela plicata*, *Molgula occidentalis*, *Didemnum* sp., *Ciona intestinalis*, *Amaroucium pellucidum*, and *Halocynthia pyriformis*, were examined for the presence of interleukin-1 (IL-1)-like activity, and it was shown that all species produce molecules with readily detectable lymphocyte activation factor (LAF) activity (Beck et al. 1989b). Two kinds of molecules were separated; one is directly mitogenic for thymocytes (molecular weight > 50 kDa) and the other is comito-genic in an IL-1 assay (molecular weight = 20 kDa). The low molecular weight fraction with LAF activity, induced an increase in vascular permeability in rabbit skin. The LAF activity was neutralized by polyclonal anti-human IL-1 antisera. Taken together with the results of identification of IL-1 in the echinoderms (Beck and Habicht 1986), these data suggest that IL-1 is an ancient and functionally conserved molecule.

From the plasma of *Styela clava,* an opsonin for yeast was found; the activity of this factor increased the overall capacity for phagocytosis and was inhibited by the carbohydrates mannan, N-acetyl-D-galactosamine, and galactose, and by the divalent cation chelator EDTA. Opsonization was also inhibited by trypsin treatment and heat denaturation (Kelly et al. 1992). Such characteristics are typical of C-type lectins. On the other hand, the opsonin showed IL-1-like activities (Kelly et al. 1993a,b). It enhances incorporation of ^{3}H-thymidine into cultures of pharyngeal tissues from *S. clava* and acts as a powerful chemoattractant for *S. clava* hemocytes. It also activates tunicate phagocytes, in addition to its original opsonization activity. The opsonin molecule is produced by hemocytes and its production is stimulated by zymosan. Molecular characterization indicated that the opsonic protein is a 17.5-kDa monomer with a pI of 7 and is not highly polymorphic. This molecular weight is the same as that of mammalian IL-1β. The functional and physicochemical properties of *S. clava* opsonin, imply that the opsonin is identical to a cytokine-like protein, designated tunIL-1β, that has been isolated from the same species (Raftos et al. 1991, 1992).

In addition to the factors mentioned above, the blood of tunicates also contains enzymes and inhibitors that may have an important role in the immuno-defense system. Two tunicate trypsin inhibitors (inhibitor I and II) were reported from the plasma of *Halocynthia roretzi* (Yokosawa et al. 1985; Kumazaki et al. 1990). Inhibitor I consists of a single polypeptide chain with 55 amino acid residues and four intramolecular disulfide bridges, whereas inhibitor II is com-posed of two polypeptide chains corresponding to a form derived from inhibitor I by cleavage at the Lys_{16}–Met_{17} bond. These humoral trypsin inhibitors possibly regulate proteolytic activity at the area of inflammation.

From the hemocytes of *H. roretzi,* release of a metallo-protease was found to be induced by stimulation with lipopolysaccharide (LPS; Azumi et al. 1991b). LPS is a major component of the cell wall of Gram-negative bacteria. Thus, this protease might be released by bacterial infection through exocytosis, but no direct evidence has been reported for the exocytosis of hemocytes containing the enzyme.

In *Ciona intestinalis,* hemocytes of a specific cell type, the morula cell, contain a factor with opsonic activity which activates phagocytosis of vacuolar and granular amoebocytes. The phagocytosis is enhanced by the stimulation of bacterial carbohydrate, LPS, and is suppressed by the serine protease inhibitor, STI (Smith and Peddie 1992). In the morula cells, phenoloxidase was also found to exist as a proenzyme, prophenoloxidase (proPO), and to be activated by protease (Smith and Söderhäll 1991). LPS and laminarin enhanced the phenoloxidase activity, but benzamidine and STI reduced it (Jackson et al. 1993). These observations strongly suggest that phagocytosis and phenoloxidase activity are well correlated with each other in this species, and thus, phenoloxidase may be involved in the defense system against bacterial infection. Phenoloxidase, the active form of proPO, is well known as the terminal component of a cascade of serine protease and other factors that bring about recognition and cellular defense in several species of crustaceans and insects. These animals belong to the Protostomia, whereas tunicates and vertebrates belong to the Deuterostomia. Therefore, the possibility that *C. intestinalis* may possess the proPO cascade in its blood is very interesting from an evolutionary point of view.

5 Humoral Factors Involved in Allogeneic Recognition

In some colonial tunicates, especially members of the family Botryllidae, a phenomenon analogous to transplantation immunity has been described and is known as "colony specificity" (Bancroft 1903; Oka and Watanabe 1957; Mukai and Watanabe 1974; Saito and Watanabe 1982; Scofield et al. 1982). Colony specificity results in fusion or nonfusion between colonies. The genetic basis for this colony specificity in botryllids resides in a single highly polymorphic gene locus (or haplotype) called Fu/HC (fusibility/histocompatibility). The rules of histocompatibility in these ascidians differ somewhat from major histocompatibility complex (MHC)-dependent graft rejection in vertebrates. Two colonies sharing at least one allele in common (such as colonies with AB and AC at Fu/HC) are fusible, while two colonies sharing no alleles in common (AB and CD colonies) are nonfusible (Oka and Watanabe 1960; Sabbadin 1962; Scofield et al. 1982). In botryllid colonies, zooids are covered with a common tunic that is gelatinous and translucent, and they are interconnected by a remified vascular network. The margin of a colony is fringed by many vascular ampullae which are terminals of vascular vessels. Fusion and rejection occur mainly in the tunic covering the ampullae and involve the ampullae themselves (Tanaka and Watanabe 1973; Katow and Watanabe 1980). The processes of fusion and nonfusion in *Botryllus primigenus* are summarized schematically in Fig. 1. Nonfusion is accompanied by an intensive rejection reaction in and around the vascular ampullae. This rejection is called "nonfusion reaction" (NFR) and has the following characteristics: (1) infiltration of blood cells, (2) aggregation of blood cells, (3) shrinking of vascular vessels, and (4) amputation or withdrawal of ampullae (Tanaka 1973). The first sign of NFR appears 3 to 5 h after contact (Stage 3'). NFR is completed about 12 h after contact (Stage 5').

It has been shown by the following experiments that it is blood components which are primarily involved in NFR. When a small AB colony fuses with a large BC colony, the blood of the AB colony is replaced by the blood of the BC colony. If, a few days after fusion, these two colonies are separated, and then the small AB colony is brought into contact with a naive CD colony, it (AB) is now fusible with the CD colony – although prior to fusion with the BC colony, it would not

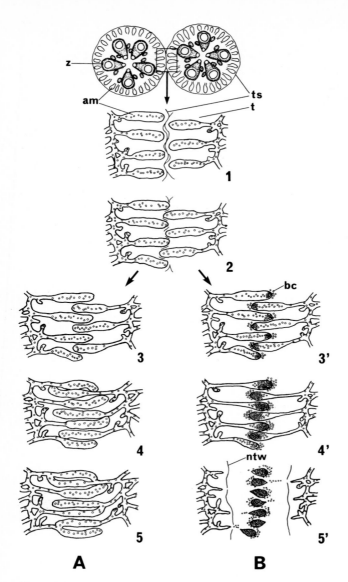

Fig. 1. Diagram summarizing the processes of fusion and nonfusion in *B. primigenus*. **A** Fusion process. Two colonies make contact with each other at the surfaces of their tunic layers (*stage 1*). The cuticle begins to dissolve at the contact area (*stage 2*). After tunic fusion, the vascular ampullae of each colony expand into the facing colony (*stage 3*). Ampullar penetration continues until the tips touch the proximal parts of ampullae of the facing colony (*stage 4*). Fusion of vascular vessels occurs between the tips and the proximal parts of ampullae, and the two colonies become a single colony with a common tunic and a common vascular system (*stage 5*). **B** Nonfusion process. The first two stages of NFR are the same as those of fusion. The first sign of NFR appears when the ampullae penetrate into the facing colony and blood cells begin to infiltrate into the tunic (*stage 3'*) Blood current becomes slow, cell aggregation occurs at the ampullar tips, and the proximal parts of ampullae gradually become thin (*stage 4'*). The proximal parts of ampullae shrink and finally become amputated. A new portion of tunic wall is formed to separate the healthy colony part from the colony part that was damaged by NFR (*stage 5'*). *am* Ampulla; *bc* blood cell; *ntw* new tunic wall; *t* tunic *ts* tunic surface; *z* zooid

fuse with a CD colony (Fig. 2; Mukai 1967). In another experiment, three colonies, AB, BC, and CD, are arranged in a line to allow contact between neighboring colonies; fusion between AB and BC and BC and CD are allowed to occur at the same time. Subsequently, NFR appears in the vascular network of the central BC colony (Fig. 3; Tanaka 1973). It has also been found that NFR is suppressed when two incompatible colonies have been irradiated with X-rays (Taneda and Watanabe 1982a). In an irradiated colony, the number of lymphocyte-like cells is remarkably reduced; therefore, lymphocyte-like cells might play an important role in NFR. Additional research has shown that in *B. primigenus,* NFR begins before fusion of vascular vessels (Fig. 1), which implies that a humoral component(s) diffusible in the tunic probably acts as an allogeneic factor(s) to the blood cells or the ampullar epithelia of the facing colony. In fact, blood plasma can induce NFR in the blood vessels of an incompatible colony by microinjection into a vessel (Taneda and Watanabe 1982b).

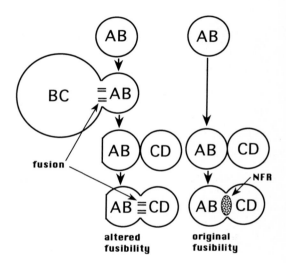

Fig. 2. Experimental alteration of fusibility. *NFR* Nonfusion reaction

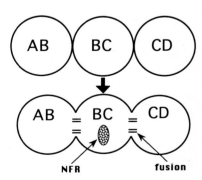

Fig. 3. Synchronous fusion among three allogeneic colonies. *NFR* Nonfusion reaction

In another botryllid, *Botrylloides simodensis,* it has also been shown, by using a plasma microinjection method, that blood plasma has a colony-specific activity coincident with colony fusibility (Saito and Watanabe 1984). Blood plasma from a colony of *B. simodensis,* when injected, induces an NFR response in a non-fusible colony, whereas it induces no response in a fusible colony such as a syngeneic or a fusible allogeneic colony. From the results of biochemical characterization of the humoral factors responsible for colony specificity, this specific activity is retained after dialysis of blood plasma against filtered seawater, but is greatly diminished by heating at temperatures higher than 55 °C. The activity is dependent on bivalent cations such as Ca^{2+} and Mg^{2+} (especially Ca^{2+}). The colony-specific activity in the blood plasma becomes nonspecific by various gentle treatments, e.g., long-term storage at 4 °C, freezing and thawing, incubation at moderate temperature (18–25 °C), and other physicochemical treatments. The blood plasma thus treated invariably induces NFR-like responses in fusible as well as nonfusible colonies. This nonspecific activity is rather stable, remaining unchanged for more than one year in the cold. Like the specific activity, the nonspecific activity is resistant to dialysis, heat-labile, and dependent on bivalent cations. Furthermore, the nonspecific activity is found in high molecular weight fractions either by ammonium sulfate fractionation or Sephadex-G75 gel filtration. The nonspecific activity is not affected by trypsin, protease, or neuraminidase (Saito and Watanabe 1984). At present, it is not known whether the observed specific and nonspecific activities belong to a single factor or to different factors in the blood plasma. The face that both activities share some characteristics (as mentioned above) suggests that the former alternative is more probable.

6 Concluding Remarks

Many kinds of humoral factors have been found in tunicates: hemagglutinins, antimicrobial substances, cytokines, opsonins, enzymes and factor(s) of allorecognition. Most of them seem to be involved in the immuno-defense system against bacterial and viral infections of tunicates, although some may have other specific functions. However, at present there is very little information on the biological functions of these factors in their host tunicates, although there are many studies focused on the similarity of physicochemical characteristics among humoral factors of tunicates and other animal groups, especially mammals.

Burnet (1968) suggested that hemagglutinins might be forerunners of vertebrate immunoglobulins. In fact, the plasma of *Halocynthia pyriformis* and *Boltenia ovifera* contain components that react with antisera made against shark immunoglobulin 7S heavy chain (Rosenshein et al. 1985). Therefore, it is possible that a protein in the plasma is the ancestral precursor of immunoglobulin. The hemagglutinin of *Didemnum candidum* reacts with polyclonal antibodies made against mammalian C-reactive protein (CRP), one of the acute phase proteins, as well as the horseshoe crab lectin, limlin (Vasta and Marchalonis 1987). The opsonin of *Styela clava* is similar to mammalian acute phase proteins in some physicochemical characteristics and functions (Kelly et al. 1992). Furthermore, cytokine-like proteins having interleukin-1-like activity were found in the plasma of several species of tunicates (Beck et al. 1989b; Raftos et al. 1991, 1992). The existence of such homologies may indicate that some innate recognition mechanisms in mammals have evolved directly from the defense system of invertebrates. On the other hand, it is notable that the agglutinin from *Styela plicata* is not a protein, but a polysaccharide or a mucopolysaccharide, and that it is similar to plant lectins rather than animal lectins (Fuke and Sugai 1972). In addition, the hemocytes of *Ciona intestinalis* probably have the proPO cascade, which is well known as a defense system in insects and some crustaceans (Jackson et al. 1993). These findings should not be ignored when discussing the evolution of the immuno-defense system.

It is clear that humoral factors always function with cellular components in the immuno-defense system. In *Halocynthia roretzi,* humoral, cellular, and cell surface agglutinins may cooperate with cellular antimicrobial substances to eliminate invasive bacteria and viruses (Azumi et al. 1991a). Allogeneic recognition in colonial tunicates is also performed by the cooperative reactions of both cellular and humoral components (Taneda and Watanabe 1982b). Thus, in the immune system, the evolution of humoral factors should be accompanied by the evolution of cellular components. Therefore, a discussion about the evolution of the immune system requires information from more detailed studies on the biological functions and mutual interaction of humoral and cellular components.

Acknowledgment. This work was supported by Grant-in-Aid for Scientific Research No. 03455008 to Y. Saito from the Ministry of Education, Science and Culture of Japan. This is Contribution No. 580 from the Shimoda Marine Research Center, University of Tsukuba.

References

Anderson RS, Good RA (1975) Naturally-occurring hemagglutinin in a tunicate *Halocynthia pyriformis.* Biol Bull 148: 357–369

Arizza V, Parrinello N, Schimmenti S (1991) In Vitro release of lectins by *Phallusia mamillata* hemocytes. Dev Comp Immunol 15: 219–226

Azumi HK, Yokosawa H, Ishii S (1987) N-Acetyl-galactosamine-specific lectin, a novel lectin in the haemolymph of the ascidian *Halocynthia roretzi:* isolation, characterization and comparison with galactose-specific lectin. Comp Biochem Physiol 88B: 375–381

Azumi K, Yokosawa H, Ishii S (1990a) Halocyamines: novel antimicrobial tetrapeptide-like substances isolated from the hemocytes of the solitary ascidians *Halocynthia roretzi.* Biochemistry 29: 159–165

Azumi K, Yoshimizu M, Ezura Y, Yokosawa H (1990b) Inhibitory effect of halocyamine, an antimicrobial substance from ascidian hemocytes, on the growth of fish viruses and marine bacteria. Experientia 46: 1066–1068

Azumi K, Ozeki S, Yokosawa H, Ishii S (1991a) A novel lipopolysaccharide-binding hemagglutinin isolated from hemocytes of the solitary ascidian, *Halocynthia roretzi:* it can agglutinate bacteria. Dev Comp Immunol 15: 9–16

Azumi K, Yokosawa H, Ishii S (1991b) Lipopolysaccharide induces release of a metallo-protease from hemocytes of the ascidian, *Halocynthia roretzi.* Dev Comp Immunol 15: 1–7

Bancroft FW (1903) Variation and fusion of colonies in compound ascidian. Proc Calif Acad Sci Ser 3: 137–186

Beck G, Habicht GS (1986) Isolation and characterization of a primitive IL-1-like protein from an invertebrate, *Asterias forbesi*. Proc Natl Acad Sci USA 83: 7429–7433

Beck G, O'Brien RF, Habicht GS (1989a) Invertebrate cytokines: the phylogenetic emergence of interleukin-1. Bio Essays 11: 62–67

Beck G, Vasta GR, Marchalonis JJ, Habicht GS (1989b) Characterization of interleukin-1 activity in tunicates. Comp Biochem Physiol 92B: 93–98

Berrill NJ (1955) The origin of vertebrates. Oxford Univ Press, London

Bretting H, Renwrantz L (1973) Untersuchungen von Invertebraten des Mittelmeeres auf ihren Gehalt an hämagglutinierenden Substanzen. Z Immun-Forsch 145: 242–249

Burnet M (1968) Evolution of the immune process in vertebrates. Nature 218: 426–430

Cammarata M, Parrinello N, Arizza V (1993) In vitro release of lectins from *Phallusia mamillata* hemocytes after their fractionation on a density gradient. J Exp Zool 266: 319–327

Coombe DR, Schluter SF, Ey PL, Jenkin CR (1982) Identification of the HA-2 agglutinin in the haemolymph of the ascidian *Botrylloides leachii* as the factor promoting adhesion of sheep erythrocytes to mouse macrophages. Dev Comp Immunol 6: 65–74

Coombe DR, Ey PL, Jenkin CR (1984) Ascidian haemagglutinins: incidence in various species, binding specificities and preliminary characterization of selected agglutinins. Comp Biochem Physiol 77B: 811–819

Drickamer K (1988) Two distinct classes of carbohydrate-recognition domains in animal lectins. J Biol Chem 263: 9557–9560

Form DM, Warr GW, Marchalonis JJ (1979) Isolation and characterization of a lectin from the hemolymph of a tunicate, *Halocynthia pyriformis*. Fed Proc 38: 934

Fuke MT (1979) Studies on the coelomic cells of some Japanese ascidians. Bull Mar Biol Stn Asamushi Tohoku Univ 16: 143–159

Fuke MT (1980) "Contact reactions" between xenogeneic or allogeneic coelomic cells of solitary ascidians. Biol Bull 158: 304–315

Fuke MT, Sugai T (1972) Studies on the naturally occurring hemagglutinin in the coelomic fluid of an ascidian. Biol Bull 143: 140–149

Ireland CM, Durso AR Jr, Newmann RA, Hacker MP (1982) Antineoplastic cyclic peptides from the marine tunicate *Lissoclinum patella*. J Org Chem 47: 1807–1811

Ishibashi M, Ohizumi Y, Sasaki T, Nakamura H, Hirata Y, Kobayashi J (1987) Pseudodistomins A and B, novel antineoplastic piperidine alkaloids with calmodulin antagonistic activity from the Okinawan tunicate *Pseudodistoma kanoko*. J Org Chem 52: 450–453

Jackson AD, Smith VJ, Peddie CM (1993) In vitro phenoloxidase activity in the blood of *Ciona intestinalis* and other ascidians. Dev Comp Immunol 17: 97–108

Katow H, Watanabe H (1980) Fine structure of fusion reaction in compound ascidian *Botryllus primigenus* Oka. Dev Biol 76: 1–14

Kawamura K, Fujiwara S, Sugino YM (1991) Budding-specific lectin induced in epithelial cells is an extracellular matrix component for stem cell aggregation in tunicates. Development 113: 905–1005

Kelly KL, Cooper EL, Raftos DA (1992) Purification and characterization of a humoral opsonin from the solitary urochordate *Styela clava*. Comp Biochem Physiol 103B: 749–753

Kelly KL, Cooper EL, Raftos DA (1993a) A humoral opsonin from the solitary urochordate *Styela clava*. Dev Comp Immunol 17: 29–39

Kelly KL, Cooper EL, Raftos DA (1993b) Cytokine-like activities of a humoral opsonin from the solitary urochordate *Styela clava*. Zool Sci 10: 57–64

Kobayshi J, Harbour GC, Gilmore J, Rinehart KL Jr (1984) Eudistomins A, D, G, H, I, J, M, N, O, P, and Q bromo-, hydroxy-, pyrrolyl-, and 1-pyrrolinyl-B-carbolines from the antiviral Caribbean tunicate *Eudistoma olivaceum*. J Am Chem Soc 106: 1526–1528

Kobayashi J, Cheng JF, Nakamura H, Ohizumi Y, Hirata Y, Sasaki T, Ohta T, Nozoe S (1988) Ascididemin, a novel pentacyclic alkaloid with potent antileukemic activity from the Okinawan tunicate *Didemnum* sp. Tetrahedron Lett 29: 1177–1180

Kumazaki T, Hoshiba N, Yokosawa H, Ishii S (1990) Primary structure of ascidian trypsin inhibitors in the hemolymph of a solitary ascidian, *Halocynthia roretzi.* J Biochem 107: 409–413

Mukai H (1967) Experimental alteration of fusibility in compound ascidians. Sci Rep Tokyo Kyoiku Daigaku 13B: 51–73

Mukai H, Watanabe H (1974) On the occurrence of colony specificity in some compound ascidians. Biol Bull 147: 411–421

Oka H, Watanabe H (1957) Colony-specificity in compound ascidians as tested by fusion experiments (A preliminary report). Proc Jpn Acad 33: 657–659

Oka H, Watanabe H (1960) Problems of colony-specificity in compound ascidians. Publ Mar Biol Stn Asamushi 10: 153–155

Parrinello N, Arizza V (1988) D-galactose binding lectins from the tunicate *Ascidia malaca:* Sub-unit characterization and hemocyte surface distribution. Dev Comp Immunol 12: 495–507

Parrinello N, Arizza V (1989) Sugar specific cellular lectins of *Phallusia mamillata* hemocytes: purification, characterization and evidence for cell surface localization. Dev Comp Immunol 13: 113–121

Parrinello N, Canicatti C (1982) Carbohydrate binding specificity and purification by biospecific affinity chromatography of *Ascidia malaca* Traust. hemagglutinins. Dev. Comp Immunol 6: 53–64

Parrinello N, Canicatti C (1983) α-Lactose binding hemagglutinins from the ascidian *Phallusia mamillata* (Cuv.) Biol Bull 164: 124–135

Parrinello N, Patricolo E (1975) Erythrocyte agglutinins in the blood of certain ascidians. Experientia 31: 1092–1093

Parrinello N, Arizza V, Cammarata M, Parrinello DM (1993) Cytotoxic activity of *Ciona intestinalis* (Tunicata) hemocytes: properties of the in vitro reaction against erythrocyte targets. Dev Comp Immunol 17: 19–27

Raftos DA, Cooper EL, Habicht G, Beck G (1991) Invertebrate cytokines: tunicate cell proliferation stimulated by an endogenous hemolymph factor. Proc Natl Acad Sci USA 88: 9518–9522

Raftos DA, Cooper EL, Stillman DL, Habicht GS, Beck G (1992) Invertebrate cytokines II: release of interleukin-1-like molecules from tunicate hemocytes stimulated with zymosan. Lymphokine Cytokine Res 11: 235–240

Rinehart KL Jr, Gloer JB, Hughes RG Jr, Renis HE, McGovern JP, Swynenberg EB, Stringfellow DA, Kuentzel SL, Li LH (1981) Didemnins: antiviral and antitumor depsipeptides from a Caribbean tunicate. Science 212: 933–935

Rinehart KL Jr, Kobayashi J, Harbour GC, Hughes RG Jr, Mizsak SA, Scahill TA (1984) Eudistomins C, E, K, and L, potent antiviral compounds containing a novel oxathiazepine ring from the Caribbean tunicate *Eudistoma olivaceum.* J Am Chem Soc 106: 1524–1526

Rosenshein IL, Schlutter SF, Vasta GR, Marchalonis JJ (1985) Phylogenetic conservation of heavy chain determinants of vertebrates and protochordates. Dev Comp Immunol 9: 783–795

Sabbadin A (1962) Le basi genetiche della capacita di fusione fra colonie in *Botryllus schlosseri* (Ascidiacea). Rend Acad Naz Lincei Ser 8 32: 1031–1035

Saito Y, Watanabe H (1982) Colony specificity in the compound ascidian, *Botryllus scalaris.* Proc Jpn Acad Ser B58: 105–108

Saito Y, Watanabe H (1984) Partial biochemical characterization of humoral factors involved in the nonfusion reaction of a botryllid ascidian, *Botrylloids simodensis.* Zool Sci 1: 229–235

Scofield VL, Schlumpberger JM, West LA, Weissman IL (1982) Protochordate allorecognition is controlled by a MHC-like gene system. Nature 295: 499–502

Sigel M, Lichter W, McCumber L, Ghaffar A, Wellham L, Hightower J (1984) A substance from the marine tunicate *Ecteinascidia turbinata* with selective action on macrophages. In: Volkman (ed) Mononuclear Phagocyte Biology. Marcel Dekker, New York, pp 451–471

Smith VJ, Peddie CM (1992) Cell cooperation during host defense in the solitary tunicate *Ciona intestinalis* (L). Biol Bull 183: 211–219

Smith VJ, Söderhäll K (1991) A comparison of phenoloxidase activity in the blood of marine invertebrates. Dev Comp Immunol 15: 251–261

Suzuki T, Takagi T, Furukohri T, Kawamura K, Nakauchi M (1990) A calcium-dependent galactose-binding lectin from the tunicate *Polyandrocarpa misakiensis.* J Biol Chem 265: 1274–1281

Tanaka K (1973) Allogeneic inhibition in a compound ascidian, *Botryllus primigenus* Oka. II. Cellular and humoral responses in "nonfusion" reaction. Cell Immunol 7: 427–443

Tanaka K, Watanabe H (1973) Allogeneic inhibition in a compound ascidian, *Botryllus primigenus* Oka. I. Processes and features of "nonfusion" reaction. Cell Immunol 7: 410–426

Taneda Y, Watanabe H (1982a) Effects of X-irradiation on colony specificity in the compound ascidian, *Botryllus primigenus* Oka. Dev Comp Immunol 6: 665–673

Taneda Y, Watanabe H (1982b) Studies on colony specificity in the compound ascidian, *Botryllus primigenus* Oka. II. In vivo bioassay for analyzing the mechanism of "nonfusion" reaction. Dev Comp Immunol 6: 243–252

Tyler A (1946) Natural heteroagglutinins in the body fluids and seminal fluids of various invertebrates. Biol Bull 90: 213–219

Vasta GR, Marchalonis JJ (1983) Lectins from tunicates and cyclostomes: a biochemical characterization. In: Bøg-Hansen Spengler (eds) Lectins: Biology, Biochemistry and Clinical Biochemistry. DeGruyter, Berlin pp 461–468

Vasta GR, Marchalonis JJ (1986) Galactosyl-binding lectins from the tunicate *Didemnum candidum*. J Biol Chem 261: 9182–9186

Vasta GR, Marchalonis JJ (1987) Lectins from protochordates as putative recognition molecules. In: Cooper EL, Kanglet C, Bierne J (eds) Developmental and comparative immunology. Alan R Liss, New York, pp 23–32

Vasta GR, Hunt JC, Marchalonis JJ, Fish WW (1986) Galactosyl-binding lectins from the tunicate *Didemnum candidum*. Purification and physicochemical characterization. J Biol Chem 261: 9174–9181

Wright RK (1974) Protochordate immunity I. Primary immune response of the tunicate *Ciona intestinalis* to vertebrate erythrocytes. J Invertebr Pathol 24: 29–36

Wright RK (1981) Urochordates. In: Ratcliffe NA, Rowley AF (eds) Invertebrate Blood Cells, vol 2. Academic Press, London, pp 565–626

Wright RK, Cooper EL (1975) Immunological maturation in the tunicate *Ciona intestinalis*. Am Zool 15: 21–27

Wright RK, Cooper EL (1984) Protochordate immunity - II. Diverse hemolymph lectins in the solitary tunicate *Styela clava*. Comp Biochem Physiol 79B: 269–277

Yokosawa H, Sawada H, Abe Y, Numakunai T, Ishii S (1982) Galactose-specific lectin in the hemolymph of solitary ascidian, *Halocynthia roretzi:* isolation and characterization. Biochem Biophys Res Commun 107: 451–457

Yokosawa H, Odajima R, Ishii S (1985) Trypsin inhibitor in the hemolymph of a solitary ascidian, *Halocynthia roretzi*. Purification and characterization. J Biochem 97: 1621–1630

Yokosawa H, Harada K, Igarashi K, Abe Y, Takahashi K, Ishii S (1986) Galactose-specific lectin in the hemolymph of solitary ascidian, *Halocynthia roretzi*. Molecular, binding and functional properties. Biochem Biophys Acta 870: 242–247

Molecular Aspects of Immune Reactions in Echinodermata

V. MATRANGA

1 Introduction

The main challenge for any living organism is its encounter with potentially pathogenic bacteria and other microorganisms. Therefore, it should posses a defense mechanism which is capable of (1) recognizing foreign material which has entered the body, and (2) finding a way to either expel it or render it inoffensive. It is always very hazardous to compare and match the host defense systems operating in invertebrate organisms to the well-known vertebrate immune system. In fact of the enormous amount of data coming from vertebrate immunology research, only a small number of reports are applicable to the defense mechanisms functioning in invertebrates.

The immune phenomena occurring in echinoderms can be schematically divided into cellular and humoral responses. Cellular responses are mediated by the cells that are circulating in the coelomic fluid, generically called coelomocytes, while humoral responses depend upon molecules that are present in the coelomic fluid.

This chapter reviews studies of the immune responses which occur after transplantation of tissues or cells into the body cavities of echinoderms, and the in vivo and in vitro clearance of bacteria and other microorganisms. These processes involve humoral molecules that recognize and attack foreign material or stimulate coelomocyte cell growth. A survey of these topics is given, with particular attention to what it is known about the molecules involved in the processes. No distinction is made between the three classes of the Echinodermata phylum, Echinoidea, Ophiuroidea and Holothuroidea, although some peculiarities of each class are reported.

This study is not intended to be a review of all the work that has been done over the years in the field of echinoderm immunology; rather it will focus on a few arguments, with the aim of reexamining some ideas and concepts. Investigations of the immune system of invertebrates, and particularly of echinoderms, will help us in understanding the complex cellular and molecular interactions that regulate host defense in higher organisms.

Istituto di Biologia dello Sviluppo del Consiglio Nazionale delle Ricerche, Via Archirafi 20, 90123 Palermo, Italy

2 Response to Allogeneic Transplants

Echinoderms possess a striking ability to reject tissue from another member of the same species. This capability has been known for a long time, ever since the pioneering experiments by Hinegardner back in the 1960s. Usually, about one month is required in order for the rejection of an allograft in *Lytechinus pictus* to be complete, although the first response is seen after only 2 days with the appearance of red cells which cover the graft throughout the rejection period. Second-set grafts are rejected in less than one-third of the time required for primary allograft rejection (Coffaro and Hinegardner 1977), suggesting that there is an adaptive immune system operating in the echinoids.

Recently Coffaro's data have been reinterpreted by Smith and Davidson (1992). They claim that no immunological memory is taking place in the graft experiments, since third-party grafts were rejected with exactly the same enhanced kinetics as second-set allografts. This has provoked some controversy among researchers in the field who believe that adaptive immunology is present in invertebrates and who claim that the sea urchin case cannot be generalized to all metazoan organisms, from which the complex vertebrate immune system has indeed evolved (see "The echinoid immune system revisited", Immunol Today 14: 91–94).

An attempt to study transplantation *sensu stricto* has been made using experiments in which the cytotoxic reaction to allogeneic mixtures of echinoid phagocytes has been demonstrated in vitro (Bertheussen 1979). The author showed that phagocytes, among other cells circulating in the coelomic fluid of the sea urchin *Strongylocentrotus droebachiensis,* have the capability of allogeneic recognition, which leads to a cytotoxic reaction after 20 h. Allogenic and xenogenic phagocyte mixtures of *Strongylocentrotus pallidus* and *Echinus esculentus,* gave cytoxicity in 70 and 90% of cases, respectively. Since other cell types did not contribute to the cytotoxic effect, the authors conclude that the phagocytic amoebocyte is probably the effector cell involved in graft rejection, and may represent the common ancestor of the vertebrate lymphocyte, macrophage and granulocyte.

In conclusion, these few contradictory studies show at least that the distinction between self and nonself developed early in evolutionary history, and therefore echinoderms do possess one important characteristic of vertebrate immune responses, namely, specificity.

3 Clearance Studies

Echinoderms posses an efficient defense mechanism against bacteria and other microorganisms which accidentally or experimentally enter their bodies. In a comprehensive study by Yui and Bayne (1983), the authors determined the clearance from the coelomic fluid of the sea urchin *Strongylocentrotus purpuratus* of injected Gram-positive and Gram-negative bacteria. All bacteria were efficiently cleared, but no differences were observed between the secondary and primary clearance rates for any of the bacteria tested, suggesting the absence of an immunological memory. An overall reduction in the number of coelomocytes was observed which paralleled the bacterial clearance; however, the decline in the number of coelomocytes could be explained by their recruitment in cellular clotting. The observed increase in the relative number of vibratile cells, a subset of coelomocytes, was explained by the authors in terms of the high motility of this cell type, which may enable it to extricate itself from the clot or to be recruited more easily from other compartments. We could also speculate that this relative abundance could be due to an active cell growth in response to cytokines (Sect. 5), as has been shown for the tunicate *Styla clava* in response to allogenic stimuli (Raftos and Cooper 1991). In any case, there was no indication for accelerated secondary clearance of bacteria.

The in vitro reaction to bacteria has been studied, with different results, by Johnson and coworkers (Johnson 1969; Johnson et al. 1970). Using the two sea urchin species *Strongylocentrotus purpuratus* and *Strongylocentrotus franciscanus* as sources of coelomocytes, they observed that phagocytosis of Gram-positive bacteria was more efficient than phagocytosis of Gram-negative bacteria.

In a recent paper by Plytycz and Seljelid (1993), the authors investigated on the bactericidal activity of a coelomocytes subpopulation, namely, the amebocytes, cultured in vitro in the presence of bacteria. Their results confirmed the Yui and Bayne in vivo experiments and demonstrated that in vitro the bactericidal activity was more efficient at 23 °C than at 4 °C. However, live sea urchins kept at 23 °C and injected with bacteria, died 3 days later. One possible explanation for these contradictory in vivo and in vitro results may be that the sea urchins had to cope with the stressing condition of elevated temperature and therefore they fail to clear the bacteria.

The clearance of bacteriophage T4 has been reported in the sea urchin *Lytechinus pictus:* in this study second injection of the host did not reduce the time of clearance (Coffaro 1978). Again, this failure in showing a secondary response contrasts with transplantation studies, in which the second grafts were rejected in less than one-third of the time required for rejection of primary grafts (Coffaro and Hinegardner 1977).

The clearance of other particles such as red blood cells, latex beeds and yeast cells (Bertheussen 1981), xenogenic cells (Reinisch and Bang 1971), and bovine serum albumin (Hilgard and Phillips 1968) have been also reported. Although it is very unlikely that an echinoderm would encounter these materials, the use of the echinoderm model system for the study of the mechanisms involved remains valid and deserves encouragement for future work.

4 Humoral Molecules

The methods that echinoderms utilize to respond to transplants of tissues and cells or to clear their bodies of bacteria and other microorganisms involve both humoral molecules and circulating coelomocytes. The humoral components presents in the coelom of echinoderms have to first recognize and later attack the foreign body. It is now believed that the specific recognition between humoral molecules and invading organisms takes place through the carbohydrate moiety present on the surfaces of host cells or bacteria. These kinds of humoral molecules, called hemagglutinins for their ability to agglutinate the red blood cells of a variety of vertebrates, are lectins. Another group of humoral molecules which are present in the perivisceral fluid and are able to lyse foreign cells or bacteria, have the properties of vertebrate perforins, cytotoxic factors released by T-lymphocytes and natural killer cells (Masson and Tschopp 1985). The molecules interact with the host-plasma membrane producing circular holes which cause the lysis of target cells. In the past they have been called hemolysins for their ability to lyse red blood cells. Over the years, many studies at different laboratories have demonstrated the presence of hemagglutinins and hemolysins in the body fluid of a variety of echinoderms (Ryoyama 1973; Parrinello et al. 1979; Bertheussen 1984; Canicattì 1990a, 1991). In particular, in the Japanese sea urchin species *Anthocidaris crassispina, Pseudocentrotus depressus* and *Hemicentrotus pulcherrimus,* the occurrence of hemagglutinins in the coelomic fluid has been shown, the same material was found to have hemolytic activity (Ryoyama 1974). Studies performed on the class Holothuroidea have shown the presence of hemagglutinating and hemolytic activities in the coelomic fluid of *Holothuria polii* (Parrinello et al. 1979; Canicattì and Parrinello 1982).

In a classic paper by Bertheussen (1983), complement-like activity is detected by the lytic action on rabbit erythrocytes and by the opsonic effect on *Strongylocentrotus droebachiensis* coelomocytes. The activity is heat unstable and calcium-dependent, and it strongly resembles the mammalian complement activity since it can be inhibited by those compounds known to inhibit human complement. Similarly, a complement-like activity has been found in the sea star *Asterias forbesi:* this effect is optimized at 25 °C and requires the presence of divalent cations. The activity is destroyed after treatment with zymosan, pronase, trypsin, and paramethylsufonilfluoride (Leonard et al. 1990). The antibacterial activity found in the coelomic fluid of the sea star *Asteria rubens* (Jolles and Jolles 1975) and in the sea urchin *Paracentrotus lividus* (Gerardi et al. 1990) is most probably exerted by lysozyme-like molecules. Other bactericidal substances are known to be involved: they comprise peroxidase (Canicattì 1990b), naphtoquinone compounds, and echinochrome-A, the pigment of red spherula coelomocytes (Wardlaw and Unkless 1978; Matthew and Wardlaw 1984; Sect. 6). Echinochrome-A was found in coloured spherula cells of the sea urchin *Echinus esculentus.*

In the serum of vertebrates some proteins are present which are specialized in binding to platelets, collagen, and other proteins, and which are involved in hemostasis by promoting the adherence of platelets at sites of injury. It has been recently shown that in the invertebrate *Caenorabditis elegans,* vitellogenin shares some sequences at the amino-terminal region with the human von Willebrand factor whose function in vertebrate blood clotting is well known (Baker 1988). Another well-known vertebrate serum protein, namely fibrinogen, reveals common ancestry with invertebrate vitellogenin (Doolittle and Riley 1990). In echinoderms, wound repair and encapsulation of foreign bodies requires the presence, in the coelomic fluid, of an adhesive activity. This activity has been found in the sea star *Asteria forbesi,* where it has been shown that a calcium- and magnesium-dependent factor is involved in the clumping of coelomocytes after injury (Kanungo 1982). Other studies on *Holothuria polii* coelomic fluid resulted in the purification of a 220-kDa factor which is heat stable and does not require divalent cations for its activity (Canicattì and Rizzo 1991). The same authors found in the coelomic fluid of *Paracentrotus lividus,* a 200-kDa protein which promotes the adhesion of autologous coelomocytes and their spreading on the substrate (Canicattì et al. 1992). Recently, we have shown that a 200-kDa protein, purified from the coelomic fluid of the sea urchin *Paracentrotus lividus,* is the precursor of a previously discovered 22S cell-adhesion molecule (Cervello and Matranga 1989) called toposome, which is responsible for the integrity of the sea urchin embryo (Noll et al. 1985; Matranga et al. 1986). Toposome is localized on the cell surface of blastula cells as well as in yolk granules of unfertilized eggs (Gratwohl et al. 1991), and it has been shown to promote the adhesion of sea urchin cells to the substrate (Matranga et al. 1991). Our results, in addition to providing a link to invertebrates for a protein involved in blood clotting, indicates a function for the toposome precursor, also referred to as vitellogenin for historical reasons, other than as mere food reservoir for the growing embryo. In support of this hypothesis is the fact that serum proteins, originating 450 millions years ago in primitive vertebrates, evolved by a combination of exon shuffling and gene duplication to form mosaic proteins with diverse function (Rogers 1985).

In conclusion, a variety of molecules are present in the coelomic fluid of echinoderms, namely lectins, perforins, complement-like molecules, and serum proteins. The coelomic fluid scenario is therefore not as simple as was first thought; many molecules are playing the game of recognition and immune response.

5 Cytokines

Cytokines are polypeptides released by activated blood cells of vertebrates, which are usually induced upon inflammation or infection. They contribute to the proliferation of immune effector cells and influence many organs and systems. A wide variety of cytokines have been discovered over the years, with different functions and a variety of targets. Most of these molecules have been cloned and a complex cross-talk among cells and systems has been described (Beck and Habicht 1991a).

Because of their great importance in the host defence mechanism, it has been proposed that cytokines have been conserved through evolution so that they were also present in invertebrate animals. In the seventies Prendergast and Suzuki (1970) accumulated data on the presence of an Interleukin-1-like molecule in the echinoderms. They found in the sea star *Asteria forbesi* a so-called SSF (sea star factor), isolated from coelomocyte lysates, with a molecular weight of 39 kDa. The molecule also caused the aggregation of coelomocytes at sites of its injection into the animal.

More recently Beck and Habicht (1986, 1991b) have purified, from the coelomic fluid of the sea star *Asteria forbesi,* a molecule that has the characteristics and functions of vertebrate Interleukin-1 (IL-1). The molecule has a molecular weight of 29.5 kDa, stimulates the proliferation of murine thymocytes and fibroblasts, stimulates protein synthesis of fibroblasts and enhances the production of prostaglandin E2, and finally, it is cytotoxic for the human cell line A375. The activity of the molecule resides in a 20-kDa peptide whose stimulatory effects on mouse thymocytes can be blocked by polyclonal antibodies specific for human IL-1. The sea star primitive IL-1 might have its precursor in the molecule described earlier by Prendergast and Suzuki (1970).

In the same laboratory two other cytokines have been isolated and characterized, namely IL-2 and TNF, from echinoderms and tunicates respectively (Beck et al. 1989). The authors propose that invertebrate host defense is regulated by a cytokine network analogous to that found in vertebrates, but not so complex and obviously easier to study.

Furthermore it has been shown that, in the sea star *Pisaster ochraceus,* the proportion of coelomocytes that phagocytate bacteria is greatly enhanced by the addition of recombinant IL-1, confirming Beck and Habich experiments (Burke and Watkins 1991). In summary, cytokines are not unique to vertebrates and, on the contrary, the presence of these molecules in echinoderms as well as in other invertebrates provides support for a continuity of cytokines in the phylogenetic tree.

6 Immune Effector Cells

In the coelomic cavity of sea urchins, sea stars, and sea cucumbers, there exists a mixed cell population: the coelomocytes. As we have already seen, they are the first line of defense against injuries and infections. Since the 19th century, echinoderm coelomocytes have been used as model systems for investigations into different phenomena ranging from invertebrate immune response to cellular morphogenesis. In a series of studies, coelomocytes have been classified into cell types on the basis of their morphology. The extensive reviews by Ratcliffe and Rowley (1979) and by Smith (1981) summarize data in the literature on cell types, to give a nomenclature based on morphological and functional criteria. According to Smith's nomenclature, at least six morphologically distinguishable cell types can be found in the coelomic fluid of echinoderms the phagocytic amoebocytes, the coloured and colourless spherula cells, the vibratile cells, the haemocytes, the crystal cells, and the progenitor cells. Not all six cell types are present in every species, and the relative abundance is very variable from species to species and even from individual to individual, depending on their physiological conditions or geographical locations. It has been suggested that these categories may correspond to functionally different cell types or that they may represent different developmental stages of the same cell line (Smith 1981).

Much work still needs to be done on the function of all these cell types. Most of the immune responses that we have examined are mediated by phagocytic amoebocytes. Many studies, reviewed by Smith (1981), have shown that they exhibit two morphologically distinct phases: the petaloid form which is actively phagocytic, and the filopodial stage which appears to be involved in the clotting. After injuries or in case of experimental grafts, it has been observed that petaloid amoebocytes migrate to the site of injury, change their shape to filopodial, and form a clot together with other coelomic cells, as well as extracellular matrix of unknown composition. A complete study of the phenomenology of the morphological transformation from petaloid to filopodial amoebocytes, involving cytoskeletal rearrangements, is given in the review by Edds (1985).

Phagocytic cells are also the first to act in the efficient clearance mechanism working in sea urchins. Bertheussen and Seljelid (1978) reported the analysis of particle recognition, ingestion, and degradation operated by phagocytic cells. They also recommend the use of in vitro phagocyte cultures in order to better analyze the mechanisms involved and to overcome the stress of manipulation. Unfortunately, this line of investigation has not been followed very much over the years, and most of the time the stressful conditions, due to artificial maintenance in the laboratory, have not been seriously taken into consideration. This is not the case for an interesting report which comes from studies by Ito et al. (1992) who found that phagocytes from the sea urchin *Strongylocentrotus nudus* cultured in vitro produce hydrogen peroxide, and that this production is enhanced after stimulation by coleomocytes. In fact phagocytic phagocytes produce more hydrogen peroxide than resting ones and the amount produced is different when they are stimulated by the injection of erythrocytes from sheep or humans implying that a specific recognition system exists in *Strongylocentrotus nudus* phagocytes.

The cellular reactions to foreign particles, instead of phagocytosis, are cell clumping and encapsulation. Both phagocytic amoebocytes and spherula cells appear to be involved in clump and capsule formation. Phagocytosis has not been observed in these cases, but it is possible that spherule cells release bactericidal substances (Johnson 1969). Despite the many morphological studies, little information is available on the molecules produced and released by the spherule cells. An arylsulfatase activity has been found in the coelomocyte populations of seven different echinoderms, comprising Asteroidea, Echinoidea and Holothuroidea (Canicattì and Miglietta 1989). Histochemically, the enzyme was detected in granules of spherula cells and therefore this cell type may be assimilated by the white series of blood cells where this enzyme has been found associated with inflammatory phenomena (Pagliara and Canicattì 1993).

It is now generally accepted that the invertebrate host defense function involves first immobilising and then encapsulating the invasive microorganism. During this process two mechanisms are operating, possibly in cooperation:

some molecules, such as perforins (Canicattì 1990a), are attacking the host, others are involved in adhesion between the cells surrounding and encapsulating the host. Candidates for the last function are the sea urchin serum toposome-precursor, also referred to as vitellogenin, (Cervello and Matranga 1989) as well as the aggregation factor found in *Holothuria polii* (Canicattì and Rizzo 1991) and sea urchins (Canicattì et al. 1992). The three molecules share many analogies, despite being different molecular species in different genera. Recently, we have shown that colourless spherula cells are the reservoirs of the so called vitellogenin: the molecule is contained in kidney-shaped granules located around the nucleus. We also showed that coelomocytes respond to stress conditions by discharging the molecule into the medium, suggesting that vitellogenin has a role in the coelomocyte clotting that occurs after injury (Cervello et al. 1994). Accordingly, a 220-kDa aggregating factor isolated from *Holothuria polii* has been shown to be released by coelomocytes and to be capable of agglutinating coelomocytes (Canicattì and Rizzo 1991).

Phagocytosis of foreign material has been attributed to amoebocytes. In *Holothuria polii* the process has been analyzed in detail (Canicattì and D'Ancona 1989). It has been shown that lysosomal enzymes including lipase, peroxidase, and serine protease, are contained intracellularly in coleomocytes (Canicattì 1990b; Canicattì and Tschopp 1990). Consequently, it seems clear that lysosomal enzymes cause the breakdown of phagocytosed material.

We have already seen that IL-1-like protein isolated from the coelomic fluid can be found intracellularly in coelomocytes and that it is secreted by coelo-mocytes into the coelom in response to stimuli (Beck and Habicht 1986). However, in this case the authors do not have evidence regarding which type of coelomocyte secretes IL-1.

Recently, a molecular approach to the echinoderm immulogy has been provided by studies of the response of sea urchin coelomocytes to injury. In brief, *Strongylocentrotus purpuratus* coelomocytes respond to minor injuries, such as the puncture of a needle into the peristomial membrane, by the increased expression of a gene (SpCoel1) coding for a newly discovered sea-urchin profilin (Smith et al. 1992). SpCoel1 encodes for a polypeptide, whose sequence has a striking similarity to the mammalian profilin, a small protein that has both actin and phosphatidyl-inositol-biphosphate binding functions (Goldschmidt-Clermont et al. 1991). The authors speculate that sea urchins can utilize a mechanism of "unspecific immunity", which works by the activation of this gene which links the cytoskeleton with the signal transduction system. From their data, Smith and Davidson (1992) proposed that sea urchins, and echinoderms in general, do not possess an adaptive immunity, but that a simple non-specific alert mechanism operates after injury, inflammation, or host invasion.

There has been some debate on the origin of coelomocytes in echinoderms, and many leukopoietic elements have been reported in several locations in the animal. However, it is now widely accepted that the main source of coelomocytes is the axial organ, which therefore may represent an ancestral primary lymphoid gland. It has also been shown that the axial organ of an echinoid releases coelomocytes after injury (Millott 1969). A recent report describes the fraction-ation from the axial organ of the sea star *Asteria rubens* of two populations of cells, namely the adherent (B-like) cells, resembling mammalian B-lymphocytes, and the non-adherent (T-like) cells, resembling mammalian T-lymphocytes (Leclerc and Bajelan 1992). Of course, it would be of great interest to analyze the molecules produced and released into the coelom by these two cell types.

In conclusion, coelomocytes can be defined as the effector cells of the echino-derm immune system. They synthesize and release upon stimulation a number of specific molecules such as cytokines, lectins, perforins, lysosomal enzymes, serum proteins, profilins, and others. Furthermore, because of their immune functions, they deserve the role of progenitor immune cells. Further studies are required for the understanding of the specific function of each cell type and the possible cross-talk among them.

7 Summary and Concluding Remarks

The study of immune reactions in echinoderms is very important, either from a phylogenetic point of view, or because it may contribute to the understanding of the complex immunology of higher animals, including humans. An attempt has been made in this review to examine the molecular aspects of the invertebrate immune system and their possible analogies with the mechanisms operating in vertebrates.

Allogenic transplantation studies have shown that two important charac-teristics of vertebrate immune response, namely specificity and to some extent memory, are retained by echinoderms (Coffaro and Hinegardner 1977). The clearance of bacteria from sea urchins resembles the equivalent process operating in mammals, with a biphase process consisting of a rapid and exponential phase of clearance (90–99% reduction in circulating bacteria) followed by a slow phase during which bacteria are removed in hours or days (Yui and Bayne 1983).

The aggregation of phagocytic amoebocytes at inflammatory sites is a feature that invertebrates and vertebrates have in common. The recruitment of cells, mainly phagocytes and red cells, is due to chemiotaxis and cellular proliferation triggered by cytokines (Prendergast et al. 1983). In most cases, the recognition and binding of nonself material are based on the presence of carbohydrate-specific molecules; this may therefore represent an ancient evolutionary binding principle (Ryoyama 1973; Canicattì et al. 1992).

Yet there still stands the fascinating idea that immunoglobulins may have evolved from lectins, as postulated in the 1960s by Burnet (1968). The fact remains that an immunoglobulin-like molecule was never found in echinoderms, although the presence of immunoglobulin molecules in very primitive vertebrates has been shown (Raison et al. 1978). A hypothesis has been presented that sets out the possibility that immunoglobulin superfamily molecules may have origi-nated from the more ubiquitous cell-recognition molecules which are common in all metazoan species (Matsunaga and Mori 1987).

No cross-talk has yet been demonstrated among coelomocytes in response to host invasion. This can be explained by (1) the self-nonself recognition system operating in vertebrates is not evolved from the echinoderm immune system which has its own unique features (Smith and Davidson 1992); or (2) multi-functional proteins are present in echinoderm coelomocytes, which may fulfill the needs of immune defense and whose domains have evolved during evolution (Doolittle 1985).

We have shown that colourless spherula cells contain the so-called vitello-genin in cytoplasmic granules and that the protein is released into the coelom under stressful conditions (Cervello et al. 1994). Our working hypothesis is that the molecule is the ancestor of those serum proteins, involved in vertebrate blood clotting, that first appeared 450 million years ago in primitive vertebrates (Doolittle 1985). Functional assays using purified molecules as well as gene cloning analysis are in progress in our laboratory in order to better elucidate the matter.

The molecular approach by Davidson's laboratory (Smith et al. 1992) seems promising for the elucidation of the molecular basis of immune response against host attack, and now the expression of the gene SpCoel1 constitutes the first molecular parameter that can be followed after coelomocytes activation.

Even if not related to immune responses, valuable data on the evolution and function of hemoglobins have been obtained from studies on invertebrate hemoglobins. For example, in the sea cucumber *Caudina arenicola,* four different globin molecules have been found, and two of these have been sequenced (McDonald et al. 1992). These data will provide new insights into the molecular evolution of globins.

An effort should be made to search for specific molecules expressed by coelomocytes. For example, it has been suggested that spherula cells may produce collagen, if not constitutively, in response to injuries to be used for deposition in wound repair (Smith 1981).

Most of the molecules that have been described here are not fully charac-terized, since they are only reported in terms of activities. More can be understood if genes for known adhesive proteins and lectins purified from echinoderms are cloned, sequenced and compared with vertebrate immune molecules. In fact, despite recent advances in molecular and cellular biology, the molecular aspects of immunity operating in the echinoderms remain obscure and poorly understood. Once again this is due to a lack of information which renders the interpretation of a small number of controversial reports very difficult. The phylogenetic position of the phylum Echinodermata, so remarkably close to vertebrates, makes them pivotal to the understanding of the evolution of verte-brate immunity. Obviously, the vertebrate immune system did not spring out of nothing and we should try to delineate the pathways followed in evolution, from the echinoderm immune reactions to the vertebrate adaptive immunity.

Acknowledgments. The author wishes to thank Drs. M. Cervello and F. Zito for stimulating dis-cussions and critical reading of the manuscript. Thanks to Dr. S. Sciarrino for assistance and encouragement in the preparation of the manuscript. Here, I wish to remember Prof. Calogero Canicattì, who has dedicated his scientific life to the study of invertebrate immunology, and who introduced me to this fascinating field.

References

Baker ME (1988) Invertebrate vitellogenin is homologous to human von Willebrand factor. Biochem J 256: 1059–1063

Beck G, Habicht GS (1986) Isolation and characterization of a primitive interleukin-1-like protein from an invertebrate, *Asteria forbesi*. Proc Natl Acad Sci USA 83: 7429–7433

Beck G, Habicht GS (1991a) Primitive cytokines: harbingers of vertebrate defense. Immunol Today 12: 180–183

Bech G, Habicht GS (1991b). Purification and biochemical characterization of an invertebrate interleukin 1. Mol. Immunol 28: 577–584

Beck G, O'Brian RF, Habicht GS (1989) Invertebrates cytokines: the phylogenetic emergence of interleukin-1. BioEssay 11: 62–67

Bertheussen K (1979) The cytotoxic reaction in allogeneic mixtures of echinoid phagocytes. Exp Cell Res 120: 373–381

Bertheussen K (1981) Endocytosis by echinoid phagocytes in vitro I. Recognition of foreign matter. Dev Comp Immunol 5: 241–250

Bertheussen K (1983) Complement-like activity in sea urchin coelomic fluid. Dev Comp Immunol 7: 21–31

Bertheussen K (1984) Complement and lysins in invertebrates. Dev Comp Immunol 3: 173–181

Bertheussen K, Seljelid R (1978) Echinoid phagocytes in vitro. Exp Cell Res 111: 401–412

Burke RD, Watkins RF (1991) Stimulation of starfish coelomocytes by interleukin-1. Biochem Biophys Res Commun 180: 579–584

Burnet FM (1968) Evolution of the immune process in vertebrates. Nature 218: 426–430

Canicattì C (1990a) Hemolysins: pore-forming proteins in invertebrates. Experientia 46: 239–244

Canicattì C (1990b) Lysosomal enzyme pattern in *Holothuria polii* coelomocytes. J Invertebr Pathol 56: 70–74

Canicattì C (1991) Binding properties of *Paracentrotus lividus* (Echinoidea) hemolysins. Comp Biochem Physiol 98A: 463–468

Canicattì C, D'Ancona G (1989) Cellular aspects of *Holothuria polii* immune response. J Invertebr Pathol 53: 152–158

Canicattì C, Miglietta A (1989) Arylsulphatase in echinoderm immunocompetent cells. Histochem J 21: 419–424

Canicattì C, Parrinello N (1982) Chromatographic separation of coelomic fluid from *Holothuria polii* (Echinodermata) and partial characterization of the fractions reacting with erithrocytes. Experientia 39: 764–766

Canicattì C, Rizzo A (1991) A 220 KDa coelomocyte aggregating factor involved in *Holothuria polii* cellular clotting. Eur J Cell Biol 56: 79–83

Canicattì C, Tschopp J (1990) Holozyme A: one of the serine proteases of *Holothuria polii* coelomocytes. Comp Biochem Physiol 96B: 739–742

Canicattì C, Pagliara P, Stabili L (1992) Sea urchin coelomic fluid agglutinin mediates coelomocyte adhesion. Eur J Cell Biol 58: 291–295

Cervello M, Matranga V (1989) Evidence of a precursor-product relationship between vitellogenin and toposome, a glycoprotein complex mediating cell adhesion. Cell Diff Dev 26: 67–76

Cervello M, Arizza V, Lattuca G, Parrinello N, Matranga V (1994) Detection of vitellogenin in a subpopulation of sea urchin coelomocytes. Eur J Cell Biol 64: 314–319

Coffaro K (1978) Clearance of bacteriophage T4 in the sea urchin *Lytechinus pictus*. J Invertebr Pathol 32: 384–385

Coffaro KA, Hinegardner RT (1977) Immune response in the sea urchin *Lytechinus pictus*. Science 197: 1389–1390

Doolittle RF (1985) The genealogy of some recently evolved vertebrate proteins. Trends Biochem Sci 10: 233–237

Doolittle RF, Riley MR (1990) The ammino terminal sequence of lobster fibrinogen reveals common ancestry with vitellogenins. Biochem Biophys Res Commun 167: 16–19

Edds KT (1985) Morphological and cytoskeletal transformation in sea urchin coelomocytes. In: Cohen WD (ed) Blood cells of marine invertebrates: experimental systems in cell biology and comparative physiology, A R Liss, New York, pp 53–74

Gerardi P, Lassegues M, Canicatti C (1990) Cellular distribution of sea urchin antibacterial activity. Biol Cell 70: 153–157

Goldschmidt-Clermont P, Machesky L, Doberstein S, Pollard T (1991) Mechanism of the interaction of human platelet profilin with actin. J Cell Biol 113: 1081–1089

Gratwohl EK, Kellenberg ME, Lorand L, Noll H (1991) Storage, ultrastructural targeting and function of toposomes and hyalin in sea urchin embryogenesis. Mech Dev 33: 127–138

Hilgard HR, Phillips JH (1968) Sea urchin response to foreign substances. Science 161: 1243–1245

Ito T, Matsutani T, Mori K, Nomura T (1992) Phagocytosis and hydrogen peroxide production by phagocytes of the sea urchin Strongylocentrotus nudus. Dev Comp Immunol 16: 287–294

Johnson PT (1969) The coelomic elements of sea urchins (Strongylocentrotus) III. In vitro reaction to bacteria. J Invertebr Pathol 13: 42–62

Johnson PT, Chien PK, Chapman FA (1970) The coelomic elements of sea urchins (Strongylocentrotus) IV. Ultrastructure of leukocytes exposed to bacteria. J Invertebr Pathol 16: 466–469

Jolles J, Jolles P (1975) The lysozyme from Asterias rubens. Eur J Biochem 54: 19–23

Kanungo K (1982) In vitro studies on the effect of cell-free coelomic fluid, calcium, and/or magnesium on clumping of coelomocytes of the sea star Asteria forbesi. Biol Bull 163: 438–452

Leclerc M, Bajelan M (1992) Homologous antigen for T cell receptor in axial organ cells from the asterid Asterias rubens. Cell Biol Int Rep 16: 487–490

Leonard LA, Strandberg JD, Winkelstein JA (1990) Complement-like activity in the sea star Asterias forbesi. Dev Comp Immunol 14: 19–30

Masson D, Tschopp J (1985) Isolation of a lytic, pore-forming protein (perforin) from cytolytic T-lymphocytes. J Biol Chem 260: 9069–9072

Matranga V, Kuwasaki B, Noll H (1986) Functional characterization of toposome from sea urchin blastula embryos by a morphogenetic cell aggregation assay. EMBO J 5: 3125–3132

Matranga V, Di Ferro D, Cervello M, Zito F, Nakano E (1991) Adhesion of sea urchin embryonic cells to substrata coated with cell adhesion molecules. Biol Cell 71: 289–191

Matsunaga T, Mori (1987) The origin of immune system. Scand J Immunol 25: 485–495

Matthew S, Wardlaw AC (1984) Echinochrome-A as a bactericidal substance in the coelomic fluid of Echinus esculentus. Comp Biochem Physiol 79B: 161–165

McDonald GD, Davidson L, Kitto GB (1992) Amino acid sequence of the coelomic C globin from the sea cucumber Caudina (Molpadia arenicola). J Protein Chem 11: 29–37

Millott N (1969) Injury and the axial organ of echinoids. Experientia 25: 756

Noll H, Matranga V, Cervello M, Humphreys T, Kuwasaki B, Adelson D (1985) Characterization of toposomes from sea urchin blastula cells: a cell organelle mediating cell adhesion and expressing positional information. Proc Natl Acad Sci USA 82: 8062–8066

Pagliara P, Canicatti C (1993) Isolation of coelomocyte granules from sea urchin amoebocytes. Eur J Cell Biol 60: 179–184

Parrinello N, Ridone D, Canicatti C (1979) Naturally occurring hemolysins in the coleomic fluid of Holothuria polii. Dev Comp Immunol 3: 45–54

Plytcz B, Seljelid R (1993) Bacterial clearance by the sea urchin Strongylocentrous droebachiensis. Dev Comp Immunol 17: 283–289

Prendergast R, Suzuki M (1970) Invertebrate protein stimulating mediators of delayed hypersensitivity. Nature 227: 277–279

Prendergast RA, Lutty GA, Scott AL (1983) Directed inflammation: the phylogeny of lymphokines. Dev Comp Immunol 7: 629–632

Raftos DA, Cooper EL (1991) Proliferation of lymphocyte-like cells from the solitary tunicate, Stlyla clava, in response to allogeneic stimuli. J Exp Zool 260: 391–400

Raison RL, Hull CJ, Hildemann WH (1978) Characterization of immunoglobulin from the Pacific hagfish, a primitive vertebrate. Proc Natl Acad Sci USA 75: 5679–5682

Ratacliffe NA, Rowley AF (1979) A comparative synopsis of the structure and function of the blood cells of insects and other invertebrates. Dev Comp Immunol 3: 189–243

Reinisch CL, Bang FB (1971) Cell recognition: reaction of the sea star (*Asterias vulgaris*) to the injection of amoebocytes of sea urchin (*Arbacia punctulata*). Cell Immunol 2: 496–503

Rogers J (1985) Exon shuffling and intron insertion in serine protease genes. Nature 315: 458–459

Ryoyama K (1973) Studies on the biological properties of coelomic fluid of sea urchin. I Natrually occurring hemolysins. Biochim Biophys Acta 320: 157–165

Ryoyama K (1974) Studies on the biological properties of coelomic fluid of sea urchin. II Naturally occurring hemagglutinin in sea urchin. Biol Bull 146: 404–414

Smith LC, Davidson EH (1992) The echinoid immune system and the phylogenetic occurence of immune mechanisms in deuterostomes. Immunol today 13: 356–362

Smith LC, Britten RJ, Davidson EH (1992) SpCoell: a sea urchin profilin gene expressed specifically in coelomocytes in response to injury. Mol Biol Cell 3: 403–414

Smith VJ (1981) The echinoderms. In: Ratcliffe NA, Rowley AT (eds) Invertebrate Blood Cells. Vol 2. Academic Press, London

Yui MA, Bayne CJ (1983) Echinoderm immunology: bacterial clearance by the sea urchin *Strongylocentrotus purpuratus*. Biol Bull 165: 473–486

Wardlaw AC, Unkless SE (1978) Bactericidal activity of coelomic fluid from the sea urchin *Echinus esculentus*. J Invertebr Pathol 32: 25–34

Subject Index

α_2-Macroglobulin   101, 173
α_2-Macroglobulin-protease complex   112
β-1,3-Glucan binding protein   57
β2-microglobulins   4, 25
Adaptive cellular response   28
Agglutination-aggregation factor   182
Agglutinin   2
Allogeneic recognition   227
Allogeneic transplants   236
Allografts   18
Allorecognition   195
Amoebocyte   11
Antibacterial responses   68
Antibacterial substances   176
Anticoagulants   171
Antilipopolysaccharide factor   177
Antimicrobial substances   224
Apidaecins   76
Attacins   72

Big defensin   178

C-reactive protein   179, 230
C-type lectin   92
Carcinoscorpin   179
Cecropins   71
Cell-adhesion proteins   55
Cell mediated cytotoxicity   27
Chimera   210
Clotting   154
Clotting cascade   158
Clotting enzyme   163
Clotting factors   90
Clotting proteins   59
Coagulin   160
Coagulogen   93, 95, 160
Coelomic fluid   14, 27
Coelomocyte   5, 11
Coelomocyte response   15
Complement   5
Complement-like proteins   58
Concanavalin A   24
Cytokine-binding activities   116
Cytokine-like substance   225

Cytokines   147, 239
Cytolytic substances   32
Cytotoxicity   12, 191

Defensins   73

Elicitor-binding proteins   52
Eleocyte   11
Elicitors   52
Encapsulation   14
Evolution   121
Evolution of immune responses   35

Factor B   92, 164
Factor C   165
Factor G   93, 167
Fasciclin   75

Graft rejection   16, 191

Halocyamine   224
Hemagglutinating systems   31
Hemagglutinins   219
Hemocyanin   135
Hemocyte granules   156
Hemocytes   89, 156, 190
Hemoglobin   134
Hemolin   75
Hemolymph   156, 222
Hemolysin families   31
Hemostasis   89
Histocompatibility   36
Honey bee venom   76
Host-parasite interactions   131
Humoral factors   218

IgG-binding proteins   25
Immune defense   154
Immune effector cells   240
Immunodefense reactions   36
Immunoglobulin joining chain   26
Immunoglobulins   25
Inducible humoral factor   79
Inflammatory response   15

Inflammatory-type reaction   2
Interleukin-1-type molecule   5
Internal defenses   133
Intracellular agglutinins   181

Lectins   69, 89, 179, 219
Limulin   179
Limulus clotting system   89
Limunectin   182
Lipofuscin   14
Lipopolysaccharide-binding lectin   70
Lombricine   34
Lysozyme   69
Lysozyme-like substances   30

Natural cytocidal response   196
Natural immunity   13
Natural killer cells   13

Opsonins   13, 70

Parasites   135
Peroxidase   14
Phagocytosis   2, 4, 12, 13, 18, 31, 70
Phylogenesis   1
Plasma-based cytolytic system   116
Polyphemin   180
Proclotting enzyme   163

Prophenoloxidase   47, 49
Prophenoloxidase activating enzyme   50
Prophenoloxidase activating
   inhibitors   50
Prophenoloxidase activating system   48
Prophenoloxidase system   2
Protease inhibitors   171
Proteolytic cascade   90

Regulators of clotting   94

Self/nonself recognition   3, 191, 212
Substrates of transglutaminase   176
Superoxide   144

Tachyplesin   177
Thy-1 gene   26
Toxic oxygen   144
Transglutaminase   93, 174
Transplant   10
Transplantation   3, 19, 194
Transplantation antigens   20
Trypsin inhibitor   173
Tumorostatic activity   34

Wound healing   18

Xenografts   16

Printing: Saladruck, Berlin
Binding: Buchbinderei Lüderitz & Bauer, Berlin